CAMBRIDGE MONOGRAPHS ON
APPLIED AND COMPUTATIONAL
MATHEMATICS

Series Editors

M. J. ABLOWITZ, S. H. DAVIS, E. J. HINCH, A. ISERLES,
J. OCKENDEN, P. J. OLVER

Learning Theory:
An Approximation Theory Viewpoint

The Cambridge Monographs on Applied and Computational Mathematics reflect the crucial role of mathematical and computational techniques in contemporary science. The series publishes expositions on all aspects of applicable and numerical mathematics, with an emphasis on new developments in this fast-moving area of research.

State-of-the-art methods and algorithms as well as modern mathematical descriptions of physical and mechanical ideas are presented in a manner suited to graduate research students and professionals alike. Sound pedagogical presentation is a prerequisite. It is intended that books in the series will serve to inform a new generation of researchers.

Within the series will be published titles in the Library of Computational Mathematics, published under the auspices of the Foundations of Computational Mathematics organisation. *Learning Theory*: An Approximation Theory View Point is the first title within this new subseries.

The Library of Computational Mathematics is edited by the following editorial board: Felipe Cucker (Managing Editor) Ron Devore, Nick Higham, Arieh Iserles, David Mumford, Allan Pinkus, Jim Renegar, Mike Shub.

Also in this series:

A practical Guide to Pseudospectral Methods, *Bengt Fornberg*

Dynamical Systems and Numerical Analysis, *A. M. Stuart and A. R. Humphries*

Level Set Methods, *J. A. Sethian*

The Numerical Solution of Integral Equations of the Second Kind,
    Kendall E. Atkinson

Orthogonal Rational Functions, *Adhemar Bultheel, Pablo González-Vera,
    Erik Hendriksen, and Olav Njåstad*

Theory of Composites, *Graeme W. Milton*

Geometry and Topology for Mesh Generation, *Herbert Edelsbrunner*

Schwarz-Christoffel Mapping, *Tobin A. Driscoll and Lloyd N. Trefethen*

High-Order Methods for Incompressible Fluid, *M.O. Deville,
    E.H. Mund and P. Fisher*

Practical Extrapolation Methods, *Avram Sidi*

Generalized Riemann Problems in Computational Fluid Dynamics,
    M. Ben-Artzi and J. Falcovtz

Radial Basis Functions, *Martin Buhmann*

Learning Theory: An Approximation Theory Viewpoint

FELIPE CUCKER
City University of Hong Kong

DING-XUAN ZHOU
City University of Hong Kong

CAMBRIDGE
UNIVERSITY PRESS

32 Avenue of the Americas, New York NY 10013-2473, USA

Cambridge University Press is part of the University of Cambridge.

It furthers the University's mission by disseminating knowledge in the pursuit of
education, learning and research at the highest international levels of excellence.

www.cambridge.org
Information on this title: www.cambridge.org/9780521865593

© Cambridge University Press 2007

First published 2007

A catalogue record for this publication is available from the British Library

Library of Congress Cataloguing in Publication data

Cucker, Felipe, 1958–
Learning theory: an approximation theory viewpoint / Felipe Cucker,
Ding-Xuan Zhou.
p. cm.
Includes bibliographical references and index.
ISBN-13: 978-0-521-86559-3 (hardback: alk. paper)
ISBN-10: 0-521-86559-X (hardback: alk. paper)

1. Computational learning theory. 2. Approximation theory. I. Zhou, Ding-Xuan. II. Title.
Q325.7.C83 2007
006.3′1–dc22
2006037012

ISBN 978-0-521-86559-3 Hardback

Contents

v

Foreword

This book by Felipe Cucker and Ding-Xuan Zhou provides solid mathematical foundations and new insights into the subject called *learning theory*.

Some years ago, Felipe and I were trying to find something about brain science and artificial intelligence starting from literature on neural nets. It was in this setting that we encountered the beautiful ideas and fast algorithms of learning theory. Eventually we were motivated to write on the mathematical foundations of this new area of science.

I have found this arena to with its new challenges and growing number of application, be exciting. For example, the unification of dynamical systems and learning theory is a major problem. Another problem is to develop a comparative study of the useful algorithms currently available and to give unity to these algorithms. How can one talk about the "best algorithm" or find the most appropriate algorithm for a particular task when there are so many desirable features, with their associated trade-offs? How can one see the working of aspects of the human brain and machine vision in the same framework?

I know both authors well. I visited Felipe in Barcelona more than 13 years ago for several months, and when I took a position in Hong Kong in 1995, I asked him to join me. There Lenore Blum, Mike Shub, Felipe, and I finished a book on real computation and complexity. I returned to the USA in 2001, but Felipe continues his job at the City University of Hong Kong. Despite the distance we have continued to write papers together. I came to know Ding-Xuan as a colleague in the math department at City University. We have written a number of papers together on various aspects of learning theory. It gives me great pleasure to continue to work with both mathematicians. I am proud of our joint accomplishments.

I leave to the authors the task of describing the contents of their book. I will give some personal perspective on and motivation for what they are doing.

Computational science demands an understanding of fast, robust algorithms. The same applies to modern theories of artificial and human intelligence. Part of this understanding is a complexity-theoretic analysis. Here I am not speaking of a literal count of arithmetic operations (although that is a by-product), but rather to the question: What sample size yields a given accuracy? Better yet, describe the error of a computed hypothesis as a function of the number of examples, the desired confidence, the complexity of the task to be learned, and variants of the algorithm. If the answer is given in terms of a mathematical theorem, the practitioner may not find the result useful. On the other hand, it is important for workers in the field or leaders in laboratories to have some background in theory, just as economists depend on knowledge of economic equilibrium theory. Most important, however, is the role of mathematical foundations and analysis of algorithms as a precursor to research into new algorithms, and into old algorithms in new and different settings.

I have great confidence that many learning-theory scientists will profit from this book. Moreover, scientists with some mathematical background will find in this account a fine introduction to the subject of learning theory.

Stephen Smale
Chicago

Preface

Broadly speaking, the goal of (mainstream) learning theory is to approximate a function (or some function features) from data samples, perhaps perturbed by noise. To attain this goal, learning theory draws on a variety of diverse subjects. It relies on statistics whose purpose is precisely to infer information from random samples. It also relies on approximation theory, since our estimate of the function must belong to a prespecified class, and therefore the ability of this class to approximate the function accurately is of the essence. And algorithmic considerations are critical because our estimate of the function is the outcome of algorithmic procedures, and the efficiency of these procedures is crucial in practice. Ideas from all these areas have blended together to form a subject whose many successful applications have triggered its rapid growth during the past two decades.

This book aims to give a general overview of the theoretical foundations of learning theory. It is not the first to do so. Yet we wish to emphasize a viewpoint that has drawn little attention in other expositions, namely, that of approximation theory. This emphasis fulfills two purposes. First, we believe it provides a balanced view of the subject. Second, we expect to attract mathematicians working on related fields who find the problems raised in learning theory close to their interests.

While writing this book, we faced a dilemma common to the writing of any book in mathematics: to strike a balance between clarity and conciseness. In particular, we faced the problem of finding a suitable degree of self-containment for a book relying on a variety of subjects. Our solution to this problem consists of a number of sections, all called "Reminders," where several basic notions and results are briefly reviewed using a unified notation.

We are indebted to several friends and colleagues who have helped us in many ways. Steve Smale deserves a special mention. We first became interested in learning theory as a result of his interest in the subject, and much of the material in this book comes from or evolved from joint papers we wrote with him. Qiang Wu, Yiming Ying, Fangyan Lu, Hongwei Sun, Di-Rong Chen, Song Li, Luoqing Li, Bingzheng Li, Lizhong Peng, and Tiangang Lei regularly attended our weekly seminars on learning theory at City University of Hong Kong, where we exposed early drafts of the contents of this book. They, and José Luis Balcázar, read preliminary versions and were very generous in their feedback. We are indebted also to David Tranah and the staff of Cambridge University Press for their patience and willingness to help. We have also been supported by the University Grants Council of Hong Kong through the grants CityU 1087/02P, 103303, and 103704.

1
The framework of learning

1.1 Introduction

We begin by describing some cases of learning, simplified to the extreme, to convey an intuition of what learning is.

Case 1.1 Among the most used instances of learning (although not necessarily with this name) is linear regression. This amounts to finding a straight line that best approximates a functional relationship presumed to be implicit in a set of data points in $\mathbb{R}^2$, $\{(x_1, y_1), (x_2, y_2), \ldots, (x_m, y_m)\}$ (Figure 1.1). The yardstick used to measure how good an approximation a given line $Y = aX + b$ is, is called *least squares*. The best line is the one that minimizes

$$Q(a, b) = \sum_{i=1}^{m} (y_i - ax_i - b)^2.$$

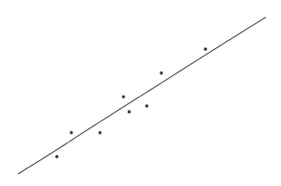

Figure 1.1

1

Case 1.2 Case 1.1 readily extends to a classical situation in science, namely, that of learning a physical law by curve fitting to data. Assume that the law at hand, an unknown function $f : \mathbb{R} \to \mathbb{R}$, has a specific form and that the space of all functions with this form can be parameterized by N real numbers. For instance, if f is assumed to be a polynomial of degree d, then $N = d + 1$ and the parameters are the unknown coefficients $w_0, \ldots, w_d$ of f. In this case, finding the *best fit* by the *least squares method* estimates the unknown f from a set of pairs $\{(x_1, y_1), \ldots, (x_m, y_m)\}$. If the measurements generating this set were exact, then y_i would be equal to $f(x_i)$. However, in general one expects the values y_i to be affected by noise. That is, $y_i = f(x_i) + \varepsilon$, where ε is a random variable (which may depend on x_i) with mean zero. One then computes the vector of coefficients w such that the value

$$\sum_{i=1}^{m} (f_w(x_i) - y_i)^2, \quad \text{with } f_w(x) = \sum_{j=0}^{d} w_j x^j$$

is minimized, where, typically, $m > N$. In general, the minimum value above is not 0. To solve this minimization problem, one uses the least squares technique, a method going back to Gauss and Legendre that is computationally efficient and relies on numerical linear algebra.

Since the values y_i are affected by noise, one might take as starting point, instead of the unknown f, a family of probability measures ε_x on $\mathbb{R}$ varying with $x \in \mathbb{R}$. The only requirement on these measures is that for all $x \in \mathbb{R}$, the mean of ε_x is $f(x)$. Then y_i is randomly drawn from ε_{x_i}. In some contexts the x_i, rather than being chosen, are also generated by a probability measure ρ_X on $\mathbb{R}$. Thus, the starting point could even be a single measure ρ on $\mathbb{R} \times \mathbb{R}$ – capturing both the measure ρ_X and the measures ε_x for $x \in \mathbb{R}$ – from which the pairs (x_i, y_i) are randomly drawn.

A more general form of the functions in our approximating class could be given by

$$f_w(x) = \sum_{i=1}^{N} w_i \phi_i(x),$$

where the ϕ_i are the elements of a basis of a specific function space, not necessarily of polynomials.

Case 1.3 The training of neural networks is an extension of Case 1.2. Roughly speaking, a neural network is a directed graph containing some input nodes, some output nodes, and some intermediate nodes where certain functions are computed. If X denotes the input space (whose elements are fed to the input nodes) and Y the output space (of possible elements returned by the output nodes), a neural network computes a function from X to Y. The literature on neural networks shows a variety of choices for X and Y, which can be continuous or discrete, as well as for the functions computed at the intermediate nodes. A common feature of all neural nets, though, is the dependence of these functions on a set of parameters, usually called *weights*, $w = \{w_j\}_{j \in J}$. This set determines the function $f_w : X \to Y$ computed by the network.

Neural networks are *trained* to learn functions. As in Case 1.2, there is a target function $f : X \to Y$, and the network is given a set of randomly chosen pairs $(x_1, y_1), \ldots, (x_m, y_m)$ in $X \times Y$. Then, training algorithms select a set of weights w attempting to minimize some distance from f_w to the target function $f : X \to Y$.

Case 1.4 A standard example of pattern recognition involves handwritten characters. Consider the problem of classifying handwritten letters of the English alphabet. Here, elements in our space X could be matrices with entries in the interval $[0, 1]$ – each entry representing a pixel in a certain gray scale of a digitized photograph of the handwritten letter or some features extracted from the letter. We may take Y to be

$$Y = \left\{ y \in \mathbb{R}^{26} \mid y = \sum_{i=1}^{26} \lambda_i e_i \text{ such that } \sum_{i=1}^{26} \lambda_i = 1 \right\}.$$

Here e_i is the ith coordinate vector in $\mathbb{R}^{26}$, each coordinate corresponding to a letter. If $\Delta \subset Y$ is the set of points y as above such that $0 \leq \lambda_i \leq 1$, for $i = 1, \ldots, 26$, one can interpret a point in Δ as a probability measure on the set $\{A, B, C, \ldots, X, Y, Z\}$. The problem is to learn the ideal function $f : X \to Y$ that associates, to a given handwritten letter x, a linear combination of the e_i with coefficients $\{\text{Prob}\{x = A\}, \text{Prob}\{x = B\}, \ldots, \text{Prob}\{x = Z\}\}$. Unambiguous letters are mapped into a coordinate vector, and in the (pure) classification problem f takes values on these e_i. "Learning f" means finding a sufficiently good approximation of f within a given prescribed class.

The approximation of f is constructed from a set of samples of handwritten letters, each of them with a label in Y. The set $\{(x_1, y_1), \ldots, (x_m, y_m)\}$ of these m samples is randomly drawn from $X \times Y$ according to a measure ρ on $X \times Y$. This measure satisfies $\rho(X \times \Delta) = 1$. In addition, in practice, it is concentrated around the set of pairs (x, y) with $y = e_i$ for some $1 \leq i \leq 26$. That is, the occurring elements $x \in X$ are handwritten letters and not, say, a digitized image of the *Mona Lisa*. The function f to be learned is the regression function f_ρ of ρ.

That is, $f_\rho(x)$ is the average of the y values of $\{x\} \times Y$ (we are more precise about ρ and the regression function in Section 1.2).

Case 1.5 A standard approach for approximating characteristic (or indicator) functions of sets is known as *PAC learning* (from "probably approximately correct"). Let T (the *target concept*) be a subset of $\mathbb{R}^n$ and ρ_X be a probability measure on $\mathbb{R}^n$ that we assume is not known in advance. Intuitively, a set $S \subset \mathbb{R}^n$ approximates T when the symmetric difference $S \Delta T = (S \setminus T) \cup (T \setminus S)$ is small, that is, has a small measure. Note that if f_S and f_T denote the characteristic functions of S and T, respectively, this measure, called the *error of* S, is $\int_{\mathbb{R}^n} |f_S - f_T| \, d\rho_X$. Note that since the functions take values in $\{0, 1\}$, only this integral coincides with $\int_{\mathbb{R}^n} (f_S - f_T)^2 \, d\rho_X$.

Let $\mathcal{C}$ be a class of subsets of $\mathbb{R}^n$ and assume that $T \in \mathcal{C}$. One strategy for constructing an approximation of T in $\mathcal{C}$ is the following. First, draw points $x_1, \ldots, x_m \in \mathbb{R}^n$ according to ρ_X and label each of them with 1 or 0 according to whether they belong to T. Second, compute any function $f_S : \mathbb{R}^n \to \{0, 1\}$, $f_S \in \mathcal{C}$, that coincides with this labeling over $\{x_1, \ldots, x_m\}$. Such a function will provide a good approximation S of T (small error with respect to ρ_X) as long as m is large enough and $\mathcal{C}$ is not too wild. Thus the measure ρ_X is used in both capacities, governing the sample drawing and measuring the error set $S \Delta T$.

A major goal in PAC learning is to estimate how large m needs to be to obtain an ε approximation of T with probability at least $1 - \delta$ as a function of ε and δ.

The situation described above is noise free since each randomly drawn point $x_i \in \mathbb{R}^n$ is correctly labeled. Extensions of PAC learning allowing for labeling mistakes with small probability exist.

Case 1.6 (Monte Carlo integration) An early instance of randomization in algorithmics appeared in numerical integration. Let $f : [0, 1]^n \to \mathbb{R}$. One way of approximating the integral $\int_{x \in [0,1]^n} f(x) \, dx$ consists of randomly drawing points $x_1, \ldots, x_m \in [0, 1]^n$ and computing

$$I_m(f) = \frac{1}{m} \sum_{i=1}^{m} f(x_i).$$

Under mild conditions on the regularity of f, $I_m(f) \to \int f$ with probability 1; that is, for all $\varepsilon > 0$,

$$\lim_{m \to \infty} \operatorname*{Prob}_{x_1, \ldots, x_m} \left\{ \left| I_m(f) - \int_{x \in [0,1]^n} f(x) \, dx \right| > \varepsilon \right\} \to 0.$$

Again we find the theme of learning an object (here a single real number, although defined in a nontrivial way through f) from a sample. In this case the measure governing the sample is known (the measure in $[0, 1]^n$ inherited from the standard Lebesgue measure on $\mathbb{R}^n$), but the same idea can be used for an unknown measure. If ρ_X is a probability measure on $X \subset \mathbb{R}^n$, a domain or manifold, $I_m(f)$ will approximate $\int_{x \in X} f(x) \, d\rho_X$ for large m with high probability as long as the points $x_1, \ldots, x_m$ are drawn from X according to the measure ρ_X. Note that no noise is involved here. An extension of this idea to include noise is, however, possible.

A common characteristic of Cases 1.2–1.5 is the existence of both an "unknown" function $f : X \to Y$ and a probability measure allowing one to randomly draw points in $X \times Y$. That measure can be on X (Case 1.5), on Y varying with $x \in X$ (Cases 1.2 and 1.3), or on the product $X \times Y$ (Case 1.4). The only requirement it satisfies is that, if for $x \in X$ a point $y \in Y$ can be randomly drawn, then the expected value of y is $f(x)$. That is, the noise is centered at zero. Case 1.6 does not follow this pattern. However, we have included it since it is a well-known algorithm and shares the flavor of learning an unknown object from random data.

The development in this book, for reasons of unity and generality, is based on a single measure on $X \times Y$. However, one should keep in mind the distinction between "inputs" $x \in X$ and "outputs" $y \in Y$.

1.2 A formal setting

Since we want to study learning from random sampling, the primary object in our development is a probability measure ρ governing the sampling that is not known in advance.

Let X be a compact metric space (e.g., a domain or a manifold in Euclidean space) and $Y = \mathbb{R}^k$. For convenience we will take $k = 1$ for the time being. Let ρ be a Borel probability measure on $Z = X \times Y$ whose regularity properties will be assumed as required. In the following we try to utilize concepts formed naturally and solely from X, Y, and ρ.

Throughout this book, if ξ is a random variable (i.e., a real-valued function on a probability space Z), we will use $\mathbf{E}(\xi)$ to denote the *expected value* (or average, or mean) of ξ and $\sigma^2(\xi)$ to denote its *variance*. Thus

$$\mathbf{E}(\xi) = \int_{z \in Z} \xi(z) \, dp \quad \text{and} \quad \sigma^2(\xi) = \mathbf{E}((\xi - \mathbf{E}(\xi))^2) = \mathbf{E}(\xi^2) - (\mathbf{E}(\xi))^2.$$

A central concept in the next few chapters is the *generalization error* (or *least squares error* or, if there is no risk of ambiguity, simply *error*) of f, for

$f : X \to Y$, defined by

$$\mathcal{E}(f) = \mathcal{E}_\rho(f) = \int_Z (f(x) - y)^2 \, d\rho.$$

For each input $x \in X$ and output $y \in Y$, $(f(x) - y)^2$ is the error incurred through the use of f as a model for the process producing y from x. This is a local error. By integrating over $X \times Y$ (w.r.t. ρ, of course) we average out this local error over all pairs (x, y). Hence the word "error" for $\mathcal{E}(f)$.

The problem posed is: *What is the f that minimizes the error $\mathcal{E}(f)$?* To answer this question we note that the error $\mathcal{E}(f)$ naturally decomposes as a sum. For every $x \in X$, let $\rho(y|x)$ be the conditional (w.r.t. x) probability measure on Y. Let also ρ_X be the marginal probability measure of ρ on X, that is, the measure on X defined by $\rho_X(S) = \rho(\pi^{-1}(S))$, where $\pi : X \times Y \to X$ is the projection. For every integrable function $\varphi : X \times Y \to \mathbb{R}$ a version of Fubini's theorem relates ρ, $\rho(y|x)$, and ρ_X as follows:

$$\int_{X \times Y} \varphi(x, y) \, d\rho = \int_X \left(\int_Y \varphi(x, y) \, d\rho(y|x) \right) d\rho_X.$$

This "breaking" of ρ into the measures $\rho(y|x)$ and ρ_X corresponds to looking at Z as a product of an input domain X and an output set Y. In what follows, unless otherwise specified, integrals are to be understood as being over ρ, $\rho(y|x)$ or ρ_X.

Define $f_\rho : X \to Y$ by

$$f_\rho(x) = \int_Y y \, d\rho(y|x).$$

The function f_ρ is called the *regression function* of ρ. For each $x \in X$, $f_\rho(x)$ is the average of the y coordinate of $\{x\} \times Y$ (in topological terms, the average of y on the fiber of x). Regularity hypotheses on ρ will induce regularity properties on f_ρ.

We will assume throughout this book that f_ρ is bounded.

Fix $x \in X$ and consider the function from Y to $\mathbb{R}$ mapping y into $(y - f_\rho(x))$. Since the expected value of this function is 0, its variance is

$$\sigma^2(x) = \int_Y (y - f_\rho(x))^2 \, d\rho(y|x).$$

Now average over X, to obtain

$$\sigma_\rho^2 = \int_X \sigma^2(x) \, d\rho_X = \mathcal{E}(f_\rho).$$

The number σ_ρ^2 is a measure of how well conditioned ρ is, analogous to the notion of condition number in numerical linear algebra.

Remark 1.7

(i) It is important to note that whereas ρ and f_ρ are generally "unknown," ρ_X is known in some situations and can even be the Lebesgue measure on X inherited from Euclidean space (as in Cases 1.2 and 1.6).
(ii) In the remainder of this book, if formulas do not make sense or ∞ appears, then the assertions where these formulas occur should be considered vacuous.

Proposition 1.8 *For every* $f : X \to Y$,

$$\mathcal{E}(f) = \int_X (f(x) - f_\rho(x))^2 \, d\rho_X + \sigma_\rho^2.$$

Proof From the definition of $f_\rho(x)$ for each $x \in X$, $\int_Y (f_\rho(x) - y) = 0$. Therefore,

$$
\begin{aligned}
\mathcal{E}(f) &= \int_Z (f(x) - f_\rho(x) + f_\rho(x) - y)^2 \\
&= \int_X (f(x) - f_\rho(x))^2 + \int_X \int_Y (f_\rho(x) - y)^2 \\
&\quad + 2 \int_X \int_Y (f(x) - f_\rho(x))(f_\rho(x) - y) \\
&= \int_X (f(x) - f_\rho(x))^2 + \sigma_\rho^2 + 2 \int_X (f(x) - f_\rho(x)) \int_Y (f_\rho(x) - y) \\
&= \int_X (f(x) - f_\rho(x))^2 + \sigma_\rho^2.
\end{aligned}
$$

∎[1]

The first term on the right-hand side of Proposition 1.8 provides an average (over X) of the error suffered from the use of f as a model for f_ρ. In addition, since σ_ρ^2 is independent of f, Proposition 1.8 implies that f_ρ has the smallest possible error among all functions $f : X \to Y$. Thus σ_ρ^2 represents a lower bound on the error $\mathcal{E}$ and it is due solely to our primary object, the measure ρ. Thus, Proposition 1.8 supports the following statement:

The goal is to "learn" (i.e., to find a good approximation of) f_ρ from random samples on Z.

---

[1] Throughout this book, the square ∎ denotes the end of a proof or the fact that no proof is given.

We now consider sampling. Let

$$\mathbf{z} \in Z^m, \quad \mathbf{z} = ((x_1, y_1), \ldots, (x_m, y_m))$$

be a *sample* in Z^m, that is, m examples independently drawn according to ρ. Here Z^m denotes the m-fold Cartesian product of Z. We define the *empirical error* of f (w.r.t. $\mathbf{z}$) to be

$$\mathcal{E}_{\mathbf{z}}(f) = \frac{1}{m} \sum_{i=1}^{m} (f(x_i) - y_i)^2.$$

If ξ is a random variable on Z, we denote the *empirical mean* of ξ (w.r.t. $\mathbf{z}$) by $\mathbf{E}_{\mathbf{z}}(\xi)$. Thus,

$$\mathbf{E}_{\mathbf{z}}(\xi) = \frac{1}{m} \sum_{i=1}^{m} \xi(z_i).$$

For any function $f : X \to Y$ we denote by f_Y the function

$$f_Y : X \times Y \to Y$$
$$(x, y) \mapsto f(x) - y.$$

With these notations we may write $\mathcal{E}(f) = \mathbf{E}(f_Y^2)$ and $\mathcal{E}_{\mathbf{z}}(f) = \mathbf{E}_{\mathbf{z}}(f_Y^2)$. We have already remarked that the expected value of $(f_\rho)_Y$ is 0; we now remark that its variance is σ_ρ^2.

Remark 1.9 Consider the PAC learning setting discussed in Case 1.5 where $X = \mathbb{R}^n$ and T is a subset of $\mathbb{R}^n$.[2] The measure ρ_X described there can be extended to a measure ρ on Z by defining, for $A \subset Z$,

$$\rho(A) = \rho_X(\{x \in X \mid (x, f_T(x)) \in A\}),$$

where, we recall, f_T is the characteristic function of the set T. The marginal measure of ρ on X is our original ρ_X. In addition, $\sigma_\rho^2 = 0$, the error $\mathcal{E}$ specializes to the error mentioned in Case 1.5, and the regression function f_ρ of ρ coincides with f_T except for a set of measure zero in X.

---

[2] Note, in this case, that X is not compact. In fact, most of the results in this book do not require compactness of X but only completeness and separability.

1.3 Hypothesis spaces and target functions

Learning processes do not take place in a vacuum. Some structure needs to be present at the beginning of the process. In our formal development, we assume that this structure takes the form of a class of functions (e.g., a space of polynomials, of splines, etc.). The goal of the learning process is thus to find the best approximation of f_ρ within this class.

Let $\mathscr{C}(X)$ be the Banach space of continuous functions on X with the norm

$$\|f\|_\infty = \sup_{x \in X} |f(x)|.$$

We consider a subset $\mathcal{H}$ of $\mathscr{C}(X)$ – in what follows called *hypothesis space* – where algorithms will work to find, as well as is possible, the best approximation for f_ρ. A main choice in this book is a compact, infinite-dimensional subset of $\mathscr{C}(X)$, but we will also consider closed balls in finite-dimensional subspaces of $\mathscr{C}(X)$ and whole linear spaces.

If $f_\rho \in \mathcal{H}$, simplifications will occur, but in general we will not even assume that $f_\rho \in \mathscr{C}(X)$ and we will have to consider a *target function* $f_\mathcal{H}$ in $\mathcal{H}$. Define $f_\mathcal{H}$ to be any function minimizing the error $\mathcal{E}(f)$ over $f \in \mathcal{H}$, namely, any optimizer of

$$\min_{f \in \mathcal{H}} \int_Z (f(x) - y)^2 \, d\rho.$$

Notice that since $\mathcal{E}(f) = \int_X (f - f_\rho)^2 + \sigma_\rho^2$, $f_\mathcal{H}$ is also an optimizer of

$$\min_{f \in \mathcal{H}} \int_X (f - f_\rho)^2 \, d\rho_X.$$

Let $\mathbf{z} \in Z^m$ be a sample. We define the *empirical target function* $f_{\mathcal{H},\mathbf{z}} = f_\mathbf{z}$ to be a function minimizing the empirical error $\mathcal{E}_\mathbf{z}(f)$ over $f \in \mathcal{H}$, that is, an optimizer of

$$\min_{f \in \mathcal{H}} \frac{1}{m} \sum_{i=1}^m (f(x_i) - y_i)^2. \tag{1.1}$$

Note that although $f_\mathbf{z}$ is not produced by an algorithm, it is close to algorithmic. The statement of the minimization problem (1.1) depends on ρ only through its dependence on $\mathbf{z}$, but once $\mathbf{z}$ is given, so is (1.1), and its solution $f_\mathbf{z}$ can be looked for without further involvement of ρ. In contrast to $f_\mathcal{H}$, $f_\mathbf{z}$ is "empirical" from its dependence on the sample $\mathbf{z}$. Note finally that $\mathcal{E}(f_\mathbf{z})$ and $\mathcal{E}_\mathbf{z}(f)$ are different objects.

We next prove that $f_\mathcal{H}$ and $f_\mathbf{z}$ exist under a mild condition on $\mathcal{H}$.

Definition 1.10 Let $f : X \to Y$ and $\mathbf{z} \in Z^m$. The *defect* of f (w.r.t. $\mathbf{z}$) is

$$L_{\mathbf{z}}(f) = L_{\rho,\mathbf{z}}(f) = \mathcal{E}(f) - \mathcal{E}_{\mathbf{z}}(f).$$

Notice that the theoretical error $\mathcal{E}(f)$ cannot be measured directly, whereas $\mathcal{E}_{\mathbf{z}}(f)$ can. A bound on $L_{\mathbf{z}}(f)$ becomes useful since it allows one to bound the actual error from an observed quantity. Such bounds are the object of Theorems 3.8 and 3.10.

Let $f_1, f_2 \in \mathscr{C}(X)$. Toward the proof of the existence of $f_{\mathcal{H}}$ and $f_{\mathbf{z}}$, we first estimate the quantity

$$|L_{\mathbf{z}}(f_1) - L_{\mathbf{z}}(f_2)|$$

linearly by $\|f_1 - f_2\|_\infty$ for almost all $\mathbf{z} \in Z^m$ (a Lipschitz estimate). We recall that a set $U \subseteq Z$ is said to be *full measure* when $Z \setminus U$ has measure zero.

Proposition 1.11 *If, for $j = 1, 2$, $|f_j(x) - y| \leq M$ on a full measure set $U \subseteq Z$ then, for all $\mathbf{z} \in U^m$,*

$$|L_{\mathbf{z}}(f_1) - L_{\mathbf{z}}(f_2)| \leq 4M \|f_1 - f_2\|_\infty.$$

Proof. First note that since

$$(f_1(x) - y)^2 - (f_2(x) - y)^2 = (f_1(x) + f_2(x) - 2y)(f_1(x) - f_2(x)),$$

we have

$$
\begin{aligned}
|\mathcal{E}(f_1) - \mathcal{E}(f_2)| &= \left| \int_Z (f_1(x) + f_2(x) - 2y)(f_1(x) - f_2(x)) \, d\rho \right| \\
&\leq \int_Z |(f_1(x) - y) + (f_2(x) - y)| \, \|f_1 - f_2\|_\infty \, d\rho \\
&\leq 2M \|f_1 - f_2\|_\infty.
\end{aligned}
$$

Also, for all $\mathbf{z} \in U^m$, we have

$$
\begin{aligned}
|\mathcal{E}_{\mathbf{z}}(f_1) - \mathcal{E}_{\mathbf{z}}(f_2)| &= \frac{1}{m} \left| \sum_{i=1}^{m} (f_1(x_i) + f_2(x_i) - 2y_i)(f_1(x_i) - f_2(x_i)) \right| \\
&\leq \frac{1}{m} \sum_{i=1}^{m} |(f_1(x_i) - y) + (f_2(x_i) - y_i)| \, \|f_1 - f_2\|_\infty \\
&\leq 2M \|f_1 - f_2\|_\infty.
\end{aligned}
$$

Thus

$$|L_z(f_1) - L_z(f_2)| = |\mathcal{E}(f_1) - \mathcal{E}_z(f_1) - \mathcal{E}(f_2) + \mathcal{E}_z(f_2)| \le 4M \|f_1 - f_2\|_\infty.$$

∎

Remark 1.12 Notice that for bounding $|\mathcal{E}_z(f_1) - \mathcal{E}_z(f_2)|$ in this proof – in contrast to the bound for $|\mathcal{E}(f_1) - \mathcal{E}(f_2)|$ – the use of the $\| \ \|_\infty$ norm is crucial. Nothing less will do.

Corollary 1.13 *Let* $\mathcal{H} \subseteq \mathscr{C}(X)$ *and* ρ *be such that, for all* $f \in \mathcal{H}$, $|f(x) - y| \le M$ *almost everywhere. Then* $\mathcal{E}, \mathcal{E}_z : \mathcal{H} \to \mathbb{R}$ *are continuous.*

Proof. The proof follows from the bounds $|\mathcal{E}(f_1) - \mathcal{E}(f_2)| \le 2M \|f_1 - f_2\|_\infty$ and $|\mathcal{E}_z(f_1) - \mathcal{E}_z(f_2)| \le 2M \|f_1 - f_2\|_\infty$ shown in the proof of Proposition 1.11. ∎

Corollary 1.14 *Let* $\mathcal{H} \subseteq \mathscr{C}(X)$ *be compact and such that for all* $f \in \mathcal{H}$, $|f(x) - y| \le M$ *almost everywhere. Then* $f_\mathcal{H}$ *and* f_z *exist.*

Proof. The proof follows from the compactness of $\mathcal{H}$ and the continuity of $\mathcal{E}, \mathcal{E}_z : \mathscr{C}(X) \to \mathbb{R}$. ∎

Remark 1.15
(i) The functions $f_\mathcal{H}$ and f_z are not necessarily unique. However, we see a uniqueness result for $f_\mathcal{H}$ in Section 3.4 when $\mathcal{H}$ is convex.
(ii) Note that the requirement of $\mathcal{H}$ to be compact is what allows Corollary 1.14 to be proved and therefore guarantees the existence of $f_\mathcal{H}$ and f_z. Other consequences (e.g., the finiteness of covering numbers) follow in subsequent chapters.

1.4 Sample, approximation, and generalization errors

For a given hypothesis space $\mathcal{H}$, the *error in* $\mathcal{H}$ of a function $f \in \mathcal{H}$ is the normalized error

$$\mathcal{E}_\mathcal{H}(f) = \mathcal{E}(f) - \mathcal{E}(f_\mathcal{H}).$$

Note that $\mathcal{E}_\mathcal{H}(f) \ge 0$ for all $f \in \mathcal{H}$ and that $\mathcal{E}_\mathcal{H}(f_\mathcal{H}) = 0$. Also note that $\mathcal{E}(f_\mathcal{H})$ and $\mathcal{E}_\mathcal{H}(f)$ are different objects.

Continuing the discussion after Proposition 1.8, it follows from our definitions and proposition that

$$\int_X (f_{\mathbf{z}} - f_\rho)^2 \, d\rho_X + \sigma_\rho^2 = \mathcal{E}(f_{\mathbf{z}}) = \mathcal{E}_\mathcal{H}(f_{\mathbf{z}}) + \mathcal{E}(f_\mathcal{H}). \qquad (1.2)$$

The quantities in (1.2) are the main characters in this book. We have already noted that σ_ρ^2 is a lower bound on the error $\mathcal{E}$ that is solely due to the measure ρ. The generalization error $\mathcal{E}(f_{\mathbf{z}})$ of $f_{\mathbf{z}}$ depends on ρ, $\mathcal{H}$, the sample $\mathbf{z}$, and the scheme (1.1) defining $f_{\mathbf{z}}$. The squared distance $\int_X (f_{\mathbf{z}} - f_\rho)^2 \, d\rho_X$ is the *excess generalization error* of $f_{\mathbf{z}}$. A goal of this book is to show that under some hypotheses on ρ and $\mathcal{H}$, this excess generalization error becomes arbitrarily small with high probability as the sample size m tends to infinity.

Now consider the sum $\mathcal{E}_\mathcal{H}(f_{\mathbf{z}}) + \mathcal{E}(f_\mathcal{H})$. The second term in this sum depends on the choice of $\mathcal{H}$ but is independent of sampling. We will call it the *approximation error*. Note that this approximation error is the sum

$$\mathcal{A}(\mathcal{H}) + \sigma_\rho^2,$$

where $\mathcal{A}(\mathcal{H}) = \int_X (f_\mathcal{H} - f_\rho)^2 \, d\rho_X$. Therefore, σ_ρ^2 is a lower bound for the approximation error.

The first term, $\mathcal{E}_\mathcal{H}(f_{\mathbf{z}})$, is called the *sample error* or *estimation error*.

Equation (1.2) thus reduces our goal above – to estimate $\int_X (f_{\mathbf{z}} - f_\rho)^2$ or, equivalently, $\mathcal{E}(f_{\mathbf{z}})$ – into two different problems corresponding to finding estimates for the sample and approximation errors. The way these problems depend on the measure ρ calls for different methods and assumptions in their analysis.

The second problem (to estimate $\mathcal{A}(\mathcal{H})$) is independent of the sample $\mathbf{z}$. But it depends heavily on the regression function f_ρ. The worse behaved f_ρ is (e.g., the more it oscillates), the more difficult it will be to approximate f_ρ well with functions in $\mathcal{H}$. Consequently, all bounds for $\mathcal{A}(\mathcal{H})$ will depend on some parameter measuring the behavior of f_ρ.

The first problem (to estimate the sample error $\mathcal{E}_\mathcal{H}(f_{\mathbf{z}})$) is posed on the space $\mathcal{H}$, and its dependence on ρ is through the sample $\mathbf{z}$. In contrast with the approximation error, it is essentially independent of f_ρ. Consequently, bounds for $\mathcal{E}_\mathcal{H}(f_{\mathbf{z}})$ will not depend on properties of f_ρ. However, due to their dependence on the random sample $\mathbf{z}$, they will hold with only a certain confidence. That is, the bound will depend on a parameter δ and will hold with a confidence of at least $1 - \delta$.

This discussion extends to some algorithmic issues. Although dependence on the behavior of f_ρ seems unavoidable in the estimates of the approximation error (and hence on the generalization error $\mathcal{E}(f_z)$ of f_z), such a dependence is undesirable in the design of the algorithmic procedures leading to f_z (e.g., the selection of $\mathcal{H}$). Ultimately, the goal is to be able, given a sample z, to select a hypothesis space $\mathcal{H}$ and compute the resulting f_z without assumptions on f_ρ and then to exhibit bounds on $\int_X (f_z - f_\rho)^2 \, d\rho_X$ that are, with high probability, reasonably good in the measure that f_ρ is well behaved. Yet, in many situations, the choice of some parameter related to the selection of $\mathcal{H}$ is performed with methods that, although satisfactory in practice, lack a proper theoretical justification. For these methods, our best theoretical results rely on information about f_ρ.

1.5 The bias–variance problem

For fixed $\mathcal{H}$ the sample error decreases when the number m of examples increases (as we see in Theorem 3.14). Fix m instead. Then, typically, the approximation error will decrease when enlarging $\mathcal{H}$, but the sample error will increase. The *bias–variance problem* consists of choosing the size of $\mathcal{H}$ when m is fixed so that the error $\mathcal{E}(f_z)$ is minimized with high probability. Roughly speaking, the "bias" of a solution f coincides with the approximation error, and its "variance" with the sample error. This is common terminology:

> A model which is too simple, or too inflexible, will have a large bias, while one which has too much flexibility in relation to the particular data set will have a large variance. Bias and variance are complementary quantities, and the best generalization [i.e. the smallest error] is obtained when we have the best compromise between the conflicting requirements of small bias and small variance.[3]

Thus, a too small space $\mathcal{H}$ will yield a large bias, whereas one that is too large will yield a large variance. Several parameters (radius of balls, dimension, etc.) determine the "size" of $\mathcal{H}$, and different instances of the bias–variance problem are obtained by fixing all of them except one and minimizing the error over this nonfixed parameter.

Failing to find a good compromise between bias and variance leads to what is called *underfitting* (large bias) or *overfitting* (large variance). As an example, consider Case 1.2 and the curve $\mathscr{C}$ in Figure 1.2(a) with the set of sample points and assume we want to approximate that curve with a polynomial of degree d (the parameter d determines in our case the dimension of $\mathcal{H}$). If d is too small, say $d = 2$, we obtain a curve as in Figure 1.2(b) which necessarily "underfits" the data points. If d is too large, we can tightly fit the data points

---

[3] [18] p. 332.

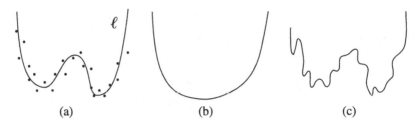

(a)   (b)   (c)

Figure 1.2

but this "overfitting" yields a curve as in Figure 1.2(c). In terms of the error decomposition (1.2) this overfitting corresponds to a small approximation error but large sample error.

As another example of overfitting, consider the PAC learning situation in Case 1.5 with $\mathscr{C}$ consisting of all subsets of $\mathbb{R}^n$. Consider also a sample $\{(x_1, 1), \ldots, (x_k, 1), (x_{k+1}, 0), \ldots, (x_m, 0)\}$. The characteristic function of the set $S = \{x_1, \ldots, x_k\}$ has zero sample error, but its approximation error is the measure (w.r.t. ρ_X) of the set $T \triangle \{x_1, \ldots, x_k\}$, which equals the measure of T as long as ρ_X has no points with positive probability mass.

1.6 The remainder of this book

In Chapter 2 we describe some common choices for the hypothesis space $\mathcal{H}$. One of them, derived from the use of reproducing kernel Hilbert spaces (RKHSs), will be systematically used in the remainder of the book.

The focus of Chapter 3 is on estimating the sample error. We want to estimate how close one may expect $f_{\mathbf{z}}$ and $f_{\mathcal{H}}$ to be, depending on the size of the sample and with a given confidence. Or, equivalently,

> *How many examples do we need to draw to assert, with a confidence greater than* $1 - \delta$, *that* $\int_X (f_{\mathbf{z}} - f_{\mathcal{H}})^2 \, d\rho_X$ *is not more than* ε?

Our main result in Chapter 3, Theorem 3.3, gives an answer.

Chapter 4 characterizes the measures ρ and some families $\{\mathcal{H}_R\}_{R>0}$ of hypothesis spaces for which $\mathcal{A}(\mathcal{H}_R)$ tends to zero with polynomial decay; that is, $\mathcal{A}(\mathcal{H}_R) = \mathcal{O}(R^{-\theta})$ for some $\theta > 0$. These families of hypothesis spaces are defined using RKHSs. Consequently, the chapter opens with several results on these spaces, including a proof of Mercer's theorem.

The bounds for the sample error in Chapter 3 are in terms of a specific measure of the size of the hypothesis space $\mathcal{H}$, namely, its covering numbers. This measure is not explicit for all of the common choices of $\mathcal{H}$. In Chapter 5 we give bounds for these covering numbers for most of the spaces $\mathcal{H}$ introduced in Chapter 2. These bounds are in terms of explicit geometric parameters of $\mathcal{H}$ (e.g., dimension, diameter, smoothness, etc.).

In Chapter 6 we continue along the lines of Chapter 4. We first show some conditions under which the approximation error can decay as $\mathcal{O}(R^{-\theta})$ only if f_ρ is $\mathscr{C}^\infty$. Then we show a polylogarithmic decay in the approximation error of hypothesis spaces defined via RKHSs for some common instances of these spaces.

Chapter 7 gives a solution to the bias–variance problem for a particular family of hypothesis spaces (and under some assumptions on f_ρ).

Chapter 8 describes a new setting, *regularization*, in which the hypothesis space is no longer required to be compact and argues some equivalence with the setting described above. In this new setting the computation of the empirical target function is algorithmically very simple. The notion of excess generalization error has a natural version, and a bound for it is exhibited.

A special case of learning is that in which Y is finite and, most particularly, when it has two elements (cf. Case 1.5). Learning problems of this kind are called *classification problems* as opposed to the ones with $Y = \mathbb{R}$, which are called *regression problems*. For classification problems it is possible to take advantage of the special structure of Y to devise learning schemes that perform better than simply specializing the schemes used for regression problems. One such scheme, known as the *support vector machine*, is described, and its error analyzed, in Chapter 9. Chapter 10 gives a detailed analysis for natural extensions of the support vector machine.

We have begun Chapters 3–10 with brief introductions. Our intention is that maybe after reading Chapter 2, a reader can form an accurate idea of the contents of this book simply by reading these introductions.

1.7 References and additional remarks

The setting described in Section 1.2 was first considered in learning theory by V. Vapnik and his collaborators. An account of Vapnik's work can be found in [134].

For the bias–variance problem in the context of learning theory see [18, 54] and the references therein.

There is a vast literature in learning theory dealing with the sample error. A pair of representative books for this topic are [6, 44].

Probably the first studies of the two terms in the error decomposition (1.2) were [96] and [36].

In this book we will not go deeper into the details of PAC learning. A standard reference for this is [67].

Other (but not all) books dealing with diverse mathematical aspects of learning theory are [7, 29, 37, 57, 59, 61, 92, 95, 107, 111, 124, 125, 132, 133, 136, 137]. In addition, a number of scientific journals publish papers on learning theory. Two devoted wholly to the theory as developed in this book are *Journal of Machine Learning Research* and *Machine Learning*.

Finally, we want to mention that the exposition and structure of this chapter largely follow [39].

2
Basic hypothesis spaces

In this chapter we describe several examples of hypothesis spaces. One of these examples (or, rather, a family of them) – a subset $\mathcal{H}$ of an RKHS – will be systematically used in the remainder of this book.

2.1 First examples of hypothesis space

Example 2.1 (Homogeneous polynomials) Let $\mathcal{H}_d = \mathcal{H}_d(\mathbb{R}^{n+1})$ be the linear space of homogeneous polynomials of degree d in $x_0, x_1, \ldots, x_n$. Let $X = S(\mathbb{R}^{n+1})$, the n-dimensional unit sphere. An element in $\mathcal{H}_d$ defines a function from X to $\mathbb{R}$ and can be written as

$$f = \sum_{|\alpha|=d} w_\alpha x^\alpha.$$

Here, $\alpha = (\alpha_0, \ldots, \alpha_n) \in \mathbb{N}^{n+1}$ is a "multi-index," $|\alpha| = \alpha_0 + \cdots + \alpha_n$, and $x^\alpha = x_0^{\alpha_0} \cdots x_n^{\alpha_n}$. Thus, $\mathcal{H}_d$ is a finite-dimensional vector space. We may consider $\mathcal{H} = \{f \in \mathcal{H}_d \mid \|f\|_\infty \leq 1\}$ to be a hypothesis space. Because of the scaling $f(\lambda x) = \lambda^d f(x)$, taking the bound $\|f\|_\infty \leq 1$ causes no loss.

Example 2.2 (Finite-dimensional function spaces) This generalizes the previous example. Let $\phi_1, \ldots, \phi_N \in \mathscr{C}(X)$ and $\mathbb{E}$ be the linear subspace of $\mathscr{C}(X)$ spanned by $\{\phi_1, \ldots, \phi_N\}$. Here we may take $\mathcal{H} = \{f \in \mathbb{E} \mid \|f\|_\infty \leq R\}$ for some $R > 0$.

The next two examples deal with infinite-dimensional linear spaces. To describe them better, we first remind the reader of a few basic notions and notations.

2.2 Reminders I

(I) We first recall some commonly used spaces of functions.

We have already defined $\mathscr{C}(X)$. Recall that this is the Banach space of bounded continuous functions on X with the norm

$$\|f\|_{\mathscr{C}(X)} = \|f\|_\infty = \sup_{x \in X} |f(x)|.$$

When $X \subseteq \mathbb{R}^n$, for $s \in \mathbb{N}$, we denote by $\mathscr{C}^s(X)$ the space of functions on X that are s times differentiable and whose sth partial derivatives $D^\alpha f$ are continuous. This is also a Banach space with the norm

$$\|f\|_{\mathscr{C}^s(X)} = \max_{|\alpha| \le s} \|D^\alpha f\|.$$

Here, $\alpha \in \mathbb{N}^n$ and $D^\alpha f$ is the partial derivative $\partial^{|\alpha|} f / \partial_1^\alpha x_1, \ldots, \partial_n^\alpha x_n$.

The space $\mathscr{C}^\infty(X)$ is the intersection of the spaces $\mathscr{C}^s(X)$ for $s \in \mathbb{N}$. We do not define any norm on $\mathscr{C}^\infty(X)$ and consider it only as a linear space.

Let ν be a Borel measure on X, $p \in [1, \infty)$, and L be the linear space of functions $f : X \to Y$ such that the integral

$$\int_X |f(x)|^p \, d\nu$$

exists. The space $\mathscr{L}_\nu^p(X)$ is defined to be the quotient of L under the equivalence relation $\equiv$ given by

$$f \equiv g \iff \int_X |f(x) - g(x)|^p \, d\nu = 0.$$

This is a Banach space with the norm

$$\|f\|_{\mathscr{L}_\nu^p(X)} = \left(\int_X |f(x)|^p \, d\nu \right)^{1/p}.$$

If $p = 2$, $\mathscr{L}_\nu^2(X)$ is actually a Hilbert space with the scalar product

$$\langle f, g \rangle_{\mathscr{L}_\nu^2(X)} = \int_X f(x)g(x) \, d\nu.$$

When there is no risk of confusion we write $\langle , \rangle_\nu$ and $\| \|_\nu$ instead of $\langle , \rangle_{\mathscr{L}_\nu^2(X)}$ and $\| \|_{\mathscr{L}_\nu^2(X)}$. In addition, when $\nu = \rho_X$ we simply write $\| \|_\rho$ instead of the more cumbersome $\| \|_{\rho_X}$.

Note that elements in $\mathscr{L}_\nu^p(X)$ are classes of functions. In general, however, one abuses language and refers to them as functions on X. For instance, we say that $f \in \mathscr{L}_\nu^p(X)$ is *continuous* when there exists a continuous function in the class of f.

The *support* of a measure ν on X is the smallest closed subset X_ν of X such that $\nu(X \setminus X_\nu) = 0$.

A function $f : X \to \mathbb{R}$ is *measurable* when, for all $\alpha \in \mathbb{R}$, the set $\{x \in X \mid f(x) \le \alpha\}$ is a Borel subset of X.

The space $\mathscr{L}_\nu^\infty(X)$ is defined to be the set of all measurable functions on X such that

$$\|f\|_{\mathscr{L}_\nu^\infty(X)} := \sup_{x \in X_\nu} |f(x)| < \infty.$$

Each element in $\mathscr{L}_\nu^\infty(X)$ is a class of functions that are identical on X_ν.

A measure ν is *finite*, when $\nu(X) < \infty$. Also, we say that ν is *nondegenerate* when, for each nonempty open subset $U \subseteq X$, $\nu(U) > 0$. Note that ν is nondegenerate if and only if $X_\nu = X$.

If ν is finite and nondegenerate, then we have a well-defined injection $\mathscr{C}(X) \hookrightarrow \mathscr{L}_\nu^p(X)$, for all $1 \le p \le \infty$.

When ν is the Lebesgue measure, we sometimes denote $\mathscr{L}_\nu^p(X)$ by $\mathscr{L}^p(X)$ or, if there is no risk of confusion, simply by $\mathscr{L}^p$.

(II) We next briefly recall some basics about the Fourier transform. The *Fourier transform* $\mathcal{F} : \mathscr{L}^1(\mathbb{R}^n) \to \mathscr{L}^1(\mathbb{R}^n)$ is defined by

$$\mathcal{F}(f)(w) = \int_{\mathbb{R}^n} e^{-iw \cdot x} f(x)\, dx.$$

The function $\mathcal{F}(f)$ is well defined and continuous on $\mathbb{R}^n$ (note, however, that $\mathcal{F}(f)$ is a complex-valued function). One major property of the Fourier transform in $\mathscr{L}^1(\mathbb{R}^n)$ is the convolution property

$$\mathcal{F}(f * g) = \mathcal{F}(f)\mathcal{F}(g),$$

where $f * g$ denotes the *convolution* of f and g defined by

$$(f * g)(x) = \int_{\mathbb{R}^n} f(x - u)g(u)\, du.$$

The extension of the Fourier transform to $\mathscr{L}^2(\mathbb{R}^n)$ requires some caution. Let $\mathscr{C}_0(\mathbb{R}^n)$ denote the space of continuous functions on $\mathbb{R}^n$ with compact support. Clearly, $\mathscr{C}_0(\mathbb{R}^n) \subset \mathscr{L}^1(\mathbb{R}^n) \cap \mathscr{L}^2(\mathbb{R}^n)$. In addition, $\mathscr{C}_0(\mathbb{R}^n)$ is dense in $\mathscr{L}^2(\mathbb{R}^n)$. Thus, for any $f \in \mathscr{L}^2(\mathbb{R}^n)$, there exists $\{\phi_k\}_{k\geq 1} \subset \mathscr{C}_0(\mathbb{R}^n)$ such that $\|\phi_k - f\| \to 0$ when $k \to \infty$. One can prove that for any such sequence (ϕ_k), the sequence $(\mathcal{F}(\phi_k))$ converges to the same element in $\mathscr{L}^2(\mathbb{R}^n)$. We denote this element by $\mathcal{F}(f)$ and we say that it is the Fourier transform of f. The notation $\widehat{f}$ instead of $\mathcal{F}(f)$ is often used.

The following result summarizes the main properties of $\mathcal{F} : \mathscr{L}^2(\mathbb{R}^n) \to \mathscr{L}^2(\mathbb{R}^n)$.

Theorem 2.3 (Plancherel's theorem) *For* $f \in \mathscr{L}^2(\mathbb{R}^n)$

(i) $\mathcal{F}(f)(w) = \lim\limits_{k\to\infty} \int_{[-k,k]^n} e^{-iw\cdot x} f(x)\, dx$, *where the convergence is for the norm in* $\mathscr{L}^2(\mathbb{R}^n)$.

(ii) $\|\mathcal{F}(f)\| = (2\pi)^{n/2}\|f\|$.

(iii) $f(x) = \lim\limits_{k\to\infty} \dfrac{1}{(2\pi)^n} \int_{[-k,k]^n} e^{iw\cdot x} \mathcal{F}(f)(w)\, dw$, *where the convergence is for the norm in* $\mathscr{L}^2(\mathbb{R}^n)$. *If* $f \in \mathscr{L}^1(\mathbb{R}^n) \cap \mathscr{L}^2(\mathbb{R}^n)$, *then the convergence holds almost everywhere.*

(iv) *The map* $\mathcal{F} : \mathscr{L}^2(\mathbb{R}^n) \to \mathscr{L}^2(\mathbb{R}^n)$ *is an isomorphism of Hilbert spaces.*

(III) Our third reminder is about compactness.

It is well known that a subset of $\mathbb{R}^n$ is compact if and only if it is closed and bounded. This is not true for subsets of $\mathscr{C}(X)$. Yet a characterization of compact subsets of $\mathscr{C}(X)$ in similar terms is still possible.

A subset S of $\mathscr{C}(X)$ is said to be *equicontinuous* at $x \in X$ when for every $\varepsilon > 0$ there exists a neighborhood V of x such that for all $y \in V$ and $f \in S$, $|f(x) - f(y)| < \varepsilon$. The set S is said to be *equicontinuous* when it is so at every x in X.

Theorem 2.4 (Arzelá–Ascoli theorem) *Let* X *be compact and* S *be a subset of* $\mathscr{C}(X)$. *Then* S *is a compact subset of* $\mathscr{C}(X)$ *if and only if* S *is closed, bounded, and equicontinuous.* ∎

The fact that every closed ball in $\mathbb{R}^n$ is compact is not true in Hilbert space. However, we will use the fact that closed balls in a Hilbert space H are *weakly compact*. That is, every sequence $\{f_n\}_{n\in\mathbb{N}}$ in a closed ball B in H has a weakly convergent subsequence $\{f_{n_k}\}_{k\in\mathbb{N}}$, or, in other words, there is some $\widetilde{f} \in B$ such that

$$\lim_{k\to\infty} \langle f_{n_k}, g \rangle = \langle \widetilde{f}, g \rangle, \quad \forall g \in H.$$

(IV) We close this section with a discussion of completely monotonic functions. This discussion is on a less general topic than the preceding contents of these reminders.

A function $f : [0, \infty) \to \mathbb{R}$ is *completely monotonic* if it is continuous on $[0, \infty)$, $\mathscr{C}^\infty$ on $(0, \infty)$, and, for all $r > 0$ and $k \geq 0$, $(-1)^k f^{(k)}(r) \geq 0$.

We will use the following characterization of completely monotonic functions.

Proposition 2.5 *A function* $f : [0, \infty) \to \mathbb{R}$ *is completely monotonic if and only if, for all* $t \in (0, \infty)$,

$$f(t) = \int_0^\infty e^{-t\sigma} d\nu(\sigma),$$

where ν is a finite Borel measure on $[0, \infty)$. ∎

2.3 Hypothesis spaces associated with Sobolev spaces

Definition 2.6 Let $J : \mathbb{E} \to \mathbb{F}$ be a linear map between the Banach spaces $\mathbb{E}$ and $\mathbb{F}$. We say that J is *bounded* when there exists $b \in \mathbb{R}$ such that for all $x \in \mathbb{E}$ with $\|x\| = 1$, $\|J(x)\| \leq b$. The *operator norm* of J is

$$\|J\| = \sup_{\|x\|=1} \|J(x)\|.$$

If J is not bounded, then we write $\|J\| = \infty$. We say that J is *compact* when the closure $\overline{J(B)}$ of $J(B)$ is compact for any bounded set $B \subset \mathbb{E}$.

Example 2.7 (Sobolev spaces) Let X be a domain in $\mathbb{R}^n$ with smooth boundary. For every $s \in \mathbb{N}$ we can define an inner product in $\mathscr{C}^\infty(X)$ by

$$\langle f, g \rangle_s = \int_X \sum_{|\alpha| \leq s} D^\alpha f D^\alpha g.$$

Here we are integrating with respect to the Lebesgue measure μ on X inherited from Euclidean space. We will denote by $\| \ \|_s$ the norm induced by $\langle \ , \ \rangle_s$. Notice that when $s = 0$, the inner product above coincides with that of $\mathscr{L}^2(X)$. That is, $\| \ \|_0 = \| \ \|$. We define the Sobolev space $H^s(X)$ to be the completion of $\mathscr{C}^\infty(X)$ with respect to the norm $\| \ \|_s$. The Sobolev embedding theorem asserts that for all $r \in \mathbb{N}$ and all $s > n/2 + r$, the inclusion

$$J_s : H^s(X) \hookrightarrow \mathscr{C}^r(X)$$

is well defined and bounded. In particular, for all $s > n/2$, the inclusion

$$J_s : H^s(X) \hookrightarrow \mathscr{C}(X)$$

is well defined and bounded. From Rellich's theorem it follows that if X is compact, this last embedding is compact as well. Thus, if B_R denotes the closed ball of radius R in $H^s(X)$ we may take $\mathcal{H}_{R,s} = \mathcal{H} = \overline{J_s(B_R)}$.

2.4 Reproducing Kernel Hilbert Spaces

Definition 2.8 Let X be a metric space. We say that $K : X \times X \to \mathbb{R}$ is *symmetric* when $K(x,t) = K(t,x)$ for all $x, t \in X$ and that it is *positive semidefinite* when for all finite sets $\mathbf{x} = \{x_1, \ldots, x_k\} \subset X$ the $k \times k$ matrix $K[\mathbf{x}]$ whose (i,j) entry is $K(x_i, x_j)$ is positive semidefinite. We say that K is a *Mercer kernel* if it is continuous, symmetric, and positive semidefinite. The matrix $K[\mathbf{x}]$ above is called the *Gramian* of K at $\mathbf{x}$.

For the remainder of this section we fix a compact metric space X and a Mercer kernel $K : X \times X \to \mathbb{R}$. Note that the positive semidefiniteness implies that $K(x,x) \geq 0$ for each $x \in X$. We define

$$\mathbf{C}_K := \sup_{x \in X} \sqrt{K(x,x)}.$$

Then

$$\mathbf{C}_K = \sup_{x,t \in X} \sqrt{|K(x,t)|}$$

since, by the positive semidefiniteness of the matrix $K[\{x,t\}]$, for all $x, t \in X$,

$$(K(x,t))^2 \leq K(x,x)K(t,t).$$

For $x \in X$, we denote by K_x the function

$$K_x : X \to \mathbb{R}$$
$$t \mapsto K(x,t).$$

The main result of this section is given in the following theorem.

Theorem 2.9 *There exists a unique Hilbert space* $(\mathcal{H}_K, \langle\,,\,\rangle_{\mathcal{H}_K})$ *of functions on X satisfying the following conditions:*

(i) *for all $x \in X$, $K_x \in \mathcal{H}_K$,*
(ii) *the span of the set $\{K_x \mid x \in X\}$ is dense in $\mathcal{H}_K$, and*
(iii) *for all $f \in \mathcal{H}_K$ and $x \in X$, $f(x) = \langle K_x, f\rangle_{\mathcal{H}_K}$.*

Moreover, $\mathcal{H}_K$ consists of continuous functions and the inclusion $I_K : \mathcal{H}_K \to \mathscr{C}(X)$ is bounded with $\|I_K\| \leq \mathbf{C}_K$.

Proof. Let H_0 be the span of the set $\{K_x \mid x \in X\}$. We define an inner product in H_0 as

$$\langle f, g\rangle = \sum_{\substack{1 \leq i \leq s \\ 1 \leq j \leq r}} \alpha_i \beta_j K(x_i, t_j), \quad \text{for} \quad f = \sum_{i=1}^{s} \alpha_i K_{x_i}, \; g = \sum_{j=1}^{r} \beta_j K_{t_j}.$$

The conditions for the inner product can be easily checked. For example, if $\langle f, f\rangle = 0$, then for each $t \in X$ the positive semidefiniteness of the Gramian of K at the subset $\{x_i\}_{i=1}^{s} \cup \{t\}$ tells us that for each $\epsilon \in \mathbb{R}$

$$\sum_{i,j=1}^{s} \alpha_i K(x_i, x_j)\alpha_j + 2\sum_{i=1}^{s} \alpha_i K(x_i, t)\epsilon + \epsilon^2 K(t, t) \geq 0.$$

However, $\sum_{i,j=1}^{s} \alpha_i K(x_i, x_j)\alpha_j = \langle f, f\rangle = 0$. By letting ϵ be arbitrarily small, we see that $f(t) = \sum_{i=1}^{s} \alpha_i K(x_i, t) = 0$. This is true for each $t \in X$; hence f is the zero function.

Let $\mathcal{H}_K$ be the completion of H_0 with the associated norm. It is easy to check that $\mathcal{H}_K$ satisfies the three conditions in the statement. We need only prove that it is unique. So, assume H is another Hilbert space of functions on X satisfying the conditions noted. We want to show that

$$H = \mathcal{H}_K \quad \text{and} \quad \langle\,,\,\rangle_H = \langle\,,\,\rangle_{\mathcal{H}_K}. \tag{2.1}$$

We first observe that $H_0 \subset H$. Also, for any $x, t \in X$, $\langle K_x, K_t\rangle_H = K(x, t) = \langle K_x, K_t\rangle_{\mathcal{H}_K}$. By linearity, for every $f, g \in H_0$, $\langle f, g\rangle_H = \langle f, g\rangle_{\mathcal{H}_K}$. Since both H and $\mathcal{H}_K$ are completions of H_0, (2.1) follows from the uniqueness of the completion.

To see the remaining assertion consider $f \in \mathcal{H}_K$ and $x \in X$. Then

$$|f(x)| = |\langle K_x, f\rangle_{\mathcal{H}_K}| \leq \|f\|_{\mathcal{H}_K}\|K_x\|_{\mathcal{H}_K} = \|f\|_{\mathcal{H}_K}\sqrt{K(x,x)}.$$

This implies that $\|f\|_\infty \leq \mathbf{C}_K\|f\|_{\mathcal{H}_K}$ and, thus, $\|I_K\| \leq \mathbf{C}_K$. Therefore, convergence in $\| \ \|_{\mathcal{H}_K}$ implies convergence in $\| \ \|_\infty$, and this shows that f is continuous since f is the limit of elements in H_0 that are continuous. ∎

In what follows, to reduce the amount of notation, we will write $\langle \, , \, \rangle_K$ instead of $\langle \, , \, \rangle_{\mathcal{H}_K}$ and $\| \ \|_K$ instead of $\| \ \|_{\mathcal{H}_K}$.

Definition 2.10 The Hilbert space $\mathcal{H}_K$ in Theorem 2.9 is said to be an *Reproducing Kernel Hilbert Space* (RKHS). Property (iii) in Theorem 2.9 is refered to as the *reproducing property*.

2.5 Some Mercer kernels

In this section we discuss some families of Mercer kernels on subsets of $\mathbb{R}^n$. In most cases, checking the symmetry and continuity of a given kernel K will be straightforward. Checking that K is positive semidefinite will be more involved.

The first family of Mercer kernels we look at is that of dot product kernels. Let $X = \{x \in \mathbb{R}^n : \|x\| \leq R\}$ be a ball of $\mathbb{R}^n$ with radius $R > 0$. A *dot product kernel* is a function $K : X \times X \to \mathbb{R}$ given by

$$K(x,y) = \sum_{d=0}^\infty a_d(x \cdot y)^d,$$

where $a_d \geq 0$ and $\sum a_d R^{2d} < \infty$.

Proposition 2.11 *Dot product kernels are Mercer kernels on X.*

Proof. The kernel K is obviously symmetric and continuous on $X \times X$. To check its positive semidefiniteness, recall that the multinomial coefficients associated with the pairs (d, α)

$$C_\alpha^d = \frac{d!}{\alpha_1! \cdots \alpha_n!}, \quad \alpha \in \mathbb{Z}_+^n, |\alpha| = d,$$

satisfy

$$(x \cdot y)^d = \sum_{|\alpha|=d} C_\alpha^d x^\alpha y^\alpha, \quad \forall x, y \in \mathbb{R}^n.$$

Let $\{x_1, \ldots, x_k\} \subset X$. Then, for all $c_1, \ldots, c_k \in \mathbb{R}$,

$$\sum_{i,j=1}^k c_i c_j K(x_i, x_j) = \sum_{d=0}^\infty a_d \sum_{|\alpha|=d} C_\alpha^d \left(\sum_{i=1}^k c_i x_i^\alpha \right)^2 \geq 0.$$

Therefore, K is a Mercer kernel. ∎

An explicit example is the linear polynomial kernel.

Example 2.12 Let X be a subset of $\mathbb{R}$ containing at least two points, and K the Mercer kernel on X given by $K(x, y) = 1 + x \cdot y$. Then $\mathcal{H}_K$ is the space of linear functions and $\{1, x\}$ forms an orthonormal basis of $\mathcal{H}_K$.

Proof. Note that for $a \in X$, K_a is the function $1 + ax$ of the variable x in $X \subseteq \mathbb{R}$. Take $a \neq b \in X$. By the definition of the inner product in $\mathcal{H}_K$,

$$\|K_a - K_b\|_K^2 = \langle K_a - K_b, K_a - K_b \rangle_K = K(a, a) - 2K(a, b) + K(b, b)$$
$$= 1 + a^2 - 2(1 + ab) + 1 + b^2 = (a - b)^2.$$

But $(K_a - K_b)(x) = (1 + ax) - (1 + bx) = (a - b)x$. So $\|K_a - K_b\|_K^2 = \|(a - b)x\|_K^2 = (a - b)^2 \|x\|_K^2$. It follows that $\|x\|_K = 1$.

In the same way,

$$\langle K_a, K_a - K_b \rangle_K = K(a, a) - K(a, b) = (1 + a^2) - (1 + ab) = a(a - b).$$

But $(K_a - K_b)(x) = (a - b)x$ and $K_a(x) = 1 + ax$. So

$$\langle K_a, K_a - K_b \rangle_K = \langle 1 + ax, (a - b)x \rangle_K = (a - b)\langle 1, x \rangle_K + a(a - b)\|x\|_K^2.$$

Since $\|x\|_K^2 = 1$, we have $(a - b)\langle 1, x \rangle_K = 0$. But $a - b \neq 0$, and hence $\langle 1, x \rangle_K = 0$.

Similarly, $\langle K_a, K_a \rangle_K = K(a, a) = 1 + a^2$ can also be written as

$$\langle 1 + ax, 1 + ax \rangle_K = \|1\|_K^2 + 2a\langle 1, x \rangle_K + a^2 \|x\|_K^2 = \|1\|_K^2 + a^2.$$

Hence $\|1\|_K = 1$. We have thus proved that $\{1, x\}$ is an orthonormal system of $\mathcal{H}_K$.

Finally, each function $K_c = 1 + cx$ with $c \in X$ is contained in span$\{1, x\}$, which is a closed subspace of $\mathcal{H}_K$. Therefore, span$\{1, x\} = \mathcal{H}_K$ and $\{1, x\}$ is an orthonormal basis of $\mathcal{H}_K$. $\blacksquare$

The above extends to the multivariate case under a slightly stronger assumption on X.

Example 2.13 Let $X \subseteq \mathbb{R}^n$ containing 0 and the coordinate vectors e_j, $j = 1, \ldots, n$. Let K be the Mercer kernel on X given by $K(x, y) = 1 + x \cdot y$. Then $\mathcal{H}_K$ is the space of linear functions and $\{1, x_1, x_2, \ldots, x_n\}$ forms an orthonormal basis of $\mathcal{H}_K$.

Proof. Note that $K(v, x) = 1 + v \cdot x = 1 + v_1 x_1 + \cdots + v_n x_n$ with $v = (v_1, \ldots, v_n) \in X \subset \mathbb{R}^n$.

We argue as in Example 2.12. For each $1 \leq j \leq n$,

$$\|K_{e_j} - K_0\|_K^2 = K(e_j, e_j) - 2K(e_j, 0) + K(0, 0) = 1.$$

But $(K_{e_j} - K_0)(x) = (1 + x_j) - 1 = x_j$, and therefore $1 = \|K_{e_j} - K_0\|_K^2 = \|x_j\|_K^2$. One can prove, similarly, that $\langle 1, x_j \rangle_K = 0$ and $\langle 1, 1 \rangle_K = 1$. Now consider $i \neq j$:

$$\langle K_{e_i}, K_{e_j} \rangle_K = K(e_i, e_j) = 1 = \langle 1 + x_i, 1 + x_j \rangle_K$$

$$= \|1\|_K^2 + \langle 1, x_i \rangle_K + \langle 1, x_j \rangle_K + \langle x_i, x_j \rangle_K.$$

Since we have shown that $\langle 1, x_i \rangle_K = \langle 1, x_j \rangle_K = 0$ and $\|1\|_K = 1$, it follows that $\langle x_i, x_j \rangle_K = 0$. We have thus proved that $\{1, x_1, x_2, \ldots, x_n\}$ is an orthonormal system of $\mathcal{H}_K$. But each function $K_v = 1 + v \cdot x$ with $v \in X$ is contained in $\mathrm{span}\{1, x_1, x_2, \ldots, x_n\}$, which is a closed subspace of $\mathcal{H}_K$. Therefore, $\mathrm{span}\{1, x_1, x_2, \ldots, x_n\} = \mathcal{H}_K$ and $\{1, x_1, x_2, \ldots, x_n\}$ is an orthonormal basis of $\mathcal{H}_K$. ∎

The second family is that of *translation invariant kernels* as given by,

$$K(x, y) = k(x - y),$$

where k is an *even* function on $\mathbb{R}^n$, that is, $k(-x) = k(x)$ for all $x \in \mathbb{R}^n$.

We say that the Fourier transform $\widehat{k}$ of k is *nonnegative* (respectively, *positive*) when it is real valued and $\widehat{k}(\xi) \geq 0$ (respectively, $\widehat{k}(\xi) > 0$) for all $\xi \in \mathbb{R}^n$.

Proposition 2.14 *Let* $k \in \mathcal{L}^2(\mathbb{R}^n)$ *be continuous and even. Suppose the Fourier transform of* k *is nonnegative. Then the kernel* $K(x, y) = k(x - y)$ *is a Mercer kernel on* $\mathbb{R}^n$ *and hence a Mercer kernel on any subset* X *of* $\mathbb{R}^n$.

Proof. We need only show the positive semidefiniteness. To do so, for any $x_1, \ldots, x_m \in \mathbb{R}^n$ and $c_1, \ldots, c_m \in \mathbb{R}$, we apply the inverse Fourier transform

$$k(x) = (2\pi)^{-n} \int_{\mathbb{R}^n} \widehat{k}(\xi) e^{ix \cdot \xi} \, d\xi$$

to get

$$\sum_{j,\ell=1}^{m} c_j c_\ell K(x_j, x_\ell) = \sum_{j,\ell=1}^{m} c_j c_\ell (2\pi)^{-n} \int_{\mathbb{R}^n} \widehat{k}(\xi) e^{ix_j \cdot \xi} e^{-ix_\ell \cdot \xi} \, d\xi$$

$$= (2\pi)^{-n} \int_{\mathbb{R}^n} \widehat{k}(\xi) \left(\sum_{j=1}^{m} c_j e^{ix_j \cdot \xi} \right) \overline{\left(\sum_{\ell=1}^{m} c_\ell e^{ix_\ell \cdot \xi} \right)} \, d\xi$$

$$= (2\pi)^{-n} \int_{\mathbb{R}^n} \widehat{k}(\xi) \left| \sum_{j=1}^{m} c_j e^{ix_j \cdot \xi} \right|^2 \, d\xi \geq 0,$$

where $| \; |$ means the module in $\mathbb{C}$ and $\bar{z}$ is the complex conjugate of z. Thus, K is a Mercer kernel on any subset of $\mathbb{R}^n$. ∎

Example 2.15 (A spline kernel) Let k be the univariate function supported on $[-2, 2]$ given by $k(x) = 1 - |x|/2$ for $-2 \leq x \leq 2$. Then the kernel K defined by $K(x, y) = k(x - y)$ is a Mercer kernel on any subset X of $\mathbb{R}$.

Proof. One can easily check that $2k(x)$ equals the convolution of the characteristic function $\chi_{[-1,1]}$ with itself. But $\widehat{\chi}_{[-1,1]}(\xi) = 2 \sin \xi / \xi$. Thus, $\widehat{k}(\xi) = 2(\sin \xi / \xi)^2 \geq 0$ and the Mercer property follows from Proposition 2.14. ∎

Remark 2.16 Note that the kernel K defined in Example 2.15 is given by

$$K(x, y) = \begin{cases} 1 - \dfrac{|x-y|}{2} & \text{if } |x - y| \leq 2 \\ 0 & \text{otherwise,} \end{cases}$$

and therefore $\mathbf{C}_K = 1$.

Multivariate splines can also be used to construct translation-invariant kernels. Take $B = [b_1 \, b_2 \, \dots \, b_q]$ to be an $n \times q$ matrix (called the *direction set*) such that $q \geq n$ and the $n \times n$ submatrix $B_0 = [b_1 \, b_2 \, \dots \, b_n]$ is invertible. Define

$$M_{B_0} = \frac{1}{|\det B_0|} \chi_{\mathrm{par}(B_0)},$$

the *normalized characteristic function* of the parallepiped

$$\mathrm{par}(B_0) := \left\{ \sum_{j=1}^{n} t_j b_j \mid |t_j| \leq \frac{1}{2}, 1 \leq j \leq n \right\}$$

spanned by the vectors $b_1, \ldots, b_n$ in $\mathbb{R}^n$. Then the (centered) *box spline* M_B can be inductively defined by

$$M_{[b_1 \, b_2 \, \ldots \, b_{n+j}]}(x) = \int_{-\frac{1}{2}}^{\frac{1}{2}} M_{[b_1 \, b_2 \, \ldots \, b_{n+j-1}]}(x - t b_{n+j}) \, dt$$

for $j = 1, \ldots, q - n$. One can check by induction that its Fourier transform satisfies

$$\widehat{M}_B(\xi) = \prod_{j=1}^{q} \frac{\sin(\xi \cdot b_j / 2)}{\xi \cdot b_j / 2}.$$

Example 2.17 (A box spline kernel) Let $B = [b_1 \, b_2 \, \ldots \, b_q]$ be an $n \times q$ matrix where $[b_1 \, b_2 \, \ldots \, b_n]$ is invertible. Choose $k(x) = (M_B * M_B)(x)$ to be the box spline with direction set $[B, B]$. Then, for all $\xi \in \mathbb{R}^n$,

$$\widehat{k}(\xi) = \prod_{j=1}^{q} \left(\frac{\sin(\xi \cdot b_j / 2)}{\xi \cdot b_j / 2} \right)^2$$

and the kernel $K(x, y) = k(x - y)$ is a Mercer kernel on any subset X of $\mathbb{R}^n$.

An interesting class of translation invariant kernels is provided by *radial basis functions*. Here the kernel takes the form $K(x, y) = f(\|x - y\|^2)$ for a univariate function f on $[0, +\infty)$. The following result allows us to verify positive semidefiniteness easily for this type of kernel.

Proposition 2.18 *Let* $X \subset \mathbb{R}^n$, $f : [0, \infty) \to \mathbb{R}$ *and* $K{:}X \times X \to \mathbb{R}$ *defined by* $K(x, y) = f(\|x - y\|^2)$. *If f is completely monotonic, then K is positive semidefinite.*

Proof. By Proposition 2.5, there is a finite Borel measure ν on $[0, \infty)$ for which

$$f(t) = \int_0^{\infty} e^{-t\sigma} \, d\nu(\sigma)$$

for all $t \in [0, \infty)$. It follows that

$$K(x, y) = f(\|x - y\|^2) = \int_0^{\infty} e^{-\sigma \|x - y\|^2} \, d\nu(\sigma).$$

Now note that for each $\sigma \in [0, \infty)$, the Fourier transform of $e^{-\sigma \|x\|^2}$ equals $(\sqrt{\pi/\sigma})^n e^{-\|\xi\|^2/4\sigma}$. Hence,

$$e^{-\sigma \|x\|^2} = (2\pi)^{-n} \int_{\mathbb{R}^n} \left(\frac{\pi}{\sigma}\right)^{\frac{n}{2}} e^{-\frac{\|\xi\|^2}{4\sigma}} e^{ix\cdot\xi} \, d\xi.$$

Therefore, reasoning as in the proof of Proposition 2.14, we have, for all $\mathbf{x} = (x_1, \ldots, x_m) \in X^m$,

$$\sum_{\ell,j=1}^{m} c_\ell c_j K(x_\ell, x_j) = \int_0^\infty (2\pi)^{-n} \int_{\mathbb{R}^n} \left(\frac{\pi}{\sigma}\right)^{\frac{n}{2}} e^{-\frac{\|\xi\|^2}{4\sigma}} \left| \sum_{j=1}^{m} c_j e^{-ix_j\cdot\xi} \right|^2 d\xi \, dv(\sigma) \geq 0.$$

∎

Corollary 2.19 *Let $c > 0$. The following functions are Mercer kernels on any subset $X \subset \mathbb{R}^n$:*

(i) *(Gaussian)* $K(x, t) = e^{-\|x-t\|^2/c^2}$.

(ii) *(Inverse multiquadrics)* $K(x, t) = (c^2 + \|x - t\|^2)^{-\alpha}$ *with $\alpha > 0$.*

Proof. Clearly, both kernels are continuous and symmetric. In (i) K is positive semidefinite by Proposition 2.18 with $f(r) = e^{-r/c^2}$. The same is true for (ii) taking $f(r) = (c^2 + r)^{-\alpha}$. ∎

Remark 2.20 The kernels of (i) and (ii) in Corollary 2.19 satisfy $C_K = 1$ and $C_K = c^{-\alpha}$, respectively.

A key example of a finite-dimensional RKHS induced by a Mercer kernel follows. Unlike in the case of the Mercer kernels of Corollary 2.19, we will not use Proposition 2.18 to show positivity.

Example 2.1 (continued) Recall that $\mathcal{H}_d = \mathcal{H}_d(\mathbb{R}^{n+1})$ is the linear space of homogeneous polynomials of degree d in $x_0, x_1, \ldots, x_n$. Its dimension (the number of coefficients of a polynomial $f \in \mathcal{H}_d$) is

$$N = \binom{n+d}{n}.$$

The number N is exponential in n and d. We notice, however, that in some situations one may consider a linear space of polynomials with a given monomial structure; that is, only a prespecified set of monomials may appear.

We can make $\mathcal{H}_d$ an inner product space by taking

$$\langle f, g \rangle_{\text{W}} = \sum_{|\alpha|=d} w_\alpha v_\alpha (C_\alpha^d)^{-1}$$

for $f, g \in \mathcal{H}_d, f = \sum w_\alpha x^\alpha, g = \sum v_\alpha x^\alpha$. This inner product, which we call the *Weyl inner product*, is natural and has an important invariance property. Let $\mathcal{O}(n+1)$ be the orthogonal group in $\mathbb{R}^{n+1}$, that is, the group of $(n+1) \times (n+1)$ real matrices whose action on $\mathbb{R}^{n+1}$ preserves the inner product on $\mathbb{R}^{n+1}$,

$$\sigma(x) \cdot \sigma(y) = x \cdot y, \quad \text{for all } x, y \in \mathbb{R}^{n+1} \text{ and all } \sigma \in \mathcal{O}(n+1).$$

The action of $\mathcal{O}(n+1)$ on $\mathbb{R}^{n+1}$ induces an action of $\mathcal{O}(n+1)$ on $\mathcal{H}_d$. For $f \in \mathcal{H}_d$ and $\sigma \in \mathcal{O}(n+1)$ we define $\sigma(f) \in \mathcal{H}_d$ by $\sigma f(x) = f(\sigma^{-1}(x))$. The invariance property of $\langle \, , \, \rangle_{\text{W}}$, called *orthogonal invariance*, is that for all $f, g \in \mathcal{H}_d$,

$$\langle \sigma(f), \sigma(g) \rangle_{\text{W}} = \langle f, g \rangle_{\text{W}}.$$

Note that if $\|f\|_{\text{W}}$ denotes the norm induced by $\langle \, , \, \rangle_{\text{W}}$, then

$$|f(x)| \le \|f\|_{\text{W}} \|x\|^d,$$

where $\|x\|$ is the standard norm of $x \in \mathbb{R}^{n+1}$. This follows from taking the action of $\sigma \in \mathcal{O}(n+1)$ such that $\sigma(x) = (\|x\|, 0, \ldots, 0)$.

Let $X = S(\mathbb{R}^{n+1})$ and

$$K : X \times X \to \mathbb{R}$$

$$(x, t) \mapsto (x \cdot t)^d.$$

Let also

$$\Phi : X \to \mathbb{R}^N$$

$$x \mapsto \left(x^\alpha (C_\alpha^d)^{1/2} \right)_{|\alpha|=d}.$$

Then, for $x, t \in X$, we have

$$\Phi(x) \cdot \Phi(t) = \sum_{|\alpha|=d} x^\alpha t^\alpha C_\alpha^d = (x \cdot t)^d = K(x, t).$$

This equality enables us to prove that K is positive semidefinite. For $t_1, \ldots, t_k \in X$, the entry in row i and column j of $K[\mathbf{t}]$ is $\Phi(t_i) \cdot \Phi(t_j)$. Therefore, if M denotes the matrix whose jth column is $\Phi(t_j)$, we have that $K[\mathbf{t}] = M^\mathsf{T}M$, from which the positivity of $K[\mathbf{t}]$ follows. Since K is clearly continuous and symmetric, we conclude that K is a Mercer kernel.

The next proposition shows the RKHS associated with K.

Proposition 2.21 $\mathcal{H}_d = \mathcal{H}_K$ *as function spaces and inner product spaces.*

Proof. We know from the proof of Theorem 2.9 that $\mathcal{H}_K$ is the completion of H_0, the span of $\{K_x \mid x \in X\}$. Since $H_0 \subseteq \mathcal{H}_d$ and $\mathcal{H}_d$ has finite dimension, the same holds for H_0. But then H_0 is complete and we deduce that

$$\mathcal{H}_K = H_0 \subseteq \mathcal{H}_d.$$

The map $\mathcal{V} : \mathbb{R}^{n+1} \to \mathbb{R}^N$ defined by $\mathcal{V}(x) = (x^\alpha)_{|\alpha|=d}$ is a well-known object in algebraic geometry, where it is called a *Veronese embedding*. We note here that the map Φ defined above is related to $\mathcal{V}$, since for every $x \in X$, $\Phi(x) = D\mathcal{V}(x)$, where D is the diagonal matrix with entries $(C_\alpha^d)^{1/2}$. The image of $\mathbb{R}^{n+1}$ by the Veronese embedding is an algebraic variety called the *Veronese variety*, which is known to be nondegenerate, that is, to span all of $\mathbb{R}^N$. This implies that $\mathcal{H}_K = \mathcal{H}_d$ as vector spaces. We now show that they are actually the same inner product space.

By definition of the inner product in H_0, for all $x, t \in X$,

$$\langle K_x, K_t \rangle_{H_0} = K(x, t) = \sum_{|\alpha|=d} C_\alpha^d x^\alpha t^\alpha.$$

On the other hand, since $K_x(w) = \sum_{|\alpha|=d} C_\alpha^d x^\alpha w^\alpha$, we know that the Weyl inner product of K_x and K_t satisfies

$$\langle K_x, K_t \rangle_\mathrm{W} = \sum_{|\alpha|=d} (C_\alpha^d)^{-1} C_\alpha^d x^\alpha C_\alpha^d t^\alpha = \sum_{|\alpha|=d} C_\alpha^d x^\alpha t^\alpha = \langle K_x, K_t \rangle_K.$$

We conclude that since the polynomials K_x span all of H_0, the inner product in $\mathcal{H}_K = H_0$ is the Weyl inner product. ∎

2.6 Hypothesis spaces associated with an RKHS

We now proceed with the last example in this chapter.

Proposition 2.22 *Let K be a Mercer kernel on a compact metric space X, and $\mathcal{H}_K$ its RKHS. For all $R > 0$, the ball $B_R := \{f \in \mathcal{H}_K : \|f\|_K \leq R\}$ is a closed subset of $\mathscr{C}(X)$.*

Proof. Suppose $\{f_n\} \subset B_R$ converges in $\mathscr{C}(X)$ to a function $f \in \mathscr{C}(X)$. Then, for all $x \in X$,

$$f(x) = \lim_{n \to \infty} f_n(x).$$

Since a closed ball of a Hilbert space is weakly compact, we have that B_R is weakly compact. Therefore, there exists a subsequence $\{f_{n_k}\}_{k \in \mathbb{N}}$ of $\{f_n\}$ and an element $\widetilde{f} \in B_R$ such that

$$\lim_{k \to \infty} \langle f_{n_k}, g \rangle_K = \langle \widetilde{f}, g \rangle_K, \quad \forall g \in \mathcal{H}_K.$$

For each $x \in X$, take $g = K_x$ to obtain

$$\lim_{k \to \infty} f_{n_k}(x) = \lim_{k \to \infty} \langle f_{n_k}, K_x \rangle_K = \langle \widetilde{f}, K_x \rangle_K = \widetilde{f}(x).$$

But $\lim_{k \to \infty} f_{n_k}(x) = f(x)$, so we have $f(x) = \widetilde{f}(x)$ for every point $x \in X$. Hence, as continuous functions on X, $f = \widetilde{f}$. Therefore, $f \in B_R$. This shows that B_R is closed as a subset of $\mathscr{C}(X)$. ∎

Proposition 2.23 *Let K be a Mercer kernel on a compact metric space X, and $\mathcal{H}_K$ be its RKHS. For all $R > 0$, the set $I_K(B_R)$ is compact.*

Proof. By the Arzelá–Ascoli theorem (Theorem 2.4) it suffices to prove that B_R is equicontinuous.

Since X is compact, so is $X \times X$. Therefore, since K is continuous on $X \times X$, K must be uniformly continuous on $X \times X$. It follows that for any $\varepsilon > 0$, there exists $\delta > 0$ such that for all $x, y, y' \in X$ with $d(y, y') \le \delta$,

$$|K(x, y) - K(x, y')| \le \varepsilon.$$

For $f \in B_R$ and $y, y' \in X$ with $d(y, y') \le \delta$, we have

$$|f(y) - f(y')| = |\langle f, K_y - K_{y'} \rangle_K| \le \|f\|_K \|K_y - K_{y'}\|_K$$
$$\le R(K(y, y) - K(y, y') + K(y', y') - K(y', y))^{1/2} \le R\sqrt{2\varepsilon}.$$

∎

Example 2.24 (Hypothesis spaces associated with an RKHS) Let X be compact and $K : X \times X \to \mathbb{R}$ be a Mercer kernel. By Proposition 2.23, for all $R > 0$ we may consider $I_K(B_R)$ to be a hypothesis space. Here and in what follows B_R denotes the closed ball of radius R centered on the origin.

2.7 Reminders II

The *general nonlinear programming problem* is the problem of finding $x \in \mathbb{R}^n$ to solve the following minimization problem:

$$\min \quad f(x)$$
$$\text{s.t.} \quad g_i(x) \leq 0, \quad i = 1, \ldots, m, \tag{2.2}$$
$$h_j(x) = 0, \quad j = 1, \ldots, p,$$

where $f, g_i, h_j : \mathbb{R}^n \to \mathbb{R}$. The function f is called the *objective function*, and the equalities and inequalities on g_i and h_j are called the *constraints*. Points $x \in \mathbb{R}^n$ satisfying the constraints are *feasible* and the subset of $\mathbb{R}^n$ of all feasible points is the *feasible set*.

Although stating this problem in all its generality leads to some conceptual clarity, it would seem that the search for an efficient algorithm to solve it is hopeless. A vast amount of research has thus focused on particular cases and the emphasis has been on those cases for which efficient algorithms exist. We do not develop here the complexity theory giving formal substance to the notion of efficiency – we do not need such a development; instead we content ourselves with understanding the notion of efficiency according to its intuitive meaning: an efficient algorithm is one that computes its outcome in a reasonably short time for reasonably long inputs. This property can be found in practice and studied in theory (via several well-developed measures of complexity available to complexity theorists).

One example of a well-studied case is *linear programming*. This is the case in which both the objective function and the constraints are linear. It is also a case in which efficient algorithms exist (and have been both used in practice and studied in theory). A much more general case for which efficient algorithms exist is that of convex programming.

A subset S of a linear space H is said to be *convex* when, for all $x, y \in S$ and all $\lambda \in [0, 1]$, $\lambda x + (1 - \lambda)y \in S$.

A function f on a convex domain S is said to be *convex* if, for all $\lambda \in [0, 1]$ and all $x, y \in S, f(\lambda x + (1 - \lambda)y) \leq \lambda f(x) + (1 - \lambda)f(y)$. If S is an interval on $\mathbb{R}$, then, for $x_0, x \in S, x_0 < x$, we have

$$f(x_0 + \lambda(x - x_0)) = f(\lambda x + (1 - \lambda)x_0) \leq \lambda f(x) + (1 - \lambda)f(x_0)$$

and

$$\frac{f(x_0 + \lambda(x - x_0)) - f(x_0)}{\lambda(x - x_0)} \leq \frac{f(x) - f(x_0)}{x - x_0}.$$

This means that the function $t \mapsto (f(t) - f(x_0))/(t - x_0)$ is increasing in the interval $[x_0, x]$. Hence, the right derivative

$$f'_+(x_0) := \lim_{t \to (x_0)_+} \frac{f(t) - f(x_0)}{t - x_0}$$

exists. In the same way we see that the left derivative

$$f'_-(x_0) := \lim_{t \to (x_0)_-} \frac{f(t) - f(x_0)}{t - x_0}$$

exists. These two derivatives, in addition, satisfy $f'_-(x_0) \le f'_+(x_0)$ whenever x_0 is a point in the interior of S. Hence, both $f'_-(x_0)$ and $f'_+(x_0)$ are nondecreasing in S.

In addition to those listed above, convex functions satisfy other properties. We highlight the fact that the addition of convex functions is convex and that if a function f is convex and $\mathscr{C}^2$ then its Hessian $D^2 f(x)$ at x is positive semidefinite for all x in its domain.

The *convex programming problem* is the problem of finding $x \in \mathbb{R}^n$ to solve (2.2) with f and g_i convex functions and h_j linear. As we have remarked, efficient algorithms for the convex programming problem exist. In particular, when f and the g_i are quadratic functions, the corresponding programming problem, called the *convex quadratic programming problem*, can be solved by even more efficient algorithms. In fact, convex quadratic programs are a particular case of second-order cone programs. And second-order cone programming today provides an example of the success of interior point methods: very large amounts of input data can be efficiently dealt with, and commercial code is available. (For references see Section 2.9).

2.8 On the computation of empirical target functions

A remarkable property of the hypothesis space $\mathcal{H} = I_K(B_R)$, where B_R is the ball of radius R in an RKHS $\mathcal{H}_K$, is the fact that the optimization problem of computing the empirical target function $f_{\mathbf{z}}$ reduces to a convex programming problem.

Let K be a Mercer kernel, and $\mathcal{H}_K$ its associated RKHS. Let $\mathbf{z} \in Z^m$. Denote by $\mathcal{H}_{K,\mathbf{z}}$ the finite-dimensional subspace of $\mathcal{H}_K$ spanned by $\{K_{x_1}, \ldots, K_{x_m}\}$ and let P be the orthogonal projection $P : \mathcal{H}_K \to \mathcal{H}_{K,\mathbf{z}}$.

Proposition 2.25 *Let $B \subseteq \mathcal{H}_K$. If $f \in \mathcal{H}_K$ is a minimizer of $\mathcal{E}_{\mathbf{z}}$ in B, then $P(f)$ is a minimizer of $\mathcal{E}_{\mathbf{z}}$ in $P(B)$, the image of B under P.*

Proof. For all $f \in B$ and all $i = 1, \ldots, m$, $\langle f, K_{x_i} \rangle_K = \langle P(f), K_{x_i} \rangle_K$. Since both f and $P(f)$ are in $\mathcal{H}_K$, the reproducing property implies that

$$f(x_i) = \langle f, K_{x_i} \rangle_K = \langle P(f), K_{x_i} \rangle_K = (P(f))(x_i).$$

It follows that $\mathcal{E}_\mathbf{z}(f) = \mathcal{E}_\mathbf{z}(P(f))$. Taking f to be a minimizer of $\mathcal{E}_\mathbf{z}$ in B proves the statement. ∎

Corollary 2.26 *Let $B \subseteq \mathcal{H}_K$ be such that $P(B) \subseteq B$. If $\mathcal{E}_\mathbf{z}$ can be minimized in B then such a minimizer can be chosen in $P(B)$.* ∎

Corollary 2.26 shows that in many situations – for example, when B is convex – the empirical target function $f_\mathbf{z}$ may be chosen in $\mathcal{H}_{K,\mathbf{z}}$. Recall from Theorem 2.9 that the norm $\| \ \|_K$ restricted to $\mathcal{H}_{K,\mathbf{z}}$ is given by

$$\left\| \sum_{i=1}^m c_i K_{x_i} \right\|^2 = \sum_{i,j=1}^m c_i K(x_i, x_j) c_j = c^\mathsf{T} K[\mathbf{x}]c.$$

Therefore, when $B = B_R$, we may take $f_\mathbf{z} = \sum_{i=1}^m c_i^* K_{x_i}$, where $c^* \in \mathbb{R}^m$ is a solution of the following problem:

$$\min \quad \frac{1}{m} \sum_{j=1}^m \left(\sum_{i=1}^m c_i K(x_i, x_j) - y_j \right)^2$$

$$\text{s.t.} \quad c^\mathsf{T} K[\mathbf{x}]c \leq R^2.$$

Note that this is a convex quadratic programming problem and, therefore, can be efficiently solved.

2.9 References and additional remarks

An exhaustive exposition of the Fourier transform can be found in [120].

For a proof of Plancherel's theorem see section 4.11 of [40]. The Arzelá–Ascoli theorem is proved, for example, in section 11.4 of [70] or section 9 of [2]. Proposition 2.5 is shown in [106] (together with a more difficult converse). For extensions to conditionally positive semidefinite kernels generated by radial basis functions see [86].

The definition of $H^s(X)$ can be extended to $s \in \mathbb{R}$, $s \geq 0$ (called *fractional Sobolev spaces*), using a Fourier transform argument [120]. We will do so in Section 5.1.

References for Sobolev space are [1, 129], and [47] for embedding theorems. A substantial amount of the theory of RKHSs was surveyed by N. Aronszajn [9]. On page 344 of this reference, Theorem 2.9, in essence, is attributed to E. H. Moore.

The special dot product kernel $K(x, y) = (c + x \cdot y)^d$ for some $c \geq 0$ and $d \in \mathbb{N}$ was introduced into the field of statistical learning theory by Vapnik (see, e.g., [134]). General dot product kernels are described in [118]; see also [101] and [79]. Spline kernels are discussed extensively in [137].

Chapter 14 of [19] is a reference for the unitary and orthogonal invariance of $\langle \, , \, \rangle_W$. A reference for the nondegeneracy of the Veronese variety mentioned in Proposition 2.21 is section 4.4 of [109].

A comprehensive introduction to convex optimization is the book [25]. For second-order cone programming see the articles [3, 119].

For more families of Mercer kernels in learning theory see [107]. More examples of box splines can be found in [41]. Reducing the computation of $f_{\mathbf{z}}$ from $\mathcal{H}_K$ to $\mathcal{H}_{K,\mathbf{z}}$ is ensured by representer theorems [137]. For a general form of these theorems see [117].

3
Estimating the sample error

The main result in this chapter provides bounds for the sample error of a compact and convex hypothesis space. We have already noted that with m fixed, the sample error increases with the size of $\mathcal{H}$. The bounds we deduce in this chapter show this behavior with respect to a particular measure for the size of $\mathcal{H}$: its capacity as measured by covering numbers.

Definition 3.1 Let S be a metric space and $\eta > 0$. We define the *covering number* $\mathcal{N}(S, \eta)$ to be the minimal $\ell \in \mathbb{N}$ such that there exist ℓ disks in S with radius η covering S. When S is compact this number is finite.

Definition 3.2 Let $M > 0$ and ρ be a probability measure on Z. We say that a set $\mathcal{H}$ of functions from X to $\mathbb{R}$ is *M-bounded* when

$$\sup_{f \in \mathcal{H}} |f(x) - y| \leq M$$

holds almost everywhere on Z.

Theorem 3.3 *Let $\mathcal{H}$ be a compact and convex subset of $\mathscr{C}(X)$. If $\mathcal{H}$ is M-bounded, then, for all $\varepsilon > 0$,*

$$\Prob_{\mathbf{z} \in Z^m} \{\mathcal{E}_{\mathcal{H}}(f_{\mathbf{z}}) \leq \varepsilon\} \geq 1 - \mathcal{N}\left(\mathcal{H}, \frac{\varepsilon}{12M}\right) \exp\left\{-\frac{m\varepsilon}{300M^2}\right\}.$$

3.1 Exponential inequalities in probability

Write the sample error $\mathcal{E}_{\mathcal{H}}(f_{\mathbf{z}}) = \mathcal{E}(f_{\mathbf{z}}) - \mathcal{E}(f_{\mathcal{H}})$ as

$$\mathcal{E}_{\mathcal{H}}(f_{\mathbf{z}}) = \mathcal{E}(f_{\mathbf{z}}) - \mathcal{E}_{\mathbf{z}}(f_{\mathbf{z}}) + \mathcal{E}_{\mathbf{z}}(f_{\mathbf{z}}) - \mathcal{E}_{\mathbf{z}}(f_{\mathcal{H}}) + \mathcal{E}_{\mathbf{z}}(f_{\mathcal{H}}) - \mathcal{E}(f_{\mathcal{H}}).$$

37

Since f_z minimizes $\mathcal{E}_z$ in $\mathcal{H}$, $\mathcal{E}_z(f_z) - \mathcal{E}_z(f_\mathcal{H}) \leq 0$. Then a bound for $\mathcal{E}_\mathcal{H}(f_z)$ follows from bounds for $\mathcal{E}(f_z) - \mathcal{E}_z(f_z)$ and $\mathcal{E}_z(f_\mathcal{H}) - \mathcal{E}(f_\mathcal{H})$. For $f : X \to \mathbb{R}$ consider a random variable ξ on Z given by $\xi(z) = (f(x) - y)^2$, where $z = (x, y) \in Z$. Then $\mathcal{E}_z(f) - \mathcal{E}(f) = \frac{1}{m} \sum_{i=1}^m \xi(z_i) - \mathbf{E}(\xi) = \mathbf{E}_z(\xi) - \mathbf{E}(\xi)$. The rate of convergence of this quantity is the subject of some well-known inequalities in probability theory.

If ξ is a nonnegative random variable and $t > 0$, then $\xi \geq \xi \chi_{\{\xi \geq t\}} \geq t \chi_{\{\xi \geq t\}}$, where χ_J denotes the characteristic function of J. Noting that $\mathrm{Prob}\{\xi \geq t\} = \mathbf{E}(\chi_{\{\xi \geq t\}})$, we obtain *Markov's inequality*,

$$\mathrm{Prob}\{\xi \geq t\} \leq \frac{\mathbf{E}(\xi)}{t}.$$

Applying Markov's inequality to $(\xi - \mathbf{E}(\xi))^2$ for an arbitrary random variable ξ yields *Chebyshev's inequality*, for any $t > 0$,

$$\mathrm{Prob}\{|\xi - \mathbf{E}(\xi)| \geq t\} = \mathrm{Prob}\{(\xi - \mathbf{E}(\xi))^2 \geq t^2\} \leq \frac{\sigma^2(\xi)}{t^2}.$$

One particular use of Chebyshev's inequality is for sums of independent random variables. If ξ is a random variable on a probability space Z with mean $\mathbf{E}(\xi) = \mu$ and variance $\sigma^2(\xi) = \sigma^2$, then, for all $\varepsilon > 0$,

$$\mathrm{Prob}_{z \in Z^m} \left\{ \left| \frac{1}{m} \sum_{i=1}^m \xi(z_i) - \mu \right| \geq \varepsilon \right\} \leq \frac{\sigma^2}{m\varepsilon^2}.$$

This inequality provides a simple form of the weak law of large numbers since it shows that when $m \to \infty$, $\frac{1}{m} \sum_{i=1}^m \xi(z_i) \to \mu$ with probability 1.

For any $0 < \delta < 1$ and by taking $\varepsilon = \sqrt{\sigma^2/(m\delta)}$ in the inequality above it follows that with confidence $1 - \delta$,

$$\left| \frac{1}{m} \sum_{i=1}^m \xi(z_i) - \mu \right| \leq \sqrt{\frac{\sigma^2}{m\delta}}. \tag{3.1}$$

The goal of this section is to extend inequality (3.1) to show a faster rate of decay. Typical bounds with confidence $1 - \delta$ will be of the form $c(\log(2/\delta)/m)^{\frac{1}{2}+\theta}$ with $0 < \theta < \frac{1}{2}$ depending on the variance of ξ. The improvement in the error is seen both in its dependence on δ – from $2/\delta$ to $\log(2/\delta)$ – and in its dependence on m – from $m^{-\frac{1}{2}}$ to $m^{-(\frac{1}{2}+\theta)}$. Note that $\{\xi_i = \xi(z_i)\}_{i=1}^m$ are independent random variables with the same mean and variance.

Proposition 3.4 (Bennett) *Let* $\{\xi_i\}_{i=1}^m$ *be independent random variables on a probability space Z with means* $\{\mu_i\}$ *and variances* $\{\sigma_i^2\}$. *Set* $\Sigma^2 := \sum_{i=1}^m \sigma_i^2$. *If for each i* $|\xi_i - \mu_i| \leq M$ *holds almost everywhere, then for every* $\varepsilon > 0$ *we have*

$$\mathrm{Prob}\left\{\sum_{i=1}^m [\xi_i - \mu_i] > \varepsilon\right\} \leq \exp\left\{-\frac{\varepsilon}{M}\left\{\left(1 + \frac{\Sigma^2}{M\varepsilon}\right)\log\left(1 + \frac{M\varepsilon}{\Sigma^2}\right) - 1\right\}\right\}.$$

Proof. Without loss of generality, we assume $\mu_i = 0$. Then the variance of ξ_i is $\sigma_i^2 = \mathbf{E}(\xi_i^2)$.

Let c be an arbitrary positive constant that will be determined later. Then

$$I := \mathrm{Prob}\left\{\sum_{i=1}^m [\xi_i - \mu_i] > \varepsilon\right\} = \mathrm{Prob}\left\{\exp\left\{\sum_{i=1}^m c\xi_i\right\} > e^{c\varepsilon}\right\}.$$

By Markov's inequality and the independence of $\{\xi_i\}$, we have

$$I \leq e^{-c\varepsilon}\mathbf{E}\left(\exp\left\{\sum_{i=1}^m c\xi_i\right\}\right) = e^{-c\varepsilon}\prod_{i=1}^m \mathbf{E}(e^{c\xi_i}). \qquad (3.2)$$

Since $|\xi_i| \leq M$ almost everywhere and $\mathbf{E}(\xi_i) = 0$, the Taylor expansion for e^x yields

$$\mathbf{E}(e^{c\xi_i}) = 1 + \sum_{\ell=2}^{+\infty} \frac{c^\ell \mathbf{E}(\xi_i^\ell)}{\ell!} \leq 1 + \sum_{\ell=2}^{+\infty} \frac{c^\ell M^{\ell-2}\sigma_i^2}{\ell!}.$$

Using $1 + t \leq e^t$, it follows that

$$\mathbf{E}(e^{c\xi_i}) \leq \exp\left\{\sum_{\ell=2}^{+\infty} \frac{c^\ell M^{\ell-2}\sigma_i^2}{\ell!}\right\} = \exp\left\{\frac{e^{cM} - 1 - cM}{M^2}\sigma_i^2\right\},$$

and therefore

$$I \leq \exp\left\{-c\varepsilon + \frac{e^{cM} - 1 - cM}{M^2}\Sigma^2\right\}.$$

Now choose the constant c to be the minimizer of the bound on the right-hand side above:

$$c = \frac{1}{M}\log\left(1 + \frac{M\varepsilon}{\Sigma^2}\right).$$

That is, $e^{cM} - 1 = M\varepsilon / \Sigma^2$. With this choice,

$$I \leq \exp\left\{-\frac{\varepsilon}{M}\left\{\left(1 + \frac{\Sigma^2}{M\varepsilon}\right)\log\left(1 + \frac{M\varepsilon}{\Sigma^2}\right) - 1\right\}\right\}.$$

This proves the desired inequality.   ∎

Let $g : [0, +\infty) \to \mathbb{R}$ be given by

$$g(\lambda) := (1 + \lambda)\log(1 + \lambda) - \lambda.$$

Then Bennett's inequality asserts that

$$\text{Prob}\left\{\sum_{i=1}^{m}[\xi_i - \mu_i] > \varepsilon\right\} \leq \exp\left\{-\frac{\Sigma^2}{M^2}g\left(\frac{M\varepsilon}{\Sigma^2}\right)\right\}. \qquad (3.3)$$

Proposition 3.5 *Let $\{\xi_i\}_{i=1}^{m}$ be independent random variables on a probability space Z with means $\{\mu_i\}$ and variances $\{\sigma_i^2\}$ and satisfying $|\xi_i(z) - \mathbf{E}(\xi_i)| \leq M$ for each i and almost all $z \in Z$. Set $\Sigma^2 := \sum_{i=1}^{m}\sigma_i^2$. Then for every $\varepsilon > 0$,*

(Generalized Bennett's inequality)

$$\text{Prob}\left\{\sum_{i=1}^{m}[\xi_i - \mu_i] > \varepsilon\right\} \leq \exp\left\{-\frac{\varepsilon}{2M}\log\left(1 + \frac{M\varepsilon}{\Sigma^2}\right)\right\}.$$

(Bernstein)

$$\text{Prob}\left\{\sum_{i=1}^{m}[\xi_i - \mu_i] > \varepsilon\right\} \leq \exp\left\{-\frac{\varepsilon^2}{2\left(\Sigma^2 + \frac{1}{3}M\varepsilon\right)}\right\}.$$

(Hoeffding)

$$\text{Prob}\left\{\sum_{i=1}^{m}[\xi_i - \mu_i] > \varepsilon\right\} \leq \exp\left\{-\frac{\varepsilon^2}{2mM^2}\right\}.$$

Proof.   The first inequality follows from (3.3) and the inequality

$$g(\lambda) \geq \frac{\lambda}{2}\log(1 + \lambda), \quad \forall \lambda \geq 0. \qquad (3.4)$$

To verify (3.4), define a $\mathscr{C}^2$ function f on $[0, \infty)$ by

$$f(\lambda) := 2\log(1 + \lambda) - 2\lambda + \lambda\log(1 + \lambda).$$

We can see that $f(0) = 0$, $f'(0) = 0$, and $f''(\lambda) = \lambda(1 + \lambda)^{-2} \geq 0$ for $\lambda \geq 0$. Hence $f(\lambda) \geq 0$ and

$$\log(1 + \lambda) - \lambda \geq -\tfrac{1}{2}\lambda\log(1 + \lambda), \quad \forall \lambda \geq 0.$$

It follows that

$$g(\lambda) = \lambda\log(1 + \lambda) + \log(1 + \lambda) - \lambda \geq \frac{\lambda}{2}\log(1 + \lambda), \quad \forall \lambda > 0.$$

This verifies (3.4) and then the generalized Bennett's inequality.

Since $g(\lambda) \geq 0$, we find that the function h defined on $[0, \infty)$ by $h(\lambda) = (6 + 2\lambda)g(\lambda) - 3\lambda^2$ satisfies similar conditions: $h(0) = h'(0) = 0$, and $h''(\lambda) = (4/(1 + \lambda))g(\lambda) \geq 0$. Hence $h(\lambda) \geq 0$ for $\lambda \geq 0$ and

$$g(\lambda) \geq \frac{3\lambda^2}{6 + 2\lambda}.$$

Applying this to (3.3), we get the proof of Bernstein's inequality.

To prove Hoeffding's inequality, we follow the proof of Proposition 3.4 and use (3.2). As the exponential function is convex and $-M \leq \xi_i \leq M$ almost surely,

$$e^{c\xi_i} \leq \frac{c\xi_i - (-cM)}{2cM}e^{cM} + \frac{cM - c\xi_i}{2cM}e^{-cM}$$

holds almost everywhere. It follows from $\mathbf{E}(\xi_i) = 0$ and the Taylor expansion for e^x that

$$\mathbf{E}\left(e^{c\xi_i}\right) \leq \frac{1}{2}e^{-cM} + \frac{1}{2}e^{cM} = \frac{1}{2}\sum_{\ell=0}^{\infty}\frac{(-cM)^\ell}{\ell!} + \frac{1}{2}\sum_{\ell=0}^{\infty}\frac{(cM)^\ell}{\ell!} = \sum_{j=0}^{\infty}\frac{(cM)^{2j}}{(2j)!}$$

$$= \sum_{j=0}^{\infty}\frac{\left((cM)^2/2\right)^j}{j!}\prod_{\ell=1}^{j}\frac{1}{2\ell - 1} \leq \sum_{j=0}^{\infty}\frac{\left((cM)^2/2\right)^j}{j!} = \exp\left\{\frac{(cM)^2}{2}\right\}.$$

This, together with (3.2), implies that $I \leq \exp\left\{-c\varepsilon + m(cM)^2/2\right\}$. Choose $c = \varepsilon/(mM^2)$. Then $I \leq \exp\left\{-\varepsilon^2/(2mM^2)\right\}$. ∎

Bounds for the distance between empirical mean and expected value follow from Proposition 3.5.

Corollary 3.6 *Let ξ be a random variable on a probability space Z with mean* $\mathbf{E}(\xi) = \mu$ *and variance* $\sigma^2(\xi) = \sigma^2$, *and satisfying* $|\xi(z) - \mathbf{E}(\xi)| \leq M$ *for almost all* $z \in Z$. *Then for all* $\varepsilon > 0$,

(Generalized Bennett)

$$\Prob_{\mathbf{z} \in Z^m} \left\{ \frac{1}{m} \sum_{i=1}^{m} \xi(z_i) - \mu \geq \varepsilon \right\} \leq \exp \left\{ -\frac{m\varepsilon}{2M} \log \left(1 + \frac{M\varepsilon}{\sigma^2} \right) \right\}.$$

(Bernstein) $\quad \Prob_{\mathbf{z} \in Z^m} \left\{ \dfrac{1}{m} \sum_{i=1}^{m} \xi(z_i) - \mu \geq \varepsilon \right\} \leq \exp \left\{ -\dfrac{m\varepsilon^2}{2 \left(\sigma^2 + \frac{1}{3} M\varepsilon \right)} \right\}.$

(Hoeffding) $\quad \Prob_{\mathbf{z} \in Z^m} \left\{ \dfrac{1}{m} \sum_{i=1}^{m} \xi(z_i) - \mu \geq \varepsilon \right\} \leq \exp \left\{ -\dfrac{m\varepsilon^2}{2M^2} \right\}.$

Proof. Apply Proposition 3.5 to the random variables $\{\xi_i = \xi(z_i)/m\}$ that satisfy $|\xi_i - \mathbf{E}(\xi_i)| \leq M/m$, $\sigma^2(\xi_i) = \sigma^2/m^2$, and $\sum \sigma_i^2 = \sigma^2/m$. ∎

Remark 3.7 Each estimate given in Corollary 3.6 is said to be a one-side probability inequality. The same bound holds true when "$\geq \varepsilon$" is replaced by "$\leq -\varepsilon$." By taking the union of these two events we obtain a two-side probability inequality stating that $\Prob_{\mathbf{z} \in Z^m} \left\{ \left| \frac{1}{m} \sum_{i=1}^{m} \xi(z_i) - \mu \right| \geq \varepsilon \right\}$ is bounded by twice the bound occurring in the corresponding one-side inequality.

Recall the definition of the defect function $L_{\mathbf{z}}(f) = \mathcal{E}(f) - \mathcal{E}_{\mathbf{z}}(f)$. Our first main result, Theorem 3.8, states a bound for $\Prob\{L_{\mathbf{z}}(f) \geq -\varepsilon\}$ for a single function $f : X \to Y$. This bound follows from Hoeffding's bound in Corollary 3.6 by taking $\xi = -f_Y^2 = -(f(x) - y)^2$ satisfying $|\xi| \leq M^2$ when f is M-bounded.

Theorem 3.8 *Let $M > 0$ and $f : X \to Y$ be M-bounded. Then, for all $\epsilon > 0$,*

$$\Prob_{\mathbf{z} \in Z^m} \{L_{\mathbf{z}}(f) \geq -\varepsilon\} \geq 1 - \exp \left\{ -\frac{m\varepsilon^2}{2M^4} \right\}. \qquad ∎$$

Remark 3.9

(i) Note that the *confidence* (i.e., the right-hand side in the inequality above) is positive and approaches 1 exponentially quickly with m.

(ii) A case implying the M-boundedness for f is the following. Define

$$M_\rho = \inf \left\{ \bar{M} \geq 0 \mid \{(x, y) \in Z : |y - f_\rho(x)| \geq \bar{M} \} \text{ has measure zero} \right\}.$$

Then take $M = P + M_\rho$, where $P \geq \|f - f_\rho\|_{\mathscr{L}_{\rho_X}^\infty} = \sup_{x \in X_{\rho_X}} |f(x) - f_\rho(x)|$.

(iii) It follows from Theorem 3.8 that for any $0 < \delta < 1$, $\mathcal{E}_{\mathbf{z}}(f) - \mathcal{E}(f) \leq M^2 \sqrt{2 \log(1/\delta)/m}$ with confidence $1 - \delta$.

3.2 Uniform estimates on the defect

The second main result in this chapter extends Theorem 3.8 to families of functions.

Theorem 3.10 *Let $\mathcal{H}$ be a compact M-bounded subset of $\mathscr{C}(X)$. Then, for all $\varepsilon > 0$,*

$$\Prob_{\mathbf{z} \in Z^m} \left\{ \sup_{f \in \mathcal{H}} L_{\mathbf{z}}(f) \leq \varepsilon \right\} \geq 1 - \mathcal{N}\left(\mathcal{H}, \frac{\varepsilon}{8M}\right) \exp\left\{ -\frac{m\varepsilon^2}{8M^4} \right\}.$$

Notice the resemblance to Theorem 3.8. The only essential difference is in the covering number, which takes into account the extension from a single f to the family $\mathcal{H}$. This has the effect of requiring the sample size m to increase accordingly to achieve the confidence level of Theorem 3.8.

Lemma 3.11 *Let $\mathcal{H} = S_1 \cup \ldots \cup S_\ell$ and $\varepsilon > 0$. Then*

$$\Prob_{\mathbf{z} \in Z^m} \left\{ \sup_{f \in \mathcal{H}} L_{\mathbf{z}}(f) \geq \varepsilon \right\} \leq \sum_{j=1}^{\ell} \Prob_{\mathbf{z} \in Z^m} \left\{ \sup_{f \in S_j} L_{\mathbf{z}}(f) \geq \varepsilon \right\}.$$

Proof. The proof follows from the equivalence

$$\sup_{f \in \mathcal{H}} L_{\mathbf{z}}(f) \geq \varepsilon \quad \Longleftrightarrow \quad \exists j \leq \ell \text{ s.t. } \sup_{f \in S_j} L_{\mathbf{z}}(f) \geq \varepsilon$$

and the fact that the probability of a union of events is bounded by the sum of the probabilities of those events. ∎

Proof of Theorem 3.10 Let $\ell = \mathcal{N}\left(\mathcal{H}, \frac{\varepsilon}{4M}\right)$ and consider $f_1, \ldots, f_\ell$ such that the disks D_j centered at f_j and with radius $\frac{\varepsilon}{4M}$ cover $\mathcal{H}$. Let U be a full measure set on which $\sup_{f \in \mathcal{H}} |f(x) - y| \leq M$. By Proposition 1.11, for all $\mathbf{z} \in U^m$ and all $f \in D_j$,

$$|L_{\mathbf{z}}(f) - L_{\mathbf{z}}(f_j)| \leq 4M \|f - f_j\|_\infty \leq 4M \frac{\varepsilon}{4M} = \varepsilon.$$

Since this holds for all $\mathbf{z} \in U^m$ and all $f \in D_j$, we get

$$\sup_{f \in D_j} L_{\mathbf{z}}(f) \geq 2\varepsilon \Rightarrow L_{\mathbf{z}}(f_j) \geq \varepsilon.$$

We conclude that for $j = 1, \ldots, \ell$,

$$\operatorname*{Prob}_{\mathbf{z} \in Z^m} \left\{ \sup_{f \in D_j} L_{\mathbf{z}}(f) \geq 2\varepsilon \right\} \leq \operatorname*{Prob}_{\mathbf{z} \in Z^m} \left\{ L_{\mathbf{z}}(f_j) \geq \varepsilon \right\} \leq \exp \left\{ -\frac{m\varepsilon^2}{2M^4} \right\},$$

where the last inequality follows from Hoeffding's bound in Corollary 3.6 for $\xi = -(f(x) - y)^2$ on Z. The statement now follows from Lemma 3.11 by replacing ε by $\frac{\varepsilon}{2}$. ∎

Remark 3.12 Hoeffding's inequality can be seen as a quantitative instance of the law of large numbers. An "abstract" uniform version of this law can be extracted from the proof of Theorem 3.10.

Proposition 3.13 *Let $\mathcal{F}$ be a family of functions from a probability space Z to $\mathbb{R}$ and d a metric on $\mathcal{F}$. Let $U \subset Z$ be of full measure and $B, \mathbf{L} > 0$ such that*

(i) $|\xi(z)| \leq B$ for all $\xi \in \mathcal{F}$ and all $z \in U$, and
(ii) $|L_{\mathbf{z}}(\xi_1) - L_{\mathbf{z}}(\xi_2)| \leq \mathbf{L}\, d(\xi_1, \xi_2)$ for all $\xi_1, \xi_2 \in \mathcal{F}$ and all $\mathbf{z} \in U^m$, where

$$L_{\mathbf{z}}(\xi) = \int_Z \xi(z) - \frac{1}{m} \sum_{i=1}^m \xi(z_i).$$

Then, for all $\varepsilon > 0$,

$$\operatorname*{Prob}_{\mathbf{z} \in Z^m} \left\{ \sup_{\xi \in \mathcal{F}} |L_{\mathbf{z}}(\xi)| \leq \varepsilon \right\} \geq 1 - \mathcal{N}\left(\mathcal{F}, \frac{\varepsilon}{2\mathbf{L}} \right) 2 \exp \left\{ -\frac{m\varepsilon^2}{8B^2} \right\}. ∎$$

3.3 Estimating the sample error

How good an approximation of $f_{\mathcal{H}}$ can we expect $f_{\mathbf{z}}$ to be? In other words, how small can we expect the sample error $\mathcal{E}_{\mathcal{H}}(f_{\mathbf{z}})$ to be? The third main result in this chapter, Theorem 3.14, gives the first answer.

Theorem 3.14 *Let $\mathcal{H}$ be a compact M-bounded subset of $\mathscr{C}(X)$. Then, for all $\varepsilon > 0$,*

$$\Prob_{\mathbf{z} \in Z^m} \{\mathcal{E}_{\mathcal{H}}(f_{\mathbf{z}}) \leq \varepsilon\} \geq 1 - \left[\mathcal{N}\left(\mathcal{H}, \frac{\varepsilon}{16M}\right) + 1 \right] \exp\left\{ -\frac{m\varepsilon^2}{32M^4} \right\}.$$

Proof. Recall that $\mathcal{E}_{\mathcal{H}}(f_{\mathbf{z}}) \leq \mathcal{E}(f_{\mathbf{z}}) - \mathcal{E}_{\mathbf{z}}(f_{\mathbf{z}}) + \mathcal{E}_{\mathbf{z}}(f_{\mathcal{H}}) - \mathcal{E}(f_{\mathcal{H}})$.

By Theorem 3.8 applied to the single function $f_{\mathcal{H}}$ and $\frac{\varepsilon}{2}$, we know that $\mathcal{E}_{\mathbf{z}}(f_{\mathcal{H}}) - \mathcal{E}(f_{\mathcal{H}}) \leq \frac{\varepsilon}{2}$ with probability at least $1 - \exp\left\{ -\frac{m\varepsilon^2}{8M^4} \right\}$.

On the other hand, Theorem 3.10 with ε replaced by $\frac{\varepsilon}{2}$ tells us that with probability at least $1 - \mathcal{N}\left(\mathcal{H}, \frac{\varepsilon}{16M}\right) \exp\left\{ -\frac{m\varepsilon^2}{32M^4} \right\}$,

$$\sup_{f \in \mathcal{H}} L_{\mathbf{z}}(f) = \sup_{f \in \mathcal{H}} \{\mathcal{E}(f) - \mathcal{E}_{\mathbf{z}}(f)\} \leq \frac{\varepsilon}{2}$$

holds, which implies in particular that $\mathcal{E}(f_{\mathbf{z}}) - \mathcal{E}_{\mathbf{z}}(f_{\mathbf{z}}) \leq \frac{\varepsilon}{2}$. Combining these two bounds, we know that $\mathcal{E}_{\mathcal{H}}(f_{\mathbf{z}}) \leq \varepsilon$ with probability at least

$$\left[1 - \mathcal{N}\left(\mathcal{H}, \frac{\varepsilon}{16M}\right) \exp\left\{ -\frac{m\varepsilon^2}{32M^4} \right\} \right] \left[1 - \exp\left\{ -\frac{m\varepsilon^2}{8M^4} \right\} \right]$$

$$\geq 1 - \left[\mathcal{N}\left(\mathcal{H}, \frac{\varepsilon}{16M}\right) + 1 \right] \exp\left\{ -\frac{m\varepsilon^2}{32M^4} \right\}.$$

This is the desired bound. ∎

Remark 3.15 Theorem 3.14 helps us deal with the question posed in Section 1.3. Given $\varepsilon, \delta > 0$, to ensure that

$$\Prob_{\mathbf{z} \in Z^m} \{\mathcal{E}_{\mathcal{H}}(f_{\mathbf{z}}) \leq \varepsilon\} \geq 1 - \delta,$$

it is sufficient that the number m of examples satisfies

$$m \geq \frac{32M^4}{\varepsilon^2} \left[\ln\left(1 + \mathcal{N}\left(\mathcal{H}, \frac{\varepsilon}{16M}\right)\right) + \ln\left(\frac{1}{\delta}\right) \right]. \tag{3.5}$$

To prove this, take $\delta = \left\{ \mathcal{N}\left(\mathcal{H}, \frac{\varepsilon}{16M}\right) + 1 \right\} \exp\left\{ -\frac{m\varepsilon^2}{32M^4} \right\}$ and solve for m. Note, furthermore, that (3.5) gives a relation between the three basic variables ε, δ, and m.

3.4 Convex hypothesis spaces

The dependency on ε in Theorem 3.14 is quadratic. Our next goal is to show that when the hypothesis space $\mathcal{H}$ is convex, this dependency is linear. This is Theorem 3.3. Its Corollary 3.17 estimates directly $\|f_{\mathbf{z}} - f_{\mathcal{H}}\|_\rho$ as well.

Toward the proof of Theorem 3.3, we show an additional property of convex hypothesis spaces. From the discussion in Section 1.3 it follows that for a convex $\mathcal{H}$, there exists a function $f_{\mathcal{H}}$ in $\mathcal{H}$ whose distance in $\mathscr{L}^2_\rho(X)$ to f_ρ is minimal. We next prove that if $\mathcal{H}$ is convex and ρ_X is nondegenerate, then $f_{\mathcal{H}}$ is unique.

Lemma 3.16 *Let $\mathcal{H}$ be a convex subset of $\mathscr{C}(X)$ such that $f_{\mathcal{H}}$ exists. Then $f_{\mathcal{H}}$ is unique as an element in $\mathscr{L}^2_\rho(X)$ and, for all $f \in \mathcal{H}$,*

$$\int_X (f_{\mathcal{H}} - f)^2 \leq \mathcal{E}_{\mathcal{H}}(f).$$

In particular, if ρ_X is not degenerate, then $f_{\mathcal{H}}$ is unique in $\mathcal{H}$.

Proof. Let $s = \overline{f_{\mathcal{H}}f}$ be the segment of line with extremities $f_{\mathcal{H}}$ and f: Since $\mathcal{H}$ is convex, $s \subset \mathcal{H}$. And, since $f_{\mathcal{H}}$ minimizes the distance in $\mathscr{L}^2_\rho(X)$ to

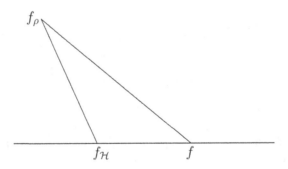

f_ρ over $\mathcal{H}$, we have that for all $g \in s$, $\|f_{\mathcal{H}} - f_\rho\|_\rho \leq \|g - f_\rho\|_\rho$. This means that for each $t \in [0, 1]$,

$$\|f_{\mathcal{H}} - f_\rho\|^2_\rho \leq \|tf + (1 - t)f_{\mathcal{H}} - f_\rho\|^2_\rho$$
$$= \|f_{\mathcal{H}} - f_\rho\|^2_\rho + 2t \left\{ \langle f - f_{\mathcal{H}}, f_{\mathcal{H}} - f_\rho \rangle_\rho + \frac{t}{2} \|f - f_{\mathcal{H}}\|^2_\rho \right\}.$$

By taking t to be small enough, we see that $\langle f - f_{\mathcal{H}}, f_{\mathcal{H}} - f_\rho \rangle_\rho \geq 0$. That is, the angle $\widehat{f_\rho f_{\mathcal{H}} f}$ is obtuse, which implies (note that the squares are crucial)

$$\|f_{\mathcal{H}} - f\|^2_\rho \leq \|f - f_\rho\|^2_\rho - \|f_{\mathcal{H}} - f_\rho\|^2_\rho;$$

that is,

$$\int_X (f_{\mathcal{H}} - f)^2 \le \mathcal{E}(f) - \mathcal{E}(f_{\mathcal{H}}) = \mathcal{E}_{\mathcal{H}}(f).$$

This proves the desired inequality. The uniqueness of $f_{\mathcal{H}}$ follows by considering the line segment joining two minimizers $f'_{\mathcal{H}}$ and $f''_{\mathcal{H}}$. Reasoning as above, one can show that both angles $\widehat{f_\rho f'_{\mathcal{H}} f''_{\mathcal{H}}}$ and $\widehat{f_\rho f''_{\mathcal{H}} f'_{\mathcal{H}}}$ are obtuse. This is possible only if $f''_{\mathcal{H}} = f'_{\mathcal{H}}$. ∎

Corollary 3.17 *With the hypotheses of Theorem 3.3, for all $\varepsilon > 0$,*

$$\Prob_{\mathbf{z} \in Z^m} \left\{ \int (f_{\mathbf{z}} - f_{\mathcal{H}})^2 \le \varepsilon \right\} \ge 1 - \mathcal{N}\left(\mathcal{H}, \frac{\varepsilon}{12M}\right) \exp\left\{-\frac{m\varepsilon}{300M^2}\right\}. \quad ∎$$

Now, in addition to convexity and M-boundedness, assume that $\mathcal{H}$ is a compact subset of $\mathscr{C}(X)$, so that the covering numbers $\mathcal{N}(\mathcal{H}, \eta)$ make sense and are finite. The main stepping stone toward the proof of Theorem 3.3 is a ratio probability inequality.

We first give such an inequality for a single random variable.

Lemma 3.18 *Suppose a random variable ξ on Z satisfies $\mathbf{E}(\xi) = \mu \ge 0$, and $|\xi - \mu| \le B$ almost everywhere. If $\mathbf{E}(\xi^2) \le c\mathbf{E}(\xi)$, then, for every $\varepsilon > 0$ and $0 < \alpha \le 1$,*

$$\Prob_{\mathbf{z} \in Z^m} \left\{ \frac{\mu - \frac{1}{m}\sum_{i=1}^m \xi(z_i)}{\sqrt{\mu + \varepsilon}} > \alpha\sqrt{\varepsilon} \right\} \le \exp\left\{ -\frac{\alpha^2 m\varepsilon}{2c + \frac{2}{3}B} \right\}$$

holds.

Proof. Since ξ satisfies $|\xi - \mu| \le B$, the one-side Bernstein inequality in Corollary 3.6 implies that

$$\Prob_{\mathbf{z} \in Z^m} \left\{ \frac{\mu - \frac{1}{m}\sum_{i=1}^m \xi(z_i)}{\sqrt{\mu + \varepsilon}} > \alpha\sqrt{\varepsilon} \right\} \le \exp\left\{ -\frac{\alpha^2 m(\mu + \varepsilon)\varepsilon}{2\left(\sigma^2(\xi) + \frac{1}{3}B\alpha\sqrt{\mu + \varepsilon}\sqrt{\varepsilon}\right)} \right\}.$$

Here $\sigma^2(\xi) \le \mathbf{E}(\xi^2) \le c\mathbf{E}(\xi) = c\mu$. Then we find that

$$\sigma^2(\xi) + \frac{1}{3}B\alpha\sqrt{\mu + \varepsilon}\sqrt{\varepsilon} \le c\mu + \frac{1}{3}B(\mu + \varepsilon) \le \left(c + \frac{B}{3}\right)(\mu + \varepsilon).$$

This yields the desired inequality. ∎

We next give a ratio probability inequality involving a set of functions.

Lemma 3.19 *Let $\mathcal{G}$ be a set of functions on Z and $c > 0$ such that for each $g \in \mathcal{G}$, $\mathbf{E}(g) \geq 0$, $\mathbf{E}(g^2) \leq c\mathbf{E}(g)$, and $|g - \mathbf{E}(g)| \leq B$ almost everywhere. Then, for every $\varepsilon > 0$ and $0 < \alpha \leq 1$, we have*

$$\mathop{\mathrm{Prob}}_{\mathbf{z} \in Z^m} \left\{ \sup_{g \in \mathcal{G}} \frac{\mathbf{E}(g) - \mathbf{E}_{\mathbf{z}}(g)}{\sqrt{\mathbf{E}(g) + \varepsilon}} > 4\alpha\sqrt{\varepsilon} \right\} \leq \mathcal{N}(\mathcal{G}, \alpha\varepsilon) \exp \left\{ -\frac{\alpha^2 m\varepsilon}{2c + \frac{2}{3}B} \right\}.$$

Proof. Let $\{g_j\}_{j=1}^{J} \subset \mathcal{G}$ with $J = \mathcal{N}(\mathcal{G}, \alpha\varepsilon)$ be such that $\mathcal{G}$ is covered by balls in $\mathscr{C}(Z)$ centered on g_j with radius $\alpha\varepsilon$.

Applying Lemma 3.18 to $\xi = g_j$ for each j, we have

$$\mathop{\mathrm{Prob}}_{\mathbf{z} \in Z^m} \left\{ \frac{\mathbf{E}(g_j) - \mathbf{E}_{\mathbf{z}}(g_j)}{\sqrt{\mathbf{E}(g_j) + \varepsilon}} \geq \alpha\sqrt{\varepsilon} \right\} \leq \exp \left\{ -\frac{\alpha^2 m\varepsilon}{2c + \frac{2}{3}B} \right\}.$$

For each $g \in \mathcal{G}$, there is some j such that $\|g - g_j\|_{\mathscr{C}(Z)} \leq \alpha\varepsilon$. Then $|\mathbf{E}_{\mathbf{z}}(g) - \mathbf{E}_{\mathbf{z}}(g_j)|$ and $|\mathbf{E}(g) - \mathbf{E}(g_j)|$ are both bounded by $\alpha\varepsilon$. Hence

$$\frac{|\mathbf{E}_{\mathbf{z}}(g) - \mathbf{E}_{\mathbf{z}}(g_j)|}{\sqrt{\mathbf{E}(g) + \varepsilon}} \leq \alpha\sqrt{\varepsilon} \quad \text{and} \quad \frac{|\mathbf{E}(g) - \mathbf{E}(g_j)|}{\sqrt{\mathbf{E}(g) + \varepsilon}} \leq \alpha\sqrt{\varepsilon}.$$

The latter implies that

$$\mathbf{E}(g_j) + \varepsilon = \mathbf{E}(g_j) - \mathbf{E}(g) + \mathbf{E}(g) + \varepsilon \leq \alpha\sqrt{\varepsilon}\sqrt{\mathbf{E}(g) + \varepsilon} + (\mathbf{E}(g) + \varepsilon)$$

$$\leq \sqrt{\varepsilon}\sqrt{\mathbf{E}(g) + \varepsilon} + (\mathbf{E}(g) + \varepsilon) \leq 2(\mathbf{E}(g) + \varepsilon).$$

It follows that $\sqrt{\mathbf{E}(g_j) + \varepsilon} \leq 2\sqrt{\mathbf{E}(g) + \varepsilon}$. We have thus seen that $(\mathbf{E}(g) - \mathbf{E}_{\mathbf{z}}(g))/\sqrt{\mathbf{E}(g) + \varepsilon} \geq 4\alpha\sqrt{\varepsilon}$ implies $(\mathbf{E}(g_j) - \mathbf{E}_{\mathbf{z}}(g_j))/\sqrt{\mathbf{E}(g) + \varepsilon} \geq 2\alpha\sqrt{\varepsilon}$ and hence $(\mathbf{E}(g_j) - \mathbf{E}_{\mathbf{z}}(g_j))/\sqrt{\mathbf{E}(g_j) + \varepsilon} \geq \alpha\sqrt{\varepsilon}$. Therefore,

$$\mathop{\mathrm{Prob}}_{\mathbf{z} \in Z^m} \left\{ \sup_{g \in \mathcal{G}} \frac{\mathbf{E}(g) - \mathbf{E}_{\mathbf{z}}(g)}{\sqrt{\mathbf{E}(g) + \varepsilon}} \geq 4\alpha\sqrt{\varepsilon} \right\} \leq \sum_{j=1}^{J} \mathop{\mathrm{Prob}}_{\mathbf{z} \in Z^m} \left\{ \frac{\mathbf{E}(g_j) - \mathbf{E}_{\mathbf{z}}(g_j)}{\sqrt{\mathbf{E}(g_j) + \varepsilon}} \geq \alpha\sqrt{\varepsilon} \right\},$$

which is bounded by $J \cdot \exp \left\{ -\alpha^2 m\varepsilon / (2c + \frac{2}{3}B) \right\}$. ∎

We are in a position to prove Theorem 3.3.

Proof of Theorem 3.3   Consider the function set

$$\mathcal{G} = \left\{ (f(x) - y)^2 - (f_{\mathcal{H}}(x) - y)^2 : f \in \mathcal{H} \right\}.$$

Each function g in $\mathcal{G}$ satisfies $\mathbf{E}(g) = \mathcal{E}_{\mathcal{H}}(f) \geq 0$. Since $\mathcal{H}$ is M-bounded, we have $-M^2 \leq g(z) \leq M^2$ almost everywhere. It follows that $|g - \mathbf{E}(g)| \leq B :=$ $2M^2$ almost everywhere. Observe that

$$g(z) = (f(x) - f_{\mathcal{H}}(x)) [(f(x) - y) + (f_{\mathcal{H}}(x) - y)], \quad z = (x, y) \in Z.$$

It follows that $|g(z)| \leq 2M |f(x) - f_{\mathcal{H}}(x)|$ and $\mathbf{E}(g^2) \leq 4M^2 \int_X (f - f_{\mathcal{H}})^2$. Taken together with Lemma 3.16, this implies $\mathbf{E}(g^2) \leq 4M^2 \mathcal{E}_{\mathcal{H}}(f) = c\mathbf{E}(g)$ with $c = 4M^2$. Thus, all the conditions in Lemma 3.19 hold true and we can draw the following conclusion from the identity $\mathbf{E}_{\mathbf{z}}(g) = \mathcal{E}_{\mathcal{H},\mathbf{z}}(f)$: for every $\varepsilon > 0$ and $0 < \alpha \leq 1$, with probability at least

$$1 - \mathcal{N}(\mathcal{G}, \alpha\varepsilon) \exp\left\{-\frac{\alpha^2 m\varepsilon}{8M^2 + \frac{2}{3}2M^2}\right\},$$

$$\sup_{f \in \mathcal{H}} \frac{\mathcal{E}_{\mathcal{H}}(f) - \mathcal{E}_{\mathcal{H},\mathbf{z}}(f)}{\sqrt{\mathcal{E}_{\mathcal{H}}(f) + \varepsilon}} \leq 4\alpha\sqrt{\varepsilon}$$

holds, and, therefore, for all $f \in \mathcal{H}$, $\mathcal{E}_{\mathcal{H}}(f) \leq \mathcal{E}_{\mathcal{H},\mathbf{z}}(f) + 4\alpha\sqrt{\varepsilon}\sqrt{\mathcal{E}_{\mathcal{H}}(f) + \varepsilon}$. Take $\alpha = \sqrt{2}/8$ and $f = f_{\mathbf{z}}$. Since $\mathcal{E}_{\mathcal{H},\mathbf{z}}(f_{\mathbf{z}}) \leq 0$ by definition of $f_{\mathbf{z}}$, we have

$$\mathcal{E}_{\mathcal{H}}(f_{\mathbf{z}}) \leq \sqrt{\varepsilon/2}\sqrt{\mathcal{E}_{\mathcal{H}}(f_{\mathbf{z}}) + \varepsilon}.$$

Solving the quadratic equation about $\sqrt{\mathcal{E}_{\mathcal{H}}(f_{\mathbf{z}})}$, we have $\mathcal{E}_{\mathcal{H}}(f_{\mathbf{z}}) \leq \varepsilon$.

Finally, by the inequality $\|g_1 - g_2\|_{\mathscr{C}(Z)} \leq \|(f_1(x) - f_2(x)) [(f_1(x) - y) + (f_2(x) - y)]\|_{\mathscr{C}(Z)} \leq 2M \|f_1 - f_2\|_{\mathscr{C}(X)}$, it follows that

$$\mathcal{N}(\mathcal{G}, \alpha\varepsilon) \leq \mathcal{N}\left(\mathcal{H}, \frac{\alpha\varepsilon}{2M}\right).$$

The desired inequality now follows by taking $\alpha = \sqrt{2}/8$. ∎

Remark 3.20 Note that to obtain Theorem 3.3, convexity was only used in the proof of Lemma 3.16. But the inequality proved in this lemma may hold true in other situations as well. A case that stands out is when $f_\rho \in \mathcal{H}$. In this case $f_{\mathcal{H}} = f_\rho$ and the inequality in Lemma 3.16 is trivial.

3.5 References and additional remarks

The exposition in this chapter largely follows [39]. The derivation of Theorem 3.3 deviates from that paper – hence the slightly different constants.

The probability inequalities given in Section 3.1 are standard in the literature on the law of large numbers or central limit theorems (e.g., [21, 103, 132]).

There is a vast literature on further extensions of the inequalities in Section 3.2 stated in terms of empirical covering numbers and other capacity measures [13, 71] (called *concentration inequalities*) that is outside the scope of this book. We mention the McDiarmid inequality [83], the Talagrand inequality [126, 22], and probability inequalities in Banach spaces [99].

The inequalities in Section 3.4 are improvements of those in the Vapnik–Chervonenkis theory [135]. In particular, Lemmas 3.18 and 3.19 are a covering number version of an inequality given by Anthony and Shawe-Taylor [8]. The convexity of the hypothesis space plays a central role in improving the sample error bounds, as in Theorem 3.3. This can be seen in [10, 12, 74].

A natural question about the sample error is whether upper bounds such as those in Theorem 3.3 are tight. In this regard, lower bounds called *minimax rates of convergence* can be obtained (see, e.g., [148]).

The ideas described in this chapter can be developed for a more general class of learning algorithms known as *empirical risk minimization* (ERM) or *structural risk minimization algorithms* [17, 60, 110]. We devote the remainder of this section to briefly describe some aspects of this development.

The greater generality of ERM comes from the fact that algorithms in this class minimize empirical errors with respect to a loss function $\psi : \mathbb{R} \to \mathbb{R}_+$. The loss function measures how the sample value y approximates the function value $f(x)$ by evaluating $\psi(y - f(x))$.

Definition 3.21 We say that $\psi : \mathbb{R} \to \mathbb{R}_+$ is a *regression loss function* if it is even, convex, and continuous and $\psi(0) = 0$.

For $(x, y) \in Z$, the value $\psi(y - f(x))$ is the local error suffered from the use of f as a model for the process producing y at x. The condition $\psi(0) = 0$ ensures a zero error when $y = f(x)$. Examples of regression loss functions include the least squares loss and Vapnik's ϵ-insensitive norm.

Example 3.22 The *least squares loss* corresponds to the loss function $\psi(t) = t^2$. For $\epsilon > 0$, the ϵ-insensitive norm is the loss function defined by

$$\psi(t) = \psi_\epsilon(t) = \begin{cases} |t| - \epsilon & \text{if } |t| > \epsilon \\ 0 & \text{otherwise.} \end{cases}$$

Given a regression loss function ψ one defines its associated *generalization error* by

$$\mathcal{E}^{\psi}(f) = \int_Z \psi(y - f(x)) \, d\rho$$

and, given $\mathbf{z} \in Z^m$ as well, its associated *empirical error* by

$$\mathcal{E}_{\mathbf{z}}^{\psi}(f) = \frac{1}{m} \sum_{i=1}^{m} \psi(y_i - f(x_i)).$$

As in our development, given a hypothesis class $\mathcal{H}$, these errors allow one to define a *target function* $f_{\mathcal{H}}^{\psi}$ and empirical target function $f_{\mathbf{z}}^{\psi}$ and to derive a decomposition bounding the excess generalization error

$$\mathcal{E}^{\psi}(f_{\mathbf{z}}^{\psi}) - \mathcal{E}^{\psi}(f_{\mathcal{H}}^{\psi}) \leq \left\{ \mathcal{E}^{\psi}(f_{\mathbf{z}}^{\psi}) - \mathcal{E}_{\mathbf{z}}^{\psi}(f_{\mathbf{z}}^{\psi}) \right\} + \left\{ \mathcal{E}_{\mathbf{z}}^{\psi}(f_{\mathcal{H}}^{\psi}) - \mathcal{E}^{\psi}(f_{\mathcal{H}}^{\psi}) \right\}.$$
$$(3.6)$$

The second term on the right-hand side of (3.6) converges to zero, with high probability when $m \to \infty$, and its convergence rate can be estimated by standard probability inequalities.

The first term on the right-hand side of (3.6) is more involved. If one writes $\xi_{\mathbf{z}}(z) = \psi(y - f_{\mathbf{z}}^{\psi}(x))$, then $\mathcal{E}^{\psi}(f_{\mathbf{z}}^{\psi}) - \mathcal{E}_{\mathbf{z}}^{\psi}(f_{\mathbf{z}}^{\psi}) = \int_Z \xi_{\mathbf{z}}(z) d\rho - \frac{1}{m} \sum_{i=1}^{m} \xi_{\mathbf{z}}(z_i)$. But $\xi_{\mathbf{z}}$ is not a single random variable; it depends on the sample $\mathbf{z}$. Therefore, the usual law of large numbers does not guarantee the convergence of this first term. One major goal of classical statistical learning theory [134] is to estimate this error term (i.e., $\mathcal{E}^{\psi}(f_{\mathbf{z}}^{\psi}) - \mathcal{E}_{\mathbf{z}}^{\psi}(f_{\mathbf{z}}^{\psi})$). The collection of ideas and techniques used to get such estimates, known as the *theory of uniform convergence*, plays the role of a uniform law of large numbers. To see why, consider the quantity

$$\sup_{f \in \mathcal{H}} |\mathcal{E}_{\mathbf{z}}^{\psi}(f) - \mathcal{E}^{\psi}(f)|,$$
$$(3.7)$$

which bounds the first term on the right-hand side of (3.6), hence providing (together with bounds for the second term) an estimate for the sample error $\mathcal{E}^{\psi}(f_{\mathbf{z}}^{\psi}) - \mathcal{E}^{\psi}(f_{\mathcal{H}}^{\psi})$. The theory of uniform convergence studies the convergence of this quantity. It characterizes those function sets $\mathcal{H}$ such that the quantity (3.7) tends to zero in probability as $m \to \infty$.

Definition 3.23 We say that a set $\mathcal{H}$ of real-valued functions on a metric space X is *uniform Glivenko–Cantelli* (UGC) if for every $\epsilon > 0$,

$$\lim_{\ell \to +\infty} \sup_{\mu} \operatorname{Prob} \left\{ \sup_{m \geq \ell} \sup_{f \in \mathcal{H}} \left| \frac{1}{m} \sum_{i=1}^{m} f(x_i) - \int_X f(x) \, d\mu \right| \geq \epsilon \right\} = 0,$$

where the supremum is taken with respect to all Borel probability distributions μ on X, and Prob denotes the probability with respect to the samples $x_1, x_2, \dots$ independently drawn according to such a distribution μ.

The UGC property can be characterized by the V_γ dimensions of $\mathcal{H}$, as has been done in [5].

Definition 3.24 Let $\mathcal{H}$ be a set of functions from X to $[0, 1]$ and $\gamma > 0$. We say that $A \subset X$ is V_γ shattered by $\mathcal{H}$ if there is a number $\alpha \in \mathbb{R}$ with the following property: for every subset E of A there exists some function $f_E \in \mathcal{H}$ such that $f_E(x) \leq \alpha - \gamma$ for every $x \in A \setminus E$, and $f_E(x) \geq \alpha + \gamma$ for every $x \in E$. The V_γ dimension of $\mathcal{H}$, $V_\gamma(\mathcal{H})$, is the maximal cardinality of a set $A \subset X$ that is V_γ shattered by $\mathcal{H}$.

The concept of V_γ dimension is related to many other quantities involving capacity of function sets studied in approximation theory or functional analysis: covering numbers, entropy numbers, VC dimensions, packing numbers, metric entropy, and others.

The following characterization of the UGC property is given in [5].

Theorem 3.25 *Let $\mathcal{H}$ be a set of functions from X to $[0, 1]$. Then $\mathcal{H}$ is UGC if and only if the V_γ dimension of $\mathcal{H}$ is finite for every $\gamma > 0$.*

Theorem 3.25 may be used to verify the convergence of ERM schemes when the hypothesis space $\mathcal{H}$ is a noncompact UGC set such as the union of unit balls of reproducing kernel Hilbert spaces associated with a set of Mercer kernels. In particular, for the Gaussian kernels with flexible variances, the UGC property holds [150].

Many fundamental problems about the UGC property remain to be solved. As an example, consider the empirical covering numbers.

Definition 3.26 For $\mathbf{x} = (x_i)_{i=1}^{m} \in X^m$ and $\mathcal{H} \subset \mathscr{C}(X)$, the ℓ^∞-*empirical covering number* $\mathcal{N}_\infty(\mathcal{H}, \mathbf{x}, \eta)$ is the covering number of $\mathcal{H}|_{\mathbf{x}} := \{(f(x_i))_{i=1}^{m} : f \in \mathcal{H}\}$ as a subset of $\mathbb{R}^m$ with the following metric. For $f, g \in \mathscr{C}(X)$ we take $d_{\mathbf{x}}(f, g) = \max_{i \leq m} |f(x_i) - g(x_i)|$. The *metric entropy* of $\mathcal{H}$ is defined as

$$H_m(\mathcal{H}, \eta) = \sup_{\mathbf{x} \in X^m} \log \mathcal{N}_\infty(\mathcal{H}, \mathbf{x}, \eta), \quad m \in \mathbb{N}, \eta > 0.$$

It is known [46] that a set $\mathcal{H}$ of functions from X to $[0, 1]$ is UGC if and only if, for every $\eta > 0$, $\lim_{m \to \infty} H_m(\mathcal{H}, \eta)/m = 0$. In this case, one has $H_m(\mathcal{H}, \eta) = \mathcal{O}(\log^2 m)$ for every $\eta > 0$. It is conjectured in [5] that $H_m(\mathcal{H}, \eta) = \mathcal{O}(\log m)$ is true for every $\eta > 0$. A weak form is, Is it true that for some $\alpha \in [1, 2)$, every UGC set $\mathcal{H}$ satisfies

$$H_m(\mathcal{H}, \eta) = \mathcal{O}(\log^\alpha m), \quad \forall \eta > 0?$$

4

Polynomial decay of the approximation error

We continue to assume that X is a compact metric space (which may be a compact subset of $\mathbb{R}^n$). Let K be a Mercer kernel on X and $\mathcal{H}_K$ be its induced RKHS. We observed in Section 2.6 that the space $\mathcal{H}_{K,R} = I_K(B_R)$ may be considered as a hypothesis space. Here I_K denotes the inclusion $I_K : \mathcal{H}_K \hookrightarrow \mathscr{C}(X)$. When R increases the quantity

$$\mathcal{A}(f_\rho, R) := \inf_{f \in \mathcal{H}_{K,R}} \mathcal{E}(f) - \mathcal{E}(f_\rho) = \inf_{\|f\|_K \leq R} \|f - f_\rho\|^2_{\mathscr{L}^2_{\rho_X}}$$

(which coincides with the approximation error modulo σ^2_ρ) decreases. The main result in this chapter characterizes the measures ρ and kernels K for which this decay is polynomial, that is, $\mathcal{A}(f_\rho, R) = \mathcal{O}(R^{-\theta})$ with $\theta > 0$.

Theorem 4.1 *Suppose ρ is a Borel probability measure on Z. Let K be a Mercer kernel on X and $L_K : \mathscr{L}^2_{\rho_X} \to \mathscr{L}^2_{\rho_X}$ be the operator given by*

$$L_K f(x) = \int_X K(x,t) f(t) d\rho_X(t), \quad x \in X.$$

Let $\theta > 0$. If $f_\rho \in \text{Range}(L_K^{\theta/(4+2\theta)})$, that is, $f_\rho = L_K^{\theta/(4+2\theta)}(g)$ for some $g \in \mathscr{L}^2_{\rho_X}$, then $\mathcal{A}(f_\rho, R) \leq 2^{2+\theta} \|g\|^{2+\theta}_{\mathscr{L}^2_{\rho_X}} R^{-\theta}$. Conversely, if ρ_X is nondegenerate and $\mathcal{A}(f_\rho, R) \leq CR^{-\theta}$ for some constants C and θ, then f_ρ lies in the range of $L_K^{\theta/(4+2\theta)-\epsilon}$ for all $\epsilon > 0$.

Although Theorem 4.1 may be applied to spline kernels (see Section 4.6), we show in Theorem 6.2 that for $\mathscr{C}^\infty$ kernels (e.g., the Gaussian kernel) and under some conditions on ρ_X, the approximation error decay cannot reach the order $\mathcal{A}(f_\rho, R) = \mathcal{O}(R^{-\theta})$ unless f_ρ is $\mathscr{C}^\infty$ itself. Instead, also in Chapter 6, we derive logarithmic orders like $\mathcal{A}(f_\rho, R) = \mathcal{O}((\log R)^{-\theta})$ for analytic kernels and Sobolev smooth regression functions.

54

4.1 Reminders III

We recall some basic facts about Hilbert spaces.

A sequence $\{\phi_n\}_{n\geq 1}$ in a Hilbert space H is said to be a *complete orthonormal system* (or an *orthonormal basis*) if the following conditions hold:

(i) for all $n \neq m \geq 1$, $\langle \phi_n, \phi_m \rangle = 0$,
(ii) for all $n \geq 1$, $\|\phi_n\| = 1$, and
(iii) for all $f \in H$, $f = \sum_{n=1}^{\infty} \langle f, \phi_n \rangle \phi_n$.

A sequence satisfying (i) and (ii) only is said to be an *orthonormal system*. The numbers $\langle f, \phi_n \rangle$ are the *Fourier coefficients* of f in the basis $\{\phi_n\}_{n\geq 1}$. It is easy to see that these coefficients are unique since, if $f = \sum a_n \phi_n$, $a_n = \langle f, \phi_n \rangle$ for all $n \geq 1$.

Theorem 4.2 (Parseval's theorem) *If $\{\phi_n\}$ is an orthonormal system of a Hilbert space H, then, for all $f \in H$, $\sum_n \langle f, \phi_n \rangle^2 \leq \|f\|^2$. Equality holds for all $f \in H$ if and only if $\{\phi_n\}$ is complete.* ■

We defined compactness of an operator in Section 2.3. We next recall some other basic properties of linear operators and a main result for operators satisfying them.

Definition 4.3 A linear operator $L: H \to H$ on a Hilbert space H is said to be *self-adjoint* if, for all $f, g \in H$, $\langle Lf, g \rangle = \langle f, Lg \rangle$. It is said to be *positive* (respectively *strictly positive*) if it is self-adjoint and, for all nontrivial $f \in H$, $\langle Lf, f \rangle \geq 0$ (respectively $\langle Lf, f \rangle > 0$).

Theorem 4.4 (Spectral theorem) *Let L be a compact self-adjoint linear operator on a Hilbert space H. Then there exists in H an orthonormal basis $\{\phi_1, \phi_2, \dots\}$ consisting of eigenvectors of L. If λ_k is the eigenvalue corresponding to ϕ_k, then either the set $\{\lambda_k\}$ is finite or $\lambda_k \to 0$ when $k \to \infty$. In addition, $\max_{k\geq 1} |\lambda_k| = \|L\|$. If, in addition, L is positive, then $\lambda_k \geq 0$ for all $k \geq 1$, and if L is strictly positive, then $\lambda_k > 0$ for all $k \geq 1$.* ■

We close this section by defining the power of a self-adjoint, positive, compact, linear operator. If L is such an operator and $\theta > 0$, then L^θ is the operator defined by

$$L^\theta \left(\sum c_k \phi_k \right) = \sum c_k \lambda_k^\theta \phi_k.$$

4.2 Operators defined by a kernel

In the remainder of this chapter we consider ν, a finite Borel measure on X, and $\mathscr{L}_\nu^2(X)$, the Hilbert space of square integrable functions on X. Note that ν can be any Borel measure. Significant particular cases are the Lebesgue measure μ and the marginal measure ρ_X of Chapter 1.

Let $K : X \times X \to \mathbb{R}$ be a continuous function. Then the linear map

$$L_K : \mathscr{L}_\nu^2(X) \to \mathscr{C}(X)$$

given by the following integral transform

$$(L_K f)(x) = \int K(x,t) f(t) \, d\nu(t), \quad x \in X,$$

is well defined. Composition with the inclusion $\mathscr{C}(X) \hookrightarrow \mathscr{L}_\nu^2(X)$ yields a linear operator $L_K : \mathscr{L}_\nu^2(X) \to \mathscr{L}_\nu^2(X)$, which, abusing notation, we also denote by L_K.

The function K is said to be the *kernel* of L_K, and several properties of L_K follow from properties of K. Recall the definitions of $\mathbf{C}_K$ and K_x introduced in Section 2.4.

Proposition 4.5 *If K is continuous, then $L_K : \mathscr{L}_\nu^2(X) \to \mathscr{C}(X)$ is well defined and compact. In addition, $\|L_K\| \leq \sqrt{\nu(X)}\mathbf{C}_K^2$. Here $\nu(X)$ denotes the measure of X.*

Proof.   To see that L_K is well defined, we need to show that $L_K f$ is continuous for every $f \in \mathscr{L}_\nu^2(X)$. To do so, we consider $f \in \mathscr{L}_\nu^2(X)$ and $x_1, x_2 \in X$. Then

$$|(L_K f)(x_1) - (L_K f)(x_2)| = \left| \int (K(x_1,t) - K(x_2,t)) f(t) \, d\nu(t) \right|$$

$$\leq \|K_{x_1} - K_{x_2}\|_{\mathscr{L}_\nu^2(X)} \|f\|_{\mathscr{L}_\nu^2(X)}$$

$$\text{(by Cauchy–Schwarz)}$$

$$\leq \sqrt{\nu(X)} \left\{ \max_{t \in X} |K(x_1,t) - K(x_2,t)| \right\} \|f\|_{\mathscr{L}_\nu^2(X)}.$$

$$(4.1)$$

Since K is continuous and X is compact, K is uniformly continuous. This implies the continuity of $L_K f$.

The assertion $\|L_K\| \leq \sqrt{\nu(X)}\mathbf{C}_K^2$ follows from the inequality

$$|(L_K f)(x)| \leq \sqrt{\nu(X)} \sup_{t \in X} |K(x,t)| \|f\|_{\mathscr{L}_\nu^2(X)},$$

which is proved as above.

Finally, to see that L_K is compact, let (f_n) be a bounded sequence in $\mathscr{L}_\nu^2(X)$. Since $\|L_K f\|_\infty \leq \sqrt{\nu(X)}\mathbf{C}_K^2 \|f\|_{\mathscr{L}_\nu^2(X)}$, we have that $(L_K f_n)$ is uniformly bounded. By (4.1) we have that the sequence $(L_K f_n)$ is equicontinuous. By the Arzelá–Ascoli theorem (Theorem 2.4), $(L_K f_n)$ contains a uniformly convergent subsequence. ∎

Two more important properties of L_K follow from properties of K. Recall that we say K is positive semidefinite if, for all finite sets $\{x_1, \ldots, x_k\} \subset X$, the $k \times k$ matrix $K[\mathbf{x}]$ whose (i,j) entry is $K(x_i, x_j)$ is positive semidefinite.

Proposition 4.6

(i) *If K is symmetric, then $L_K : \mathscr{L}_\nu^2(X) \to \mathscr{L}_\nu^2(X)$ is self-adjoint.*
(ii) *If, in addition, K is positive semidefinite, then L_K is positive.*

Proof. Part (i) follows easily from Fubini's theorem and the symmetry of K. For Part (ii), just note that

$$\int_X \int_X K(x,t)f(x)f(t)\, d\nu(x)\, d\nu(t) = \lim_{k \to \infty} \frac{(\nu(X))^2}{k^2} \sum_{i,j=1}^k K(x_i,x_j)f(x_i)f(x_j)$$

$$= \lim_{k \to \infty} \frac{(\nu(X))^2}{k^2} f_{\mathbf{x}}^{\mathsf{T}} K[\mathbf{x}] f_{\mathbf{x}},$$

where, for all $k \geq 1$, $x_1, \ldots, x_k \in X$ is a set of points conveniently chosen and $f_{\mathbf{x}} = (f(x_1), \ldots, f(x_k))^{\mathsf{T}}$. Since $K[\mathbf{x}]$ is positive semidefinite the result follows. ∎

Theorem 4.7 *Let $K : X \times X \to \mathbb{R}$ be a Mercer kernel. There exists an orthonormal basis $\{\phi_1, \phi_2, \ldots\}$ of $\mathscr{L}_\nu^2(X)$ consisting of eigenfunctions of L_K. If λ_k is the eigenvalue corresponding to ϕ_k, then either the set $\{\lambda_k\}$ is finite or $\lambda_k \to 0$ when $k \to \infty$. In addition, $\lambda_k \geq 0$ for all $k \geq 1$, $\max_{k \geq 1} \lambda_k = \|L_K\|$, and if $\lambda_k \neq 0$, then ϕ_k can be chosen to be continuous on X.*

Proof. By Propositions 4.5 and 4.6 $L_K : \mathscr{L}_\nu^2(X) \to \mathscr{L}_\nu^2(X)$ is a self-adjoint, positive, compact operator. Theorem 4.4 yields all the statements except the continuity of the ϕ_k.

To prove this last fact, use the fact that $\phi_k = (1/\lambda_k)/L_K(\phi_k)$ in $\mathscr{L}_\nu^2(X)$. Then we can choose the eigenfunction to be $(1/\lambda_k)L_K(\phi_k)$, which is a continuous function. ∎

In what follows we fix a Mercer kernel K and let $\{\phi_k \in \mathscr{L}_\nu^2(X)\}$ be an orthonormal basis of $\mathscr{L}_\nu^2(X)$ consisting of eigenfunctions of L_K. We call the ϕ_k *orthonormal eigenfunctions*. Denote by λ_k, $k \geq 1$, the eigenvalue of L_K corresponding to ϕ_k. If $\lambda_k > 0$, the function ϕ_k is continuous. In addition, it lies in the RKHS $\mathcal{H}_K$. This is so since

$$\phi_k(x) = \frac{1}{\lambda_k} L_K(\phi_k)(x) = \frac{1}{\lambda_k} \int K(x,t)\phi_k(t) \, d\nu(t),$$

and, thus, ϕ_k can be approximated by elements in the span of $\{K_x \mid x \in X\}$. Theorem 2.9 then shows that $\phi_k \in \mathcal{H}_K$. In fact,

$$\|\phi_k\|_K = \left\| \frac{1}{\lambda_k} \int_X K_t \phi_k(t) \, d\nu(t) \right\|_K \leq \frac{1}{\lambda_k} \int_X \|K_t\|_K |\phi_k(t)| \, d\nu(t)$$
$$\leq \frac{\mathbf{C}_K}{\lambda_k} \sqrt{\nu(X)} < \infty.$$

We shall assume, without loss of generality, that $\lambda_k \geq \lambda_{k+1}$ for all $k \geq 1$.

Using the eigenfunctions $\{\phi_k\}$, we can find an orthonormal system of the RKHS $\mathcal{H}_K$.

Theorem 4.8 *Let ν be a Borel measure on X, and $K : X \times X \to \mathbb{R}$ a Mercer kernel. Let λ_k be the kth eigenvalue of L_K, and ϕ_k the corresponding orthonormal eigenfunction. Then $\{\sqrt{\lambda_k}\phi_k : \lambda_k > 0\}$ forms an orthonormal system in $\mathcal{H}_K$.*

Proof. We apply the reproducing property stated in Theorem 2.9. Assume $\lambda_i, \lambda_j > 0$; we have

$$\langle \sqrt{\lambda_i}\phi_i, \sqrt{\lambda_j}\phi_j \rangle_K = \left\langle \frac{1}{\sqrt{\lambda_i}} \int_X K(\cdot, y)\phi_i(y) \, d\nu(y), \sqrt{\lambda_j}\phi_j \right\rangle_K$$
$$= \frac{\sqrt{\lambda_j}}{\sqrt{\lambda_i}} \int_X \phi_i(y) \langle K_y, \phi_j \rangle_K \, d\nu(y)$$
$$= \frac{\sqrt{\lambda_j}}{\sqrt{\lambda_i}} \int_X \phi_i(y)\phi_j(y) \, d\nu(y) = \frac{\sqrt{\lambda_j}}{\sqrt{\lambda_i}} \langle \phi_i, \phi_j \rangle_{\mathscr{L}_\nu^2(X)} = \delta_{ij}.$$

It follows that $\{\sqrt{\lambda_k}\phi_k : \lambda_k > 0\}$ forms an orthonormal system in $\mathcal{H}_K$. ∎

Remark 4.9 When ν is nondegenerate, one can easily see from the definition of the integral operator that L_K has no eigenvalue 0 if and only if $\mathcal{H}_K$ is dense in $\mathscr{L}_\nu^2(X)$.

In fact, the orthonormal system above forms an orthonormal basis of $\mathcal{H}_K$ when ρ_X is nondegenerate. This will be proved in Section 4.4. Toward this end, we next prove Mercer's theorem.

4.3 Mercer's theorem

If $f \in \mathscr{L}_\nu^2(X)$ and $\{\phi_1, \phi_2, \ldots\}$ is an orthonormal basis of $\mathscr{L}_\nu^2(X)$, f can be uniquely written as $f = \sum_{k \geq 1} a_k \phi_k$, and, when the basis has infinitely many functions, the partial sums $\sum_{k=1}^N a_k \phi_k$ converge to f in $\mathscr{L}_\nu^2(X)$. If this convergence also holds in $\mathscr{C}(X)$, we say that the series *converges uniformly* to f. Also, we say that a series $\sum a_k$ *converges absolutely* if the series $\sum |a_k|$ is convergent.

When L_K has only finitely many positive eigenvalues $\{\lambda_k\}_{k=1}^m$, $K(x,t) = \sum_{k=1}^m \lambda_k \phi_k(x)\phi_k(t)$.

Theorem 4.10 *Let ν be a Borel, nondegenerate measure on X, and $K : X \times X \to \mathbb{R}$ a Mercer kernel. Let λ_k be the kth positive eigenvalue of L_K, and ϕ_k the corresponding continuous orthonormal eigenfunction. For all $x, t \in X$,*

$$K(x,t) = \sum_{k \geq 1} \lambda_k \phi_k(x)\phi_k(t),$$

where the convergence is absolute (for each $x, t \in X \times X$) and uniform (on $X \times X$).

Proof. By Theorem 4.8, the sequence $\{\sqrt{\lambda_k}\phi_k\}_{k \geq 1}$ is an orthonormal system of $\mathcal{H}_K$. Let $x \in X$. The Fourier coefficients of the function $K_x \in \mathcal{H}_K$ with respect to this system are

$$\langle \sqrt{\lambda_k}\phi_k, K_x \rangle_K = \sqrt{\lambda_k}\phi_k(x),$$

where Theorem 2.9(iii) is used. Then, by Parseval's theorem, we know that

$$\sum_{k \geq 1} |\sqrt{\lambda_k}\phi_k(x)|^2 = \sum_{k \geq 1} \lambda_k |\phi_k(x)|^2 \leq \|K_x\|_K^2 = K(x,x) \leq \mathbf{C}_K^2.$$

Hence the series $\sum_{k \geq 1} \lambda_k |\phi_k(x)|^2$ converges. This is true for each point $x \in X$.

Now we fix a point $x \in X$. When the basis $\{\phi_k\}_{k \geq 1}$ has infinitely many functions, the estimate above, together with the Cauchy–Schwarz inequality, tells us that for each $t \in X$,

$$
\left| \sum_{k=m}^{m+\ell} \lambda_k \phi_k(x) \phi_k(t) \right| \leq \left(\sum_{k=m}^{m+\ell} \lambda_k |\phi_k(t)|^2 \right)^{1/2} \left(\sum_{k=m}^{m+\ell} \lambda_k |\phi_k(x)|^2 \right)^{1/2}
$$

$$
\leq \mathbf{C}_K \left(\sum_{k=m}^{m+\ell} \lambda_k |\phi_k(x)|^2 \right)^{1/2},
$$

which tends to zero uniformly (for $t \in X$). Hence the series $\sum_{k \geq 1} \lambda_k \phi_k(x) \phi_k(t)$ (as a function of t) converges absolutely and uniformly on X to a continuous function g_x. On the other hand, as a function in $\mathcal{L}_\nu^2(X)$, K_x can be expanded by means of an orthonormal basis consisting of $\{\phi_k\}$ and an orthonormal basis Ψ of the nullspace of L_K. For $f \in \Psi$,

$$
\langle K_x, f \rangle_{\mathcal{L}_\nu^2(X)} = \int_X K(x,y) f(y) \, d\nu = 0.
$$

Hence the expansion of K_x is

$$
K_x = \sum_{k \geq 1} \langle K_x, \phi_k \rangle_{\mathcal{L}_\nu^2(X)} \phi_k = \sum_{k \geq 1} L_K(\phi_k)(x) \phi_k = \sum_{k \geq 1} \lambda_k \phi_k(x) \phi_k.
$$

Thus, as functions in $\mathcal{L}_\nu^2(X)$, $K_x = g_x$. Since ν is nondegenerate, K_x and g_x are equal on a dense subset of X. But both are continuous functions and, therefore, must be equal for each $t \in X$. It follows that for any $x, t \in X$, the series $\sum_{k \geq 1} \lambda_k \phi_k(x) \phi_k(t)$ converges to $K(x, t)$. Since the limit function $K(x, t)$ is continuous, we know that the series must converge uniformly on $X \times X$. ∎

Corollary 4.11 *The sum $\sum \lambda_k$ is convergent and*

$$
\sum_{k \geq 1} \lambda_k = \int_X K(x,x) \leq \nu(X) \mathbf{C}_K^2.
$$

Moreover, for all $k \geq 1$, $\lambda_k \leq \nu(X) \mathbf{C}_K^2 / k$.

Proof. Taking $x = t$ in Theorem 4.10, we get $K(x,x) = \sum_{k \geq 1} \lambda_k \phi_k(x)^2$. Integrating on both sides of this equality gives

$$
\sum_{k \geq 1} \lambda_k \int_X \phi_k(x)^2 \, d\nu = \int_X K(x,x) \, d\nu \leq \nu(X) \mathbf{C}_K^2.
$$

However, since $\{\phi_1, \phi_2, \ldots\}$ is an orthonormal basis, $\int \phi_k^2 = 1$ for all $k \geq 1$ and the first statement follows. The second statement holds true because the assumption $\lambda_k \geq \lambda_j$ for $j > k$ tells us that $k\lambda_k \leq \sum_{j=1}^{k} \lambda_j \leq \nu(X) \mathbf{C}_K^2$. ∎

4.4 RKHSs revisited

In this section we show that the RKHS $\mathcal{H}_K$ has an orthonormal basis $\{\sqrt{\lambda_k}\phi_k\}$ derived from the integral operator L_K (and thus dependent on the measure ν).

Theorem 4.12 *Let ν be a Borel, nondegenerate measure on X, and $K : X \times X \to \mathbb{R}$ a Mercer kernel. Let λ_k be the kth positive eigenvalue of L_K, and ϕ_k the corresponding continuous orthonormal eigenfunction. Then $\{\sqrt{\lambda_k}\phi_k : \lambda_k > 0\}$ is an orthonormal basis of $\mathcal{H}_K$.*

Proof. By Theorem 4.8, $\{\sqrt{\lambda_k}\phi_k : \lambda_k > 0\}$ is an orthonormal system in $\mathcal{H}_K$. To prove the completeness we need only show that for each $x \in X$, K_x lies in the closed span of this orthonormal system. Complete this system to form an orthonormal basis $\{\sqrt{\lambda_k}\phi_k : \lambda_k > 0\} \cup \{\psi_j : j \geq 0\}$ of $\mathcal{H}_K$. By Parseval's theorem,

$$\|K_x\|_K^2 = \sum_k \left\langle K_x, \sqrt{\lambda_k}\phi_k \right\rangle_K^2 + \sum_j \langle K_x, \psi_j \rangle_K^2.$$

Therefore, to show that K_x lies in the closed span of $\{\sqrt{\lambda_k}\phi_k : \lambda_k > 0\}$, it is enough to require that

$$\|K_x\|_K^2 = \sum_k \langle K_x, \sqrt{\lambda_k}\phi_k \rangle_K^2,$$

that is, that

$$K(x, x) = \sum_k \lambda_k \langle K_x, \phi_k \rangle_K^2 = \sum_k \lambda_k \phi_k(x)^2.$$

Theorem 4.10 with $x = t$ yields this identity for each $x \in X$. ∎

Since the RKHS $\mathcal{H}_K$ is independent of the measure ν, it follows that when ν is nondegenerate and $\dim \mathcal{H}_K = \infty$, L_K has infinitely many positive eigenvalues $\lambda_k, k \geq 1$, and

$$\mathcal{H}_K = \left\{ f = \sum_{k=1}^{\infty} a_k \sqrt{\lambda_k}\phi_k : \{a_k\}_{k=1}^{\infty} \in \ell^2 \right\}.$$

When dim $\mathcal{H}_K = m < \infty$, L_K has only m positive (repeated) eigenvalues. In this case,

$$\mathcal{H}_K = \left\{ f = \sum_{k=1}^{m} a_k \sqrt{\lambda_k} \phi_k : (a_1, \ldots, a_m) \in \mathbb{R}^m \right\}.$$

In both cases, the map

$$L_K^{1/2} : \mathscr{L}_\nu^2(X) \to \mathcal{H}_K$$

$$\sum a_k \phi_k \mapsto \sum a_k \sqrt{\lambda_k} \phi_k$$

defines an isomorphism of Hilbert spaces between the closed span of $\{\phi_k : \lambda_k > 0\}$ in $\mathscr{L}_\nu^2(X)$ and $\mathcal{H}_K$. In addition, considered as an operator on $\mathscr{L}_\nu^2(X)$, $L_K^{1/2}$ is the square root of L_k in the sense that $L_K = L_K^{1/2} \circ L_K^{1/2}$ (hence the notation $L_K^{1/2}$). This yields the following corollary.

Corollary 4.13 *Let ν be a Borel, nondegenerate measure on X, and $K : X \times X \to \mathbb{R}$ a Mercer kernel. Then $\mathcal{H}_K = L_K^{1/2}(\mathscr{L}_\nu^2(X))$. That is, every function $f \in \mathcal{H}_K$ can be written as $f = L_K^{1/2} g$ for some $g \in \mathscr{L}_\nu^2(X)$ with $\|f\|_K = \|g\|_{\mathscr{L}_\nu^2(X)}$.* ∎

A different approach to the orthonormal basis $\{\sqrt{\lambda_k} \phi_k\}$ is to regard it as a function on X with values in ℓ^2.

Theorem 4.14 *The map*

$$\Phi : X \to \ell^2$$

$$x \mapsto \left(\sqrt{\lambda_k} \phi_k(x) \right)_{k \geq 1}$$

is well defined and continuous, and satisfies

$$K(x, t) = \langle \Phi(x), \Phi(t) \rangle.$$

Proof. For every $x \in X$, by Mercer's theorem $\sum \lambda_k \phi_k^2(x)$ converges to $K(x, x)$. This shows that $\Phi(x) \in \ell^2$. Also by Mercer's theorem, for every $x, t \in X$,

$$K(x, t) = \sum_{k=1}^{\infty} \lambda_k \phi_k(x) \phi_k(t) = \langle \Phi(x), \Phi(t) \rangle_{\ell^2}.$$

It remains only to prove that $\Phi : X \to \ell^2$ is continuous. For any $x, t \in X$,

$$\|\Phi(x) - \Phi(t)\|_{\ell^2}^2 = \langle \Phi(x), \Phi(x) \rangle_{\ell^2} + \langle \Phi(t), \Phi(t) \rangle_{\ell^2} - 2\langle \Phi(x), \Phi(t) \rangle_{\ell^2}$$
$$= K(x, x) + K(t, t) - 2K(x, t),$$

which tends to zero when x tends to t by the continuity of K. ∎

4.5 Characterizing the approximation error in RKHSs

In this section we prove Theorem 4.1. It actually follows from a more general characterization of the decay of the approximation error given using interpolation spaces.

Definition 4.15 Let $(B, \| \cdot \|)$ and $(\mathcal{H}, \| \cdot \|_{\mathcal{H}})$ be Banach spaces and assume $\mathcal{H}$ is a subspace of B. The $\mathbb{K}$-*functional* $\mathbb{K} : B \times (0, \infty) \to \mathbb{R}$ of the pair $(B, \mathcal{H})$ is defined, for $a \in B$ and $t > 0$, by

$$\mathbb{K}(a, t) := \inf_{b \in \mathcal{H}} \{ \|a - b\| + t\|b\|_{\mathcal{H}} \}. \tag{4.2}$$

It can easily be seen that for fixed $a \in B$, the function $\mathbb{K}(a, t)$ of t is continuous, nondecreasing, and bounded by $\|a\|$ (take $b = 0$ in (4.2)). When $\mathcal{H}$ is dense in B, $\mathbb{K}(a, t)$ tends to zero as $t \to 0$. The interpolation spaces for the pair $(B, \mathcal{H})$ are defined in terms of the convergence rate of this function.

For $0 < r < 1$, the *interpolation space* $(B, \mathcal{H})_r$ consists of all the elements $a \in B$ such that the norm

$$\|a\|_r := \sup_{t > 0} \{ \mathbb{K}(a, t)/t^r \}$$

is finite.

Theorem 4.16 *Let* $(B, \| \|)$ *be a Banach space, and* $(\mathcal{H}, \| \|_{\mathcal{H}})$ *a subspace, such that* $\|b\| \leq C_0 \|b\|_{\mathcal{H}}$ *for all* $b \in \mathcal{H}$ *and a constant* $C_0 > 0$. *Let* $0 < r < 1$. *If* $a \in (B, \mathcal{H})_r$, *then, for all* $R > 0$,

$$\mathcal{A}(a, R) := \inf_{\|b\|_{\mathcal{H}} \leq R} \{ \|a - b\|^2 \} \leq \|a\|_r^{2/(1-r)} R^{-2r/(1-r)}.$$

Conversely, if $\mathcal{A}(a, R) \leq CR^{-2r/(1-r)}$ *for all* $R > 0$, *then* $a \in (B, \mathcal{H})_r$ *and* $\|a\|_r \leq 2C^{(1-r)/2}$.

Proof.   Consider the function $f(t) := \mathbb{K}(a,t)/t$. It is continuous on $(0,+\infty)$. Since $\mathbb{K}(a,t) \leq \|a\|$, $\inf_{t>0}\{f(t)\} = 0$.

Fix $R > 0$. If $\sup_{t>0}\{f(t)\} \geq R$, then, for any $0 < \epsilon < 1$, there exists some $t_{R,\epsilon} \in (0,+\infty)$ such that

$$f(t_{R,\epsilon}) = \frac{\mathbb{K}(a,t_{R,\epsilon})}{t_{R,\epsilon}} = (1-\epsilon)R.$$

By the definition of the $\mathbb{K}$-functional, we can find $b_\epsilon \in \mathcal{H}$ such that

$$\|a - b_\epsilon\| + t_{R,\epsilon}\|b_\epsilon\|_{\mathcal{H}} \leq \mathbb{K}(a,t_{R,\epsilon})/(1-\epsilon).$$

It follows that

$$\|b_\epsilon\|_{\mathcal{H}} \leq \frac{\mathbb{K}(a,t_{R,\epsilon})}{(1-\epsilon)t_{R,\epsilon}} = R$$

and

$$\|a - b_\epsilon\| \leq \frac{\mathbb{K}(a,t_{R,\epsilon})}{1-\epsilon}.$$

But the definition of the norm $\|a\|_r$ implies that

$$\frac{\mathbb{K}(a,t_{R,\epsilon})}{t_{R,\epsilon}^r} \leq \|a\|_r.$$

Therefore

$$\|a - b_\epsilon\| \leq \left[\frac{\mathbb{K}(a,t_{R,\epsilon})}{(1-\epsilon)t_{R,\epsilon}}\right]^{-r/(1-r)} \left[\frac{\mathbb{K}(a,t_{R,\epsilon})}{(1-\epsilon)t_{R,\epsilon}^r}\right]^{1/(1-r)}$$

$$\leq R^{-r/(1-r)}\left(\frac{1}{1-\epsilon}\right)^{1/(1-r)}(\|a\|_r)^{1/(1-r)}.$$

Thus,

$$\mathcal{A}(a,R) \leq \inf_{0<\epsilon<1}\left\{\|a - b_\epsilon\|^2\right\} \leq \|a\|_r^{2/(1-r)}R^{-2r/(1-r)};$$

that is, the desired error estimate holds in this case.

Turn now to the case where $\sup_{t>0}\{f(t)\} < R$. Then, for any $0 < \epsilon < 1 - \sup_{u>0}\{f(u)\}/R$ and any $t > 0$, there exists some $b_{t,\epsilon} \in \mathcal{H}$ such that

$$\|a - b_{t,\epsilon}\| + t\|b_{t,\epsilon}\|_{\mathcal{H}} \leq \mathbb{K}(a,t)/(1-\epsilon).$$

This implies that

$$\|b_{t,\epsilon}\|_{\mathcal{H}} \leq \frac{\mathbb{K}(a,t)}{(1-\epsilon)t} \leq \frac{1}{1-\epsilon} \sup_{u>0}\{f(u)\} < R$$

and

$$\|a - b_{t,\epsilon}\| \leq \frac{\mathbb{K}(a,t)}{1-\epsilon}.$$

Hence

$$\mathcal{A}(a,R) \leq \inf_{t>0}\left\{\|a - b_{t,\epsilon}\|^2\right\} \leq \left(\inf_{t>0}\{\mathbb{K}(a,t)\}/(1-\epsilon)\right)^2$$

$$\leq \left(\inf_{t>0}\left\{\|a\|_r t^r\right\}/(1-\epsilon)\right)^2 = 0.$$

This again proves the desired error estimate. Hence the first statement of the theorem holds.

Conversely, suppose that $\mathcal{A}(a,R) \leq CR^{-2r/(1-r)}$ for all $R > 0$. Let $t > 0$. Choose $R_t = (\sqrt{C}/t)^{1-r}$. Then, for any $\epsilon > 0$, we can find $b_{t,\epsilon} \in \mathcal{H}$ such that

$$\|b_{t,\epsilon}\|_{\mathcal{H}} \leq R_t \quad \text{and} \quad \|a - b_{t,\epsilon}\|^2 \leq CR_t^{-2r/(1-r)}(1+\epsilon)^2.$$

It follows that

$$\mathbb{K}(a,t) \leq \|a - b_{t,\epsilon}\| + t\|b_{t,\epsilon}\|_{\mathcal{H}} \leq \sqrt{C}R_t^{-r/(1-r)}(1+\epsilon)$$

$$+ tR_t \leq 2(1+\epsilon)C^{(1-r)/2}t^r.$$

Since ϵ can be arbitrarily small, we have

$$\mathbb{K}(a,t) \leq 2C^{(1-r)/2}t^r.$$

Thus, $\|a\|_r = \sup_{t>0}\{\mathbb{K}(a,t)/t^r\} \leq 2C^{(1-r)/2} < \infty.$ ∎

The proof shows that if $a \in \mathcal{H}$, then $\mathcal{A}(a,R) = 0$ for $R > \|a\|_{\mathcal{H}}$. In addtion, $\mathcal{A}(a,R) = \mathcal{O}(R^{-2r/(1-r)})$ if and only if $a \in (B,\mathcal{H})_r$. A special case of Theorem 4.16 characterizes the decay of $\mathcal{A}$ for RKHSs.

Corollary 4.17 *Suppose ρ is a Borel probability measure on Z. Let $\theta > 0$. Then $\mathcal{A}(f_\rho, R) = \mathcal{O}(R^{-\theta})$ if and only if $f_\rho \in (\mathscr{L}^2_{\rho_X}, \mathcal{H}_K^+)_{\theta/(2+\theta)}$, where $\mathcal{H}_K^+$ is the closed subspace of $\mathcal{H}_K$ spanned by the orthonormal system $\{\sqrt{\lambda_k}\phi_k : \lambda_k > 0\}$ given in Theorem 4.8 with the measure ρ_X.*

Proof. Take $B = \mathscr{L}^2_{\rho_X}$ and $\mathcal{H} = \mathcal{H}^+_K$ with the norm inherited from $\mathcal{H}_K$. Then $\|b\| \leq \sqrt{\lambda_1}\|b\|_{\mathcal{H}}$ for all $b \in \mathcal{H}$. The statement now follows from Theorem 4.16 taking $r = \theta/(2 + \theta)$. ∎

Remark 4.18 When ρ_X is nondegenerate, $\mathcal{H}^+_K = \mathcal{H}_K$ by Theorem 4.12.

Recall that the space $L^r_K(\mathscr{L}^2_\nu(X))$ is $\{\sum_{\lambda_k>0} a_k\lambda^r_k\phi_k : a_k\} \in \ell^2\}$ with the norm

$$\left\| \sum_{\lambda_k>0} a_k\lambda^r_k\phi_k \right\|_{L^r_K(\mathscr{L}^2_\nu(X))} = \left(\sum_{\lambda_k>0} |a_k|^2 \right)^{1/2}.$$

Proof of Theorem 4.1   Take $\mathcal{H}^+_K$ as in Corollary 4.17 and $r = \theta/(2 + \theta)$.

If $f_\rho \in \text{Range}(L^{\theta/(4+2\theta)}_K)$, then $f_\rho = L^{\theta/(4+2\theta)}_K g$ for some $g \in \mathscr{L}^2_{\rho_X}$. Without loss of generality, we may take $g = \sum_{\lambda_k>0} a_k\phi_k$. Then $\|g\|^2 = \sum_{\lambda_k>0} a^2_k < \infty$ and $f_\rho = \sum_{\lambda_k>0} a_k\lambda^{\theta/(4+2\theta)}_k\phi_k$.

We show that $f_\rho \in (\mathscr{L}^2_{\rho_X}, \mathcal{H}^+_K)_r$. Indeed, for every $t \leq \sqrt{\lambda_1}$, there exists some $N \in \mathbb{N}$ such that

$$\lambda_{N+1} < t^2 \leq \lambda_N.$$

Choose $f = \sum^N_{k=1} a_k\lambda^{\theta/(4+2\theta)}_k\phi_k \in \mathcal{H}^+_K$. We can see from Theorem 4.8 that

$$\|f\|^2_K = \left\| \sum^N_{k=1} a_k\lambda^{(\theta/(4+2\theta))-1/2}_k\sqrt{\lambda_k}\phi_k \right\|^2_K = \sum^N_{k=1} a^2_k\lambda^{-2/(2+\theta)}_k \leq \lambda^{-2/(2+\theta)}_N\|g\|^2.$$

In addition,

$$\|f_\rho - f\|^2_{\mathscr{L}^2_{\rho_X}} = \left\| \sum_{k>N} a_k\lambda^{\theta/(4+2\theta)}_k\phi_k \right\|^2_{\mathscr{L}^2_{\rho_X}} = \sum_{k>N} a^2_k\lambda^{\theta/(2+\theta)}_k \leq \lambda^{\theta/(2+\theta)}_{N+1}\|g\|^2.$$

Let $\mathbb{K}$ be the $\mathbb{K}$-functional for the pair $(\mathscr{L}^2_{\rho_X}, \mathcal{H}^+_K)$. Then

$$\mathbb{K}(f_\rho, t) \leq \|f_\rho - f\|_{\mathscr{L}^2_{\rho_X}} + t\|f\|_K \leq \lambda^{\theta/(4+2\theta)}_{N+1}\|g\| + t\lambda^{-1/(2+\theta)}_N\|g\|.$$

By the choice of N, we have

$$\mathbb{K}(f_\rho, t) \leq \|g\| 2t^{\theta/(2+\theta)} = 2\|g\| t^r.$$

Since $\mathbb{K}(f_\rho, t) \leq \|f_\rho\|_{\mathscr{L}^2_{\rho_X}} \leq \lambda_1^{r/2} \|g\|$, we can also see that for $t > \sqrt{\lambda_1}$, $\mathbb{K}(f_\rho, t)/t^r \leq \|g\|$ holds. Therefore, $f_\rho \in (\mathscr{L}^2_{\rho_X}, \mathcal{H}_K^+)_r$ and $\|f_\rho\|_r \leq 2\|g\|$. It follows from Theorem 4.16 that

$$\mathcal{A}(f_\rho, R) \leq \inf_{f \in \mathcal{H}_K^+, \|f\|_K \leq R} \|f - f_\rho\|^2_{\mathscr{L}^2_{\rho_X}} \leq \left(2\|g\|\right)^{2/(1-r)} R^{-2r/(1-r)}$$

$$= 2^{2+\theta} \|g\|^{2+\theta} R^{-\theta}.$$

Conversely, if ρ_X is nondegenerate and $\mathcal{A}(f_\rho, R) \leq CR^{-\theta}$ for some constant C and all $R > 0$, then Theorem 4.12 states that $\mathcal{H}_K^+ = \mathcal{H}_K$. This, together with Theorem 4.16 and the polynomial decay of $\mathcal{A}(f_\rho, R)$, implies that $f_\rho \in (\mathscr{L}^2_{\rho_X}, \mathcal{H}_K)_r$ and $\|f_\rho\|_r \leq 2C^{1/(2+\theta)}$.

Let $m \in \mathbb{N}$. There exists a function $f_m \in \mathcal{H}_K$ such that

$$\|f_\rho - f_m\|_{\mathscr{L}^2_{\rho_X}} + 2^{-m}\|f_m\|_K \leq 4C^{1/(2+\theta)}2^{-mr}.$$

Then

$$\|f_\rho - f_m\|_{\mathscr{L}^2_{\rho_X}} \leq 4C^{1/(2+\theta)}2^{-mr} \quad \text{and} \quad \|f_m\|_K \leq 4C^{1/(2+\theta)}2^{m(1-r)}.$$

Write $f_\rho = \sum_k c_k \phi_k$ and $f_m = \sum_k b_k^{(m)} \phi_k$. Then, for all $0 < \epsilon < r$,

$$\sum_{2^{-2m} \leq \lambda_k < 2^{-2(m-1)}} \frac{c_k^2}{\lambda_k^{r-2\epsilon}} \leq 2 \sum_{2^{-2m} \leq \lambda_k < 2^{-2(m-1)}} \frac{\left(c_k - b_k^{(m)}\right)^2}{\lambda_k^{r-2\epsilon}}$$

$$+ 2 \sum_{2^{-2m} \leq \lambda_k < 2^{-2(m-1)}} \frac{\left(b_k^{(m)}\right)^2}{\lambda_k^{r-2\epsilon}},$$

which can be bounded by

$$2^{1+2m(r-2\epsilon)} \|f_\rho - f_m\|^2_{\mathscr{L}^2_{\rho_X}} + 2^{1+2(1-m)(1-r+2\epsilon)} \|f_m\|^2_K \leq C^{2/(2+\theta)}2^{5-4m\epsilon}$$

$$+ C^{2/(2+\theta)}2^{5+2(1-r)+4\epsilon(1-m)}.$$

Therefore,

$$\sum_{\lambda_k < 1} \frac{c_k^2}{\lambda_k^{r-2\epsilon}} \leq \frac{160}{16^\epsilon - 1} C^{2/(2+\theta)} < \infty.$$

This means that $f_\rho \in \text{Range}(L_K^{\theta/(4+2\theta)-\epsilon})$. ■

4.6 An example

In this section we describe a simple example for the approximation error in RKHSs.

Example 4.19 Let $X = [-1, 1]$, and let K be the spline kernel given in Example 2.15, that is, $K(x,y) = \max\{1 - |x - y|/2, 0\}$. We claim that $\mathcal{H}_K$ is the Sobolev space $H^1(X)$ with the following equivalent inner product:

$$\langle f, g \rangle_K = \langle f', g' \rangle_{\mathscr{L}^2[-1,1]} + \tfrac{1}{2}(f(-1) + f(1)) \cdot (g(-1) + g(1)). \quad (4.3)$$

Assume now that ρ_X is the Lebesgue measure. For $\theta > 0$ and a function $f_\rho \in \mathscr{L}^2[-1, 1]$, we also claim that $\mathcal{A}(f_\rho, R) = \mathcal{O}(R^{-\theta})$ if and only if $\|f_\rho(x + t) - f_\rho(x)\|_{\mathscr{L}^2[-1,1-t]} = \mathcal{O}(t^{\theta/(2+\theta)})$.

To prove the first claim, note that we know from Example 2.15 that K is a Mercer kernel. Also, $K_x \in H^1(X)$ for any $x \in X$. To show that (4.3) is the inner product in $\mathcal{H}_K$, it is sufficient to prove that $\langle f, K_x \rangle_K = f(x)$ for any $f \in H^1(X)$ and $x \in X$. To see this, note that $K_x' = \tfrac{1}{2}\chi_{[-1,x)} - \tfrac{1}{2}\chi_{(x,1]}$ and $K_x(-1) + K_x(1) = 1$. Then,

$$\langle f, K_x \rangle_K = \tfrac{1}{2} \int_{-1}^{x} f'(y)\, dy - \tfrac{1}{2} \int_{x}^{1} f'(y)\, dy + \tfrac{1}{2}(f(-1) + f(1)) = f(x).$$

To prove the second claim, we use Theorem 4.16 with $B = \mathscr{L}^2(X)$ and $\mathcal{H} = \mathcal{H}_K = H^1(X)$. Our conclusion follows from the following statement: for $0 < r < 1$ and $f \in \mathscr{L}^2$, $\mathbb{K}(f, t) = \mathcal{O}(t^r)$ if and only if $\|f(x+t) - f(x)\|_{\mathscr{L}^2[-1,1-t]} = \mathcal{O}(t^r)$.

To verify the sufficiency of this statement, define the function $f_t : X \to \mathbb{R}$ by

$$f_t(x) = \frac{1}{t} \int_0^t f(x + h)\, dh.$$

Taking norms of functions on the variable x, we can see that

$$\|f - f_t\|_{\mathscr{L}^2} = \left\| \frac{1}{t} \int_0^t f(x+h) - f(x)\, dh \right\|_{\mathscr{L}^2} \le \frac{1}{t} \int_0^t \|f(x+h) - f(x)\|_{\mathscr{L}^2}\, dh$$

$$\le \frac{1}{t} \int_0^t Ch^r\, dh = \frac{C}{r+1} t^r$$

and

$$\|f_t\|_{H^1(X)} = \left\| \frac{1}{t}(f(x+t) - f(x)) \right\|_{\mathscr{L}^2} \le Ct^{r-1}.$$

Hence

$$\mathbb{K}(f,t) \le \|f - f_t\|_{\mathscr{L}^2} + t\|f_t\|_{H^1(X)} \le (C/(r+1) + C)t^r.$$

Conversely, if $\mathbb{K}(f,t) \le Ct^r$, then, for any $g \in H^1(X)$, we have

$$\|f(x+t) - f(x)\|_{\mathscr{L}^2} = \|f(x+t) - g(x+t) + g(x+t) - g(x)$$
$$+ g(x) - f(x)\|_{\mathscr{L}^2},$$

which can be bounded by

$$2\|f - g\|_{\mathscr{L}^2} + \left\| \int_0^t g'(x+h)\, dh \right\|_{\mathscr{L}^2} \le 2\|f - g\|_{\mathscr{L}^2} + \int_0^t \|g'(x+h)\|_{\mathscr{L}^2}\, dh$$

$$\le 2\|f - g\|_{\mathscr{L}^2} + t\|g\|_{H^1(X)}.$$

Taking the infimum over $g \in H^1(X)$, we see that

$$\|f(x+t) - f(x)\|_{\mathscr{L}^2} \le 2\mathbb{K}(f,t) \le 2Ct^r.$$

This proves the statement and, with it, the second claim.

4.7 References and additional remarks

For a proof of the spectral theorem for compact operators see, for example, [73] and Section 4.10 of [40].

Mercer's theorem was originally proved [85] for $X = [0,1]$ and ν, the Lebesgue measure. Proofs for this simple case can also be found in [63, 73].

Theorems 4.10 and 4.12 are for general nondegenerate measures ν on a compact space X. For an extension to a noncompact space X see [123].

The map Φ in Theorem 4.14 is called the *feature map* in the literature on learning theory [37, 107, 134]. More general characterizations for the decay of the approximation error being of type $\mathcal{O}(\varphi(R))$ with φ decreasing on $(0, +\infty)$ can be derived from the literature on approximation theory e.g., ([87, 94]) by means of $\mathbb{K}$-functionals and moduli of smoothness. For interpolation spaces see [16].

RKHSs generated by general spline kernels are described in [137]. In the proof of Example 4.19 we have used a standard technique in approximation theory (see [78]). Here the function needs to be extended outside $[-1, 1]$ for defining f_t, or the norm $\mathscr{L}^2$ should be taken on $[-1, 1 - t]$. For simplicity, we have omitted this discussion.

The characterization of the approximation error described in Section 4.5 is taken from [113].

Consider the approximation for the ERM scheme with a general loss function ψ in Section 3.5. The target function $f_{\mathcal{H}}^{\psi}$ minimizes the generalization error $\mathcal{E}^{\psi}$ over $\mathcal{H}$. If we minimize instead over the set of all measurable functions we obtain a version (w.r.t. ψ) of the regression function.

Definition 4.20 Given the regression loss function ψ the *ψ-regression function* is given by

$$f_{\rho}^{\psi}(x) = \underset{t \in \mathbb{R}}{\operatorname{argmin}} \int_Y \psi(y - t) \, d\rho(y|x), \quad x \in X.$$

The *approximation error* (w.r.t. ψ) associated with the hypothesis space $\mathcal{H}$ is defined as

$$\mathcal{E}^{\psi}(f_{\mathcal{H}}^{\psi}) - \mathcal{E}^{\psi}(f_{\rho}^{\psi}) = \min_{f \in \mathcal{H}} \mathcal{E}^{\psi}(f) - \mathcal{E}^{\psi}(f_{\rho}^{\psi}).$$

Proposition 4.21 *Let ψ be a regression loss function. If $\epsilon \geq 0$ is the largest zero of ψ and $|y| \leq M$ almost surely, then, for all $x \in X$, $f_{\rho}^{\psi}(x) \in [-M - \epsilon, M + \epsilon]$.* ∎

The approximation error can be estimated as follows [141].

Theorem 4.22 *Assume $|y - f(x)| \leq M$ and $|y - f_{\rho}^{\psi}(x)| \leq M$ almost surely.*

(i) *If ψ is a regression loss function satisfying, for some $0 < s \leq 1$,*

$$\sup_{t, t' \in [-M, M]} \frac{|\psi(t) - \psi(t')|}{|t - t'|^s} = C < \infty, \tag{4.4}$$

then

$$\mathcal{E}^{\psi}(f) - \mathcal{E}^{\psi}(f_{\rho}^{\psi}) \leq C \int_{X} |f(x) - f_{\rho}^{\psi}(x)|^{s} \, d\rho_{X}$$

$$\leq C \|f - f_{\rho}^{\psi}\|_{\mathscr{L}_{\rho_{X}}^{1}}^{s} \leq C \|f - f_{\rho}^{\psi}\|_{\mathscr{L}_{\rho_{X}}^{2}}^{s/2} .$$

(ii) *If ψ is $\mathscr{C}^{1}$ on $[-M, M]$ and its derivative satisfies (4.4), then*

$$\mathcal{E}^{\psi}(f) - \mathcal{E}^{\psi}(f_{\rho}^{\psi}) \leq C \|f - f_{\rho}^{\psi}\|_{\mathscr{L}_{\rho_{X}}^{1+s}}^{1+s} \leq C \|f - f_{\rho}^{\psi}\|_{\mathscr{L}_{\rho_{X}}^{2}}^{(1+s)/2} .$$

(iii) *If $\psi''(u) \geq c > 0$ for every $u \in [-M, M]$, then*

$$\mathcal{E}^{\psi}(f) - \mathcal{E}^{\psi}(f_{\rho}^{\psi}) \geq \tfrac{c}{2} \|f - f_{\rho}^{\psi}\|_{\mathscr{L}_{\rho_{X}}^{2}}^{2} . \qquad \blacksquare$$

5

Estimating covering numbers

The bounds for the sample error described in Chapter 3 are in terms of, among other quantities, some covering numbers. In this chapter, we provide estimates for these covering numbers when we take a ball in an RKHS as a hypothesis space. Our estimates are given in terms of the regularity of the kernel. As a particular case, we obtain the following.

Theorem 5.1 *Let X be a compact subset of $\mathbb{R}^n$, and $\mathrm{Diam}(X) := \max_{x,y \in X} \|x - y\|$ its diameter.*

(i) *If $K \in \mathscr{C}^s(X \times X)$ for some $s > 0$ and X has piecewise smooth boundary, then there is $C > 0$ depending on X and s only such that*

$$\ln \mathcal{N}\left(I_K(B_R), \eta\right) \le C(\mathrm{Diam}(X))^n \|K\|_{\mathscr{C}^s(X \times X)}^{n/s} \left(\frac{R}{\eta}\right)^{2n/s}, \quad \forall 0 < \eta \le R/2.$$

(ii) *If $K(x, y) = \exp\left\{-\|x - y\|^2/\sigma^2\right\}$ for some $\sigma > 0$, then, for all $0 < \eta \le R/2$,*

$$\ln \mathcal{N}\left(I_K(B_R), \eta\right) \le n \left(32 + \frac{640n(\mathrm{Diam}(X))^2}{\sigma^2}\right)^{n+1} \left(\ln \frac{R}{\eta}\right)^{n+1}.$$

If, moreover, X contains a cube in the sense that $X \supseteq x^* + [-\frac{\Delta}{2}, \frac{\Delta}{2}]^n$ for some $x^* \in X$ and $\Delta > 0$, then, for all $0 < \eta \le R/2$,

$$\ln \mathcal{N}\left(I_K(B_R), \eta\right) \ge C_1 \left(\ln \frac{R}{\eta}\right)^{n/2}.$$

Here C_1 is a positive constant depending only on σ and Δ.

Part (i) of Theorem 5.1 follows from Theorem 5.5 and Lemma 5.6. It shows how the covering number decreases as the index s of the Sobolev smooth kernel increases. A case where the hypothesis of Part (ii) applies is that of the box spline kernels described in Example 2.17. We show this is so in Proposition 5.25.

When the kernel is analytic, better than Sobolev smoothness for any index $s > 0$, one can see from Part (i) that $\ln \mathcal{N}(I_K(B_R), \eta)$ decays at a rate faster than $(R/\eta)^\epsilon$ for any $\epsilon > 0$. Hence one would expect a decay rate such as $(\ln(R/\eta))^s$ for some s. This is exactly what Part (ii) of Theorem 5.1 shows for Gaussian kernels. The lower bound stated in Part (ii) also tells us that the upper bound is almost sharp. The proof for Part (ii) is given in Corollaries 5.14 and 5.24 together with Proposition 5.13, where an explicit formula for the constant C_1 can be found.

5.1 Reminders IV

To prove the main results of this chapter, we use some basic knowledge from function spaces and approximation theory.

Approximation theory studies the approximation of functions by functions in some "good" family – for example, polynomials, splines, wavelets, radial basis functions, ridge functions. The quality of the approximation usually depends on, in addition to the size of the approximating family, the regularity of the approximated function. In this section, we describe some common measures of regularity for functions.

(I) Consider functions on an arbitrary metric space (X, d). Let $0 < s \le 1$. We say that a continuous function f on X is *Lipschitz-s* when there exists a constant $C > 0$ such that for all $x, y \in X$,

$$|f(x) - f(y)| \le C(d(x, y))^s.$$

We denote by $\mathsf{Lip}(s)$ the space of all Lipschitz-s functions with the norm

$$\|f\|_{\mathsf{Lip}(s)} := |f|_{\mathsf{Lip}(s)} + \|f\|_{\mathscr{C}(X)},$$

where $|\ |_{\mathsf{Lip}(s)}$ is the seminorm

$$|f|_{\mathsf{Lip}(s)} := \sup_{x \ne y \in X} \frac{|f(x) - f(y)|}{(d(x, y))^s}.$$

This is a Banach space. The regularity of a function $f \in \mathsf{Lip}(s)$ is measured by the index s. The bigger the index s, the higher the regularity of f.

(II) When X is a subset of a Euclidean space $\mathbb{R}^n$, we can consider a more general measure of regularity. This can be done by means of various orders of divided differences.

Let X be a closed subset of $\mathbb{R}^n$ and $f : X \to \mathbb{R}$. For $r \in \mathbb{N}$, $t \in \mathbb{R}^n$, and $x \in X$ such that $x, x + t, \ldots, x + rt \in X$, define the *divided difference*

$$\Delta_t^r f(x) := \sum_{j=0}^{r} \binom{r}{j} (-1)^{r-j} f(x + jt).$$

In particular, when $r = 1$, $\Delta_t^1 f(x) = f(x + t) - f(x)$. Divided differences can be used to characterize various types of function spaces. Let

$$X_{r,t} = \{x \in X \mid x, x + t, \ldots, x + rt \in X\}.$$

For $0 < s < r$ and $1 \le p < \infty$, the *generalized Lipschitz space* $\mathrm{Lip}^*(s, \mathscr{L}^p(X))$ consists of functions f in $\mathscr{L}^p(X)$ for which the seminorm

$$|f|_{\mathrm{Lip}^*(s, \mathscr{L}^p(X))} := \sup_{t \in \mathbb{R}^n} \|t\|^{-s} \left\{ \int_{X_{r,t}} |\Delta_t^r f(x)|^p \, dx \right\}^{1/p}$$

is finite. This is a Banach space with the norm

$$\|f\|_{\mathrm{Lip}^*(s, \mathscr{L}^p(X))} := |f|_{\mathrm{Lip}^*(s, \mathscr{L}^p(X))} + \|f\|_{\mathscr{L}^p(X)}.$$

The space $\mathrm{Lip}^*(s, \mathscr{C}(X))$ is defined in a similar way, taking

$$|f|_{\mathrm{Lip}^*(s, \mathscr{C}(X))} := \sup_{t \in \mathbb{R}^n} \|t\|^{-s} \sup_{x \in X_{r,t}} |\Delta_t^r f(x)|$$

and

$$\|f\|_{\mathrm{Lip}^*(s, \mathscr{C}(X))} := |f|_{\mathrm{Lip}^*(s, \mathscr{C}(X))} + \|f\|_{\mathscr{C}(X)}.$$

Clearly, $\|f\|_{\mathrm{Lip}^*(s, \mathscr{L}^p(X))} \le (\mu(X))^{1/p} \|f\|_{\mathrm{Lip}^*(s, \mathscr{C}(X))}$ for all $p < \infty$ when X is compact.

When X has piecewise smooth boundary, each function $f \in \mathrm{Lip}^*(s, \mathscr{L}^p(X))$ can be extended to $\mathbb{R}^n$. If we still denote this extension by f, then there exists a constant $C_{X,s,p}$ depending on X, s, and p such that for all $f \in \mathrm{Lip}^*(s, \mathscr{L}^p(X))$,

$$\|f\|_{\mathrm{Lip}^*(s, \mathscr{L}^p(\mathbb{R}^n))} \leq C_{X,s,p} \|f\|_{\mathrm{Lip}^*(s, \mathscr{L}^p(X))}.$$

When $p = \infty$, we write $\|f\|_{\mathrm{Lip}^*(s)}$ instead of $\|f\|_{\mathrm{Lip}^*(s, \mathscr{C}(X))}$. In this case, under some mild regularity condition for X (e.g., when the boundary of X is piecewise smooth), for s not an integer, $s = \ell + s_0$ with $\ell \in \mathbb{N}$, and $0 < s_0 < 1$, $\mathrm{Lip}^*(s, \mathscr{C}(X))$ consists of continuous functions on X such that $D^\alpha f \in \mathrm{Lip}(s_0)$, for any $\alpha = (\alpha_1, \ldots, \alpha_n) \in \mathbb{N}^n$ with $|\alpha| \leq \ell$. In particular, if $X = [0, 1]^n$ or $\mathbb{R}^n$. In this case, $\mathrm{Lip}^*(s, \mathscr{C}(X)) = \mathscr{C}^s(X)$ when s is not an integer, and $\mathscr{C}^s(X) \subset \mathrm{Lip}^*(s, \mathscr{C}(X))$ when s is an integer. Also, $\mathscr{C}^1(X) \subset \mathrm{Lip}(1) \subset \mathrm{Lip}^*(1, \mathscr{C}(X))$, the last being known as the *Zygmund class*.

Again, the regularity of a function $f \in \mathrm{Lip}^*(s, \mathscr{L}^p(X))$ is measured by the index s. The bigger the index s, the higher the regularity of f.

(III) When $p = 2$ and $X = \mathbb{R}^n$, it is also natural to measure the regularity of functions in $\mathscr{L}^2(\mathbb{R}^n)$ by means of the Fourier transform. Let $s > 0$. The *fractional Sobolev space* $H^s(\mathbb{R}^n)$ consists of functions in $\mathscr{L}^2(\mathbb{R}^n)$ such that the following norm is finite:

$$\|f\|_{H^s(\mathbb{R}^n)} := \left\{ \frac{1}{(2\pi)^n} \int_{\mathbb{R}^n} (1 + \|\xi\|^2)^s |\widehat{f}(\xi)|^2 \, d\xi \right\}^{1/2}.$$

When $s \in \mathbb{N}$, $H^s(\mathbb{R}^n)$ coincides with the Sobolev space defined in Section 2.3.

Note that for a function $f \in H^s(\mathbb{R}^n)$, its regularity s is tied to the decay of $\widehat{f}$. The larger s is, the faster the decay of $\widehat{f}$ is. These subspaces of $\mathscr{L}^2(\mathbb{R}^n)$ and those described in (II) are related as follows. For any $\epsilon > 0$,

$$\mathrm{Lip}^*(s, \mathscr{L}^2(\mathbb{R}^n)) \subset H^s(\mathbb{R}^n) \subset \mathrm{Lip}^*(s - \epsilon, \mathscr{L}^2(\mathbb{R}^n)).$$

For any integer $d < s - \frac{n}{2}, f \in \mathscr{C}^d(\mathbb{R}^n)$ and $\|f\|_{\mathscr{C}^d} \leq C_d \|f\|_{H^s}$. In particular, if $s > \frac{n}{2}$, it follows by taking $d = 0$ that $H^s(\mathbb{R}^n) \subseteq \mathscr{C}(\mathbb{R}^n)$. Note that this is the Sobolev embedding theorem mentioned in Section 2.3 for $X = \mathbb{R}^n$. (These facts can be easily shown using the inverse Fourier transform when $s > n$ and $d < s - n$. When $n \geq s > \frac{n}{2}$ and $s - n \leq d < s - \frac{n}{2}$, the proofs are more involved.) Thus, if $X \subseteq \mathbb{R}^n$ has piecewise smooth boundary and $d < s - \frac{n}{2}$, each function $f \in \mathrm{Lip}^*(s, \mathscr{L}^2(X))$ can be extended to a $\mathscr{C}^d$ function on $\mathbb{R}^n$ and there exists a constant $C_{X,s,d}$ such that for all $f \in \mathrm{Lip}^*(s, \mathscr{L}^2(X))$,

$$\|f\|_{\mathscr{C}^d(\mathbb{R}^n)} \leq C_{X,s,d} \|f\|_{\mathrm{Lip}^*(s, \mathscr{L}^2(X))}. \tag{5.1}$$

5.2 Covering numbers for Sobolev smooth kernels

Recall that if $\mathbb{E}$ is a Banach space and $R > 0$, we denote

$$B_R(\mathbb{E}) = \{x \in \mathbb{E} : \|x\| \leq R\}.$$

If the space $\mathbb{E}$ is clear from the context, we simply write B_R.

Lemma 5.2 *Let* $\mathbb{E} \subseteq \mathscr{C}(X)$ *be a Banach space. For all* $\eta, R > 0$,

$$\mathcal{N}(B_R, \eta) = \mathcal{N}\left(B_1, \frac{\eta}{R}\right).$$

Proof. The proof follows from the fact that $\{B(f_1, \eta), \ldots, B(f_k, \eta)\}$ is a covering of B_R if and only if $\{B(f_1/R, \eta/R), \ldots, B(f_k/R, \eta/R)\}$ is a covering of B_1. ∎

It follows from Lemma 5.2 that it is enough to estimate covering numbers of the unit ball. We start with balls in finite-dimensional spaces.

Theorem 5.3 *Let E be a finite-dimensional Banach space, $N = \dim E$, and $R > 0$. For $0 < \eta < R$,*

$$\mathcal{N}(B_R, \eta) \leq \left(\frac{2R}{\eta} + 1\right)^N$$

and, for $\eta \geq R$, $\mathcal{N}(B_R, \eta) = 1$.

Proof. Choose a basis $\{e_j\}_{j=1}^N$ of E. Define a norm $|\ |$ on $\mathbb{R}^N$ by

$$|x| := \left\|\sum_{j=1}^N x_j e_j\right\|, \quad x = (x_1, \ldots, x_N) \in \mathbb{R}^N.$$

Let $\eta > 0$. Suppose that $\mathcal{N}(B_R, \eta) > ((2R/\eta)+1)^N$. Then B_R cannot be covered by $((2R/\eta)+1)^N$ balls with radius η. Hence we can find elements $f^{(1)}, \ldots, f^{(\ell)}$ in B_R such that

$$\ell > \left(\frac{2R}{\eta} + 1\right)^N \quad \text{and} \quad f^{(j)} \notin \bigcup_{i=1}^{j-1} B(f^{(i)}, \eta), \quad \forall j \in \{2, \ldots, \ell\}.$$

Therefore, for $i \neq j$, $\|f^{(i)} - f^{(j)}\| > \eta$.

Set $f^{(j)} = \sum_{m=1}^N x_m^{(j)} e_m \in B_R$ and $x^{(j)} = (x_1^{(j)}, \ldots, x_N^{(j)}) \in \mathbb{R}^N$. Then, for $i \neq j$,

$$|x^{(i)} - x^{(j)}| > \eta.$$

Also, $|x^{(j)}| \le R$.

Denote by $\widetilde{B}_r$ the ball of radius $r > 0$ centered on the origin in $(\mathbb{R}^N, |\ |)$. Then

$$\bigcup_{j=1}^{\ell} \left\{ x^{(j)} + \frac{\eta}{2} \widetilde{B}_1 \right\} \subseteq \widetilde{B_{R+\frac{\eta}{2}}},$$

and the sets in this union are disjoint. Therefore, if μ denotes the Lebesgue measure on $\mathbb{R}^N$,

$$\mu \left(\bigcup_{j=1}^{\ell} \left\{ x^{(j)} + \frac{\eta}{2} \widetilde{B}_1 \right\} \right) = \sum_{j=1}^{\ell} \mu \left(x^{(j)} + \frac{\eta}{2} \widetilde{B}_1 \right) \le \mu \left(\widetilde{B_{R+\frac{\eta}{2}}} \right).$$

It follows that

$$\ell \left(\frac{\eta}{2} \right)^N \mu \left(\widetilde{B}_1 \right) \le \left(R + \frac{\eta}{2} \right)^N \mu \left(\widetilde{B}_1 \right),$$

and thereby

$$\ell \le \left(\frac{R + (\eta/2)}{\eta/2} \right)^N = \left(\frac{2R}{\eta} + 1 \right)^N.$$

This is a contradiction. Therefore we must have, for all $\eta > 0$,

$$\mathcal{N}(B_R, \eta) \le \left(\frac{2R}{\eta} + 1 \right)^N.$$

If, in addition, $\eta \ge R$, then B_R can be covered by the ball with radius η centered on the origin and hence $\mathcal{N}(B_R, \eta) = 1$.   ■

The study of covering numbers is a standard topic in the field of function spaces. The asymptotic behavior of the covering numbers for Sobolev spaces is a well-known result. For example, the ball $B_R(\mathsf{Lip}^*(s, \mathscr{C}([0,1]^n)))$ in the generalized Lipschitz space on $[0,1]^n$ satisfies

$$C_s \left(\frac{R}{\eta} \right)^{n/s} \le \ln \mathcal{N} \left(B_R(\mathsf{Lip}^*(s, \mathscr{C}([0,1]^n))), \eta \right) \le C_s' \left(\frac{R}{\eta} \right)^{n/s}, \qquad (5.2)$$

where the positive constants C_s and C_s' depend only on s and n (i.e., they are independent of R and η).

We will not prove the bound (5.2) here in all its generality (references for a proof can be found in Section 5.6). However, to give an idea of the methods involved in such a proof, we deal next with the special case $0 < s < 1$ and $n = 1$. Recall that in this case, $\mathrm{Lip}^*(s, \mathscr{C}([0, 1])) = \mathrm{Lip}(s)$. Since $B_1(\mathrm{Lip}(s)) \subset B_1(\mathscr{C}([0, 1]))$, we have $\mathscr{N}(B_1(\mathrm{Lip}(s)), \eta) = 1$ for all $\eta \geq 1$.

Proposition 5.4 *Let $0 < s < 1$ and $X = [0, 1]$. Then, for all $0 < \eta \leq \frac{1}{4}$,*

$$\frac{1}{8}\left(\frac{1}{2\eta}\right)^{1/s} \leq \ln \mathscr{N}(B_1(\mathrm{Lip}(s)), \eta) \leq 4\left(\frac{4}{\eta}\right)^{1/s}.$$

The restriction $\eta \leq \frac{1}{4}$ is required only for the lower bound.

Proof. We first deal with the upper bound. Set $\varepsilon = (\eta/4)^{1/s}$. Define $\mathbf{x} = \{x_i = i\varepsilon\}_{i=1}^d$, where $d = \lfloor \frac{1}{\varepsilon} \rfloor$ denotes the integer part of $1/\varepsilon$. Then $\mathbf{x}$ is an ε-net of $X = [0, 1]$ (i.e., for all $x \in X$, the distance from x to $\mathbf{x}$ is at most ε). If $f \in B_1(\mathrm{Lip}(s))$, then $\|f\|_{\mathscr{C}(X)} \leq 1$ and $-1 \leq f(x_i) \leq 1$ for all $i = 1, \ldots, d$. Hence, $(v_i - 1)\frac{\eta}{2} \leq f(x_i) \leq v_i\frac{\eta}{2}$ for some $v_i \in J := \{-m + 1, \ldots, m\}$, where m is the smallest integer greater than $\frac{2}{\eta}$. For $v = (v_1, \ldots, v_d) \in J^d$ define

$$V_v := \left\{ f \in B_1(\mathrm{Lip}(s)) \mid (v_i - 1)\frac{\eta}{2} \leq f(x_i) \leq v_i\frac{\eta}{2} \quad \text{for } i = 1, \ldots, d \right\}.$$

Then $B_1(\mathrm{Lip}(s)) \subseteq \bigcup_{v \in J^d} V_v$. If $f, g \in V_v$, then, for each $i \in \{1, \ldots, d\}$,

$$\max_{|x - x_i| \leq \varepsilon} |f(x) - g(x)| \leq |f(x_i) - g(x_i)| + \max_{|x - x_i| \leq \varepsilon} |f(x) - f(x_i)|$$

$$+ \max_{|x - x_i| \leq \varepsilon} |g(x) - g(x_i)|$$

$$\leq \frac{\eta}{2} + 2\varepsilon^s = \eta.$$

Therefore, V_v has diameter at most η as a subset of $\mathscr{C}(X)$. That is, $\{V_v\}_{v \in J^d}$ is an η-covering of $B_1(\mathrm{Lip}(s))$. What is left is to count nonempty sets V_v.

If V_v is nonempty, then V_v contains some function $f \in B_1(\mathrm{Lip}(s))$. Since $-\frac{\eta}{2} \leq f(x_{i+1}) - v_{i+1}\frac{\eta}{2} \leq 0$ and $-\frac{\eta}{2} \leq f(x_i) - v_i\frac{\eta}{2} \leq 0$, we have

$$-\frac{\eta}{2} \leq f(x_{i+1}) - f(x_i) - (v_{i+1} - v_i)\frac{\eta}{2} \leq \frac{\eta}{2}.$$

It follows that for each $i = 1, \ldots, d - 1$,

$$|v_{i+1} - v_i|\frac{\eta}{2} - \frac{\eta}{2} \leq |f(x_{i+1}) - f(x_i)| \leq |x_{i+1} - x_i|^s \leq \varepsilon^s.$$

This yields $|v_{i+1} - v_i| \le \frac{2}{\eta}(\varepsilon^s + \frac{\eta}{2}) = \frac{3}{2}$ and then $v_{i+1} \in \{v_i - 1, v_i, v_i + 1\}$. Since v_1 has $2m$ possible values, the number of nonempty V_v is at most $2m \cdot 3^{d-1}$. Therefore,

$$\ln \mathcal{N}(B_1(\mathsf{Lip}(s)), \eta) \le \ln(2m \cdot 3^{d-1})$$
$$= \ln 2 + (d-1)\ln 3 + \ln m$$
$$\le \ln 2 + \frac{1}{\varepsilon}\ln 3 + \ln\left(\frac{2}{\eta} + 1\right).$$

But $\ln(1 + t) \le t$ for all $t \ge 0$, so we have

$$\ln\left(\frac{2}{\eta} + 1\right) \le \frac{2}{\eta} \le 2\left(\frac{1}{\eta}\right)^{1/s}$$

and hence

$$\ln \mathcal{N}(B_1(\mathsf{Lip}(s)), \eta) \le \ln 2 + \left(\frac{4}{\eta}\right)^{1/s}\ln 3 + 2\left(\frac{1}{\eta}\right)^{1/s} \le 4\left(\frac{4}{\eta}\right)^{1/s}.$$

We now prove the lower bound. Set $\varepsilon = (2\eta)^{1/s}$ and $\mathbf{x}$ as above. For $i = 1, \ldots, d - 1$, define f_i to be the hat function of height $\frac{\eta}{2}$ on the interval $[x_i - \varepsilon, x_i + \varepsilon]$; that is,

$$f_i(x_i + t) = \begin{cases} \frac{\eta}{2} - \frac{\eta}{2\varepsilon}t & \text{if } 0 \le t \le \varepsilon \\ \frac{\eta}{2} + \frac{\eta}{2\varepsilon}t & \text{if } -\varepsilon \le t < 0 \\ 0 & \text{if } t \notin [-\varepsilon, \varepsilon]. \end{cases}$$

Note that $f_i(x_j) = \frac{\eta}{2}\delta_{ij}$.

For every nonempty subset I of $\{1, \ldots, d - 1\}$ we define

$$f_I(x) = \sum_{i \in I} f_i(x).$$

If $I_1 \ne I_2$, there is some $i \in (I_1 \setminus I_2) \cup (I_2 \setminus I_1)$ and

$$\|f_{I_1} - f_{I_2}\|_{\mathscr{C}(X)} \ge |f_{I_1}(x_i) - f_{I_2}(x_i)| = \frac{\eta}{2}.$$

It follows that $\mathcal{N}(B_1(\mathsf{Lip}(s)), \eta)$ is at least the number of nonempty subsets of $\{1, \ldots, d - 1\}$ (i.e., $2^{d-1} - 1$), provided that each f_I lies in $B_1(\mathsf{Lip}(s))$. Let us prove that this is the case.

Observe that f_I is piecewise linear on each $[x_i, x_{i+1}]$ and its values on x_i are either $\frac{\eta}{2}$ or 0. Hence $\|f_I\|_{\mathscr{C}(X)} \leq \frac{\eta}{2} \leq \frac{1}{2}$. To evaluate the Lipschitz-s seminorm of f_I we take $x, x + t \in X$ with $t > 0$. If $t \geq \varepsilon$, then

$$\frac{|f_I(x+t) - f_I(x)|}{t^s} \leq \frac{2\|f_I\|_{\mathscr{C}(X)}}{\varepsilon^s} \leq \frac{\eta}{\varepsilon^s} = \frac{1}{2}.$$

If $t < \varepsilon$ and $x_i \notin (x, x + t)$ for all $i \leq d - 1$, then f_I is linear on $[x, x + t]$ with slope at most $\frac{\eta}{2\varepsilon}$ and hence

$$\frac{|f_I(x+t) - f_I(x)|}{t^s} \leq \frac{(\eta/2\varepsilon)t}{t^s} = \frac{\eta}{2\varepsilon}t^{1-s} < \frac{\eta}{2\varepsilon}\varepsilon^{1-s} = \frac{1}{4}.$$

If $t < \varepsilon$ and $x_i \in (x, x + t)$ for some $i \leq d - 1$, then

$$|f_I(x+t) - f_I(x)| \leq |f_I(x+t) - f_I(x_i)| + |f_I(x_i) - f_I(x)|$$

$$\leq \frac{\eta}{2\varepsilon}(x+t-x_i) + \frac{\eta}{2\varepsilon}(x_i - x) = \frac{\eta}{2\varepsilon}t$$

and hence

$$\frac{|f_I(x+t) - f_I(x)|}{t^s} \leq \frac{\eta}{2\varepsilon}t^{1-s} \leq \frac{\eta}{2}\varepsilon^{-s} = \frac{1}{4}.$$

Thus, in all three cases, $|f_I|_{\mathsf{Lip}(s)} \leq \frac{1}{2}$, and therefore $\|f_I\|_{\mathsf{Lip}(s)} \leq 1$. This shows $f_I \in B_1(\mathsf{Lip}(s))$.

Finally, since $\eta \leq \frac{1}{4}$, we have $\varepsilon \leq \frac{1}{2}$ and $d \geq 2$. It follows that

$$\mathcal{N}(B_1(\mathsf{Lip}(s)), \eta) \geq 2^{d-1} - 1 \geq 2^{d-2} \geq 2^{(1/2\varepsilon)-2},$$

which implies

$$\ln \mathcal{N}(B_1(\mathsf{Lip}(s)), \eta) \geq \frac{1}{2}\ln 2 \left(\frac{1}{2\eta}\right)^{1/s} - \ln 4 \geq \frac{1}{8}\left(\frac{1}{2\eta}\right)^{1/s}. \qquad \blacksquare$$

Now we can give some upper bounds for the covering number of balls in RKHSs. The bounds depend on the regularity of the Mercer kernel. When the kernel K has Sobolev or generalized Lipschitz regularity, we can show that the RKHS $\mathcal{H}_K$ can be embedded into a generalized Lipschitz space. Then an estimate for the covering number follows.

Theorem 5.5 *Let X be a closed subset of $\mathbb{R}^n$, and $K : X \times X \to \mathbb{R}$ be a Mercer kernel. If $s > 0$ and $K \in \mathsf{Lip}^*(s, \mathscr{C}(X \times X))$, then $\mathcal{H}_K \subset \mathsf{Lip}^*(\frac{s}{2}, \mathscr{C}(X))$ and, for all $r \in \mathbb{N}$, $r > s$,*

$$\|f\|_{\mathsf{Lip}^*(s/2)} \leq \sqrt{2^{r+1}\|K\|_{\mathsf{Lip}^*(s)}}\|f\|_K, \quad \forall f \in \mathcal{H}_K.$$

Proof. Let $s < r \in \mathbb{N}$ and $f \in \mathcal{H}_K$. Let $x, t \in \mathbb{R}^n$ such that $x, x+t, \ldots, x+rt \in X$. By Theorem 2.9(iii),

$$\Delta_t^r f(x) = \sum_{j=0}^r \binom{r}{j}(-1)^{r-j}\langle K_{x+jt}, f\rangle_K = \left\langle \sum_{j=0}^r \binom{r}{j}(-1)^{r-j}K_{x+jt}, f\right\rangle_K.$$

It follows from the Cauchy–Schwarz inequality that

$$|\Delta_t^r f(x)| \leq \|f\|_K \left(\sum_{j=0}^r \binom{r}{j}(-1)^{r-j}\sum_{i=0}^r \binom{r}{i}(-1)^{r-i}K(x+jt, x+it)\right)^{1/2}$$

$$= \|f\|_K \left(\sum_{j=0}^r \binom{r}{j}(-1)^{r-j}\Delta_{(0,t)}^r K(x+jt, x)\right)^{1/2}.$$

Here $(0, t)$ denotes the vector in $\mathbb{R}^{2n}$ where the first n components are zero. By hypothesis, $K \in \mathsf{Lip}^*(s, \mathscr{C}(X \times X))$. Hence

$$\left|\Delta_{(0,t)}^r K(x+jt, x)\right| \leq |K|_{\mathsf{Lip}^*(s)}\|t\|^s.$$

This yields

$$|\Delta_t^r f(x)| \leq \|f\|_K \left(\sum_{j=0}^r \binom{r}{j}|K|_{\mathsf{Lip}^*(s)}\|t\|^s\right)^{1/2} \leq \sqrt{2^r|K|_{\mathsf{Lip}^*(s)}}\|f\|_K\|t\|^{s/2}.$$

Therefore,

$$|f|_{\mathsf{Lip}^*(\frac{s}{2})} \leq \sqrt{2^r|K|_{\mathsf{Lip}^*(s)}}\|f\|_K.$$

Combining this inequality with the fact (cf. Theorem 2.9) that

$$\|f\|_\infty \leq \sqrt{\|K\|_\infty}\|f\|_K,$$

we conclude that $f \in \mathsf{Lip}^*(\frac{s}{2}, \mathscr{C}(X))$ and

$$\|f\|_{\mathsf{Lip}^*(s/2)} \leq \sqrt{2^{r+1} \|K\|_{\mathsf{Lip}^*(s)}} \, \|f\|_K.$$

The proof of the theorem is complete.   ∎

Lemma 5.6 *Let* $n \in \mathbb{N}$, $s > 0$, $D \geq 1$, *and* $x^* \in \mathbb{R}^n$. *If* $X \subseteq x^* + D[0,1]^n$ *and* X *has piecewise smooth boundary then*

$$\mathcal{N}(B_R(\mathsf{Lip}^*(s, \mathscr{C}(X))), \eta) \leq \mathcal{N}(B_{C_{X,s}D^sR}(\mathsf{Lip}^*(s, \mathscr{C}([0,1]^n))), \eta/2),$$

where $C_{X,s}$ *is a constant depending only on* X *and* s.

Proof. For $f \in \mathscr{C}([0,1]^n)$ define $f^* \in \mathscr{C}(x^* + D[0,1]^n)$ by

$$f^*(x) = f\left(\frac{x - x^*}{D}\right).$$

Then $f^* \in \mathsf{Lip}^*(s, \mathscr{C}(x^* + D[0,1]^n))$ if and only if $f \in \mathsf{Lip}^*(s, \mathscr{C}([0,1]^n))$. Moreover,

$$D^{-s}\|f\|_{\mathsf{Lip}^*(s,\mathscr{C}([0,1]^n))} \leq \|f^*\|_{\mathsf{Lip}^*(s,\mathscr{C}(x^*+D[0,1]^n))} \leq \|f\|_{\mathsf{Lip}^*(s,\mathscr{C}([0,1]^n))}.$$

Since $X \subseteq x^* + D[0,1]^n$ and X has piecewise smooth boundary, there is a constant $C_{X,s}$ depending only on X and s such that

$$B_R(\mathsf{Lip}^*(s, \mathscr{C}(X))) \subseteq \{f^*|_X \mid f^* \in B_{C_{X,s}R}(\mathsf{Lip}^*(s, \mathscr{C}(x^* + D[0,1]^n)))\}$$

$$\subseteq \{f^*|_X \mid f \in B_{C_{X,s}D^sR}(\mathsf{Lip}^*(s, \mathscr{C}([0,1]^n)))\}.$$

Let $\{f_1, \ldots, f_N\}$ be an $\frac{\eta}{2}$-net of $B_{C_{X,s}D^sR}(\mathsf{Lip}^*(s, \mathscr{C}([0,1]^n)))$ and N its covering number. For each $j = 1, \ldots, N$ take a function $g_j^*|_X \in B_R(\mathsf{Lip}^*(s, \mathscr{C}(X)))$ with $g_j \in B_{C_{X,s}D^sR}(\mathsf{Lip}^*(s, \mathscr{C}([0,1]^n)))$ and $\|g_j - f_j\|_{\mathsf{Lip}^*(s,\mathscr{C}([0,1]^n))} \leq \frac{\eta}{2}$ if it exists. Then $\{g_j^*|_X \mid j = 1, \ldots, N\}$ provides an η-net of $B_R(\mathsf{Lip}^*(s, \mathscr{C}(X)))$: each $f \in B_R(\mathsf{Lip}^*(s, \mathscr{C}(X)))$ can be written as the restriction $g^*|_X$ to X of some function $g \in B_{C_{X,s}D^sR}(\mathsf{Lip}^*(s, \mathscr{C}([0,1]^n)))$, so there is some j such that $\|g - f_j\|_{\mathsf{Lip}^*(s,\mathscr{C}([0,1]^n))} \leq \frac{\eta}{2}$. This implies that

$$\|f - g_j^*|_X\|_{\mathsf{Lip}^*(s,\mathscr{C}(X))} \leq \|g^* - g_j^*\|_{\mathsf{Lip}^*(s,\mathscr{C}(x^*+D[0,1]^n))}$$

$$\leq \|g - g_j\|_{\mathsf{Lip}^*(s,\mathscr{C}([0,1]^n))} \leq \eta.$$

This proves the statement.   ∎

Proof of Theorem 5.1(i)   Recall that $\mathscr{C}^s(X) \subset \mathsf{Lip}^*(s, \mathscr{C}(X))$ for any $s > 0$. Then, by Theorem 5.5 with $s < r \leq s + 1$, the assumption $K \in \mathscr{C}^s(X \times X)$ implies that $I_K(B_R) \subseteq B_{\sqrt{2^{s+2}\|K\|_{\mathsf{Lip}^*_{(s)}}}R}(\mathsf{Lip}^*(s/2, \mathscr{C}(X)))$. This, together with (5.2) and Lemma 5.6 with $D \geq \mathsf{Diam}(X)$, shows Theorem 5.1(i). ∎

When s is not an integer, and the boundary of X is piecewise smooth, $\mathscr{C}^s(X) = \mathsf{Lip}^*(s, \mathscr{C}(X))$. As a corollary of Theorem 5.5, we have the following.

Proposition 5.7 *Let X be a closed subset of $\mathbb{R}^n$ with piecewise smooth boundary, and $K : X \times X \to \mathbb{R}$ a Mercer kernel. If $s > 0$ is not an even integer and $K \in \mathscr{C}^s(X \times X)$, then $\mathcal{H}_K \subset \mathscr{C}^{s/2}(X)$ and*

$$\|f\|_{\mathsf{Lip}^*(s/2)} \leq \sqrt{2^{r+1}\|K\|_{\mathsf{Lip}^*_{(s)}}}\|f\|_K, \quad \forall f \in \mathcal{H}_K.$$ ∎

Theorem 5.5 and the upper bound in (5.2) yield upper-bound estimates for the covering numbers of RKHSs when the Mercer kernel has Sobolev regularity.

Theorem 5.8 *Let X be a closed subset of $\mathbb{R}^n$ with piecewise smooth boundary, and $K : X \times X \to \mathbb{R}$ a Mercer kernel. Let $s > 0$ such that K belongs to $\mathsf{Lip}^*(s, \mathscr{C}(X \times X))$. Then, for all $0 < \eta \leq R$,*

$$\ln \mathcal{N}(I_K(B_R), \eta) \leq C \left(\frac{R}{\eta}\right)^{2n/s},$$

where C is a constant independent of R and η. ∎

It is natural to expect covering numbers to have smaller upper bounds when the kernel is analytic, a regularity stronger than Sobolev smoothness. Proving this is our next step.

5.3 Covering numbers for analytic kernels

In this section we continue our discussion of the covering numbers of balls of RKHSs and provide better estimates for analytic kernels. We consider a convolution kernel K given by $K(x, t) = k(x - t)$, where k is an even function in $\mathscr{L}^2(\mathbb{R}^n)$ and $\widehat{k}(\xi) > 0$ almost everywhere on $\mathbb{R}^n$. Let $X = [0, 1]^n$. Then K is a Mercer kernel on X. Our purpose here is to bound the covering number $\mathcal{N}(I_K(B_R), \eta)$ when k is analytic.

We will use the Lagrange interpolation polynomials. Denote by $\Pi_s(\mathbb{R})$ the space of real polynomials in one variable of degree at most s. Let

$t_0, \ldots, t_s \in \mathbb{R}$ be different. We say that $w_{l,s} \in \Pi_s(\mathbb{R})$, $l = 0, \ldots, s$, are the *Lagrange interpolation polynomials* with interpolating points $\{t_0, \ldots, t_s\}$ when $\sum_{l=0}^{s} w_{l,s}(t) \equiv 1$ and

$$w_{l,s}(t_m) = \delta_{l,m}, \quad l, m \in \{0, 1, \ldots, s\}.$$

It is easy to check that

$$w_{l,s}(t) = \prod_{j \in \{0,1,\ldots,s\} \setminus \{l\}} \frac{t - t_j}{t_l - t_j}$$

satisfy these conditions.

We consider the set of interpolating points $\{0, \frac{1}{s}, \frac{2}{s}, \ldots, 1\}$ and univariate functions $\{w_{l,s}(t)\}_{l=0}^{s}$ defined by

$$w_{l,s}(t) := \sum_{j=l}^{s} \frac{st(st-1)\cdots(st-j+1)}{j!} \binom{j}{l} (-1)^{j-l}. \tag{5.3}$$

Since

$$\sum_{l=0}^{j} \binom{j}{l} (-1)^{j-l} z^l = (z-1)^j,$$

the following functions of the variable z are equal:

$$\sum_{l=0}^{s} w_{l,s}(t) z^l = \sum_{j=0}^{s} \frac{st(st-1)\cdots(st-j+1)}{j!} (z-1)^j. \tag{5.4}$$

In particular, $\sum_{l=0}^{s} w_{l,s}(t) \equiv 1$. In addition, it can be easily checked that

$$w_{l,s}\left(\frac{m}{s}\right) = \delta_{l,m}, \quad l, m \in \{0, 1, \ldots, s\}. \tag{5.5}$$

This means that the $w_{l,s}$ are the Lagrange interpolation polynomials, and hence

$$w_{l,s}(t) = \prod_{j \in \{0,1,\ldots,s\} \setminus \{l\}} \frac{t - j/s}{l/s - j/s} = \prod_{j \in \{0,1,\ldots,s\} \setminus \{l\}} \frac{st - j}{l - j}.$$

The norm of these polynomials (as elements in $\mathscr{C}([0,1])$) can be estimated as follows:

Lemma 5.9 *Let $s \in \mathbb{N}, l \in \{0, 1, \ldots, s\}$. Then, for all $t \in [0, 1]$,*

$$|w_{l,s}(t)| \leq s \begin{pmatrix} s \\ l \end{pmatrix}.$$

Proof. Let $m \in \{0, 1, \ldots, s-1\}$ and $st \in (m, m+1)$. Then for $l \in \{0, 1, \ldots, m-1\}$,

$$|w_{l,s}(t)| = \frac{\left| \prod_{j=0}^{l-1}(st-j) \prod_{j=l+1}^{m}(st-j) \prod_{j=m+1}^{s}(st-j) \right|}{l!(s-l)!}$$

$$\leq \frac{(m+1)!(s-m)!}{(st-l)l!(s-l)!} \leq s \begin{pmatrix} s \\ l \end{pmatrix}.$$

When $l \in \{m+1, \ldots, s\}$,

$$|w_{l,s}(t)| = \frac{\left| \prod_{j=0}^{m}(st-j) \prod_{j=m+1}^{l-1}(st-j) \prod_{j=l+1}^{s}(st-j) \right|}{l!(s-l)!}$$

$$\leq \frac{(m+1)!(s-m)!}{(l-m)l!(s-l)!} \leq s \begin{pmatrix} s \\ l \end{pmatrix}.$$

The case $l = m$ can be dealt with in the same way. ∎

We now turn to the multivariate case. Denote $X_N := \{0, 1, \ldots, N\}^n$. The multivariate polynomials $\{w_{\alpha,N}(x)\}_{\alpha \in X_N}$ are defined as

$$w_{\alpha,N}(x) = \prod_{j=1}^{n} w_{\alpha_j,N}(x_j), \quad x = (x_1, \ldots, x_n), \alpha = (\alpha_1, \ldots, \alpha_n). \quad (5.6)$$

We use the polynomials in (5.6) as a family of multivariate polynomials, not as interpolation polynomials any more. For these polynomials, we have the following result.

Lemma 5.10 *Let $x \in [0, 1]^n$ and $N \in \mathbb{N}$. Then*

$$\sum_{\alpha \in X_N} |w_{\alpha,N}(x)| \leq (N2^N)^n \quad (5.7)$$

and, for $\theta \in [-\frac{1}{2}, \frac{1}{2}]^n$,

$$\left| e^{-i\theta \cdot Nx} - \sum_{\alpha \in X_N} w_{\alpha,N}(x) e^{-i\theta \cdot \alpha} \right| \le n \left(1 + \frac{1}{2^N} \right)^{n-1} \left(\max_{1 \le j \le n} |\theta_j| \right)^N \quad (5.8)$$

holds.

Proof.   The bound (5.7) follows directly from Lemma 5.9.

To derive the second bound (5.8), we first consider the univariate case. Let $t \in [0, 1]$. Then the univariate function z^{Nt} is analytic on the region $|z - 1| \le \frac{1}{2}$. On this region,

$$z^{Nt} = (1 + (z - 1))^{Nt} = \sum_{j=0}^{\infty} \frac{Nt(Nt - 1) \cdots (Nt - j + 1)}{j!} (z - 1)^j.$$

It follows that for $\eta \in [-\frac{1}{2}, \frac{1}{2}]$ and $z = e^{-i\eta}$,

$$\left| e^{-i\eta \cdot Nt} - \sum_{j=0}^{N} \frac{Nt(Nt - 1) \cdots (Nt - j + 1)}{j!} (e^{-i\eta} - 1)^j \right|$$

$$\le \sum_{j=N+1}^{\infty} \left| \frac{Nt(Nt - 1) \cdots (Nt - j + 1)}{j!} \right| |\eta|^j \le |\eta|^N.$$

This, together with (5.4) for $z = e^{-i\eta}$, implies

$$\left| \sum_{l=0}^{N} w_{l,N}(t) e^{-i\eta \cdot l} \right| = \left| \sum_{j=0}^{N} \frac{Nt(Nt - 1) \cdots (Nt - j + 1)}{j!} (e^{-i\eta} - 1)^j \right|$$

$$\le 1 + |\eta|^N \le 1 + \frac{1}{2^N} \quad (5.9)$$

and

$$\left| e^{-i\eta \cdot Nt} - \sum_{l=0}^{N} w_{l,N}(t) e^{-i\eta \cdot l} \right| \le |\eta|^N. \quad (5.10)$$

Now we can derive the bound in the multivariate case. Let $\theta \in [-\frac{1}{2}, \frac{1}{2}]^n$. Then $e^{-i\theta \cdot Nx} = \prod_{m=1}^{n} e^{-i\theta_m \cdot Nx_m}$. We approximate $e^{-i\theta_m \cdot Nx_m}$ by

$\sum_{\alpha_m=0}^{N} w_{\alpha_m,N}(x_m)e^{-i\theta_m\cdot\alpha_m}$ for $m = 1, 2, \ldots, n$. We have

$$\left| e^{-i\theta\cdot Nx} - \sum_{\alpha\in X_N} w_{\alpha,N}(x)e^{-i\theta\cdot\alpha} \right| = \left| \sum_{m=1}^{n} \left[\prod_{s=1}^{m-1} e^{-i\theta_s\cdot Nx_s} \right. \right.$$

$$\times \left[e^{-i\theta_m\cdot Nx_m} - \sum_{\alpha_m=0}^{N} w_{\alpha_m,N}(x_m)e^{-i\theta_m\cdot\alpha_m} \right] \prod_{s=m+1}^{n} \left[\sum_{\alpha_s=0}^{N} w_{\alpha_s,N}(x_s)e^{-i\theta_s\cdot\alpha_s} \right] \right] \right|.$$

Applying (5.9) to the last term and (5.10) to the middle term, we see that this expression can be bounded by

$$\sum_{m=1}^{n} \left(\max_{1\le j\le n} |\theta_j| \right)^N \left(1 + \frac{1}{2^N} \right)^{n-m} \le n \left(1 + \frac{1}{2^N} \right)^{n-1} \left(\max_{1\le j\le n} |\theta_j| \right)^N.$$

Thus, bound (5.8) holds. ∎

We can now state estimates on the covering number $\mathcal{N}(I_K(B_R), \eta)$ for a convolution-type kernel $K(x,t) = k(x - t)$. The following function measures the regularity of the kernel function k:

$$\Upsilon_k(N) = n^3 \left(1 + \frac{1}{2^N} \right)^{2n-2} (2\pi)^{-n} \max_{1\le j\le n} \left\{ \int_{\xi\in[-N/2,N/2]^n} \widehat{k}(\xi) \left(\frac{|\xi_j|}{N} \right)^{2N} d\xi \right\}$$

$$+ \left(1 + (N2^N)^n \right)^2 (2\pi)^{-n} \int_{\xi\notin[-N/2,N/2]^n} \widehat{k}(\xi)\, d\xi.$$

The domain of this function is split into two parts. In the first part, $\xi \in [-N/2, N/2]^n$, and therefore, for $j = 1, \ldots, n$, $\left(|\xi_j|/N \right)^N \le 2^{-N}$; hence this first part decays exponentially quickly as N becomes large. In the second part, $\xi \notin [-N/2, N/2]^n$, and therefore ξ is large when N is large. The decay of $\widehat{k}$ (which is equivalent to the regularity of k; see Part (III) in Section 5.1) yields the fast decay of Υ_k on this second part. For more details and examples of bounding $\Upsilon_k(N)$ by means of the decay of $\widehat{k}$ (or, equivalently, the regularity of k), see Corollaries 5.12 and 5.16.

Theorem 5.11 *Assume that k is an even function in $\mathscr{L}^2(\mathbb{R}^n)$ and $\widehat{k}(\xi) > 0$ almost everywhere on $\mathbb{R}^n$. Let $K(x,t) = k(x - t)$ for $x, t \in [0, 1]^n$. Suppose $\lim_{N\to\infty} \Upsilon_k(N) = 0$. Then, for $0 < \eta < \frac{R}{2}$,*

$$\ln \mathcal{N}(I_K(B_R), \eta) \le (N + 1)^n \ln \left(8\sqrt{k(0)}(N + 1)^{n/2}(N2^N)^n \frac{R}{\eta} \right) \tag{5.11}$$

holds, where N is any integer satisfying

$$\Upsilon_k(N) \leq \left(\frac{\eta}{2R}\right)^2. \tag{5.12}$$

Proof. By Proposition 2.14, K is a Mercer kernel on $\mathbb{R}^n$. Let $f \in B_R$. Then, by reproducing that property of K, $f(x) = \langle f, K_x \rangle_K$. Recall that $X_N = \{0, 1, \ldots, N\}^n$. For $x \in [0, 1]^n$, we have

$$\left| f(x) - \sum_{\alpha \in X_N} f\left(\frac{\alpha}{N}\right) w_{\alpha,N}(x) \right| = \left| \left\langle f, K_x - \sum_{\alpha \in X_N} w_{\alpha,N}(x) K_{\frac{\alpha}{N}} \right\rangle_K \right|$$

$$\leq \|f\|_K \{Q_N(x)\}^{1/2},$$

where $\{Q_N(x)\}^{1/2}$ is the $\mathcal{H}_K$-norm of the function $K_x - \sum_{\alpha \in X_N} w_{\alpha,N}(x) K_{\alpha/N}$. It is explicitly given by

$$Q_N(x) := k(0) - 2 \sum_{\alpha \in X_N} w_{\alpha,N}(x) k\left(x - \frac{\alpha}{N}\right)$$

$$+ \sum_{\alpha,\beta \in X_N} w_{\alpha,N}(x) k\left(\frac{\alpha - \beta}{N}\right) w_{\beta,N}(x). \tag{5.13}$$

By the evenness of k and the inverse Fourier transform,

$$k\left(x - \frac{\alpha}{N}\right) = (2\pi)^{-n} \int_{\mathbb{R}^n} \widehat{k}(\xi) e^{i\xi \cdot (x - \frac{\alpha}{N})} \, d\xi,$$

we obtain

$$Q_N(x) = (2\pi)^{-n} \int_{\mathbb{R}^n} \widehat{k}(\xi) \left| 1 - \sum_{\alpha \in X_N} w_{\alpha,N}(x) e^{i\xi \cdot (x - \frac{\alpha}{N})} \right|^2 d\xi$$

$$= (2\pi)^{-n} \int_{\mathbb{R}^n} \widehat{k}(\xi) \left| e^{-i\frac{\xi}{N} \cdot Nx} - \sum_{\alpha \in X_N} w_{\alpha,N}(x) e^{-i\frac{\xi}{N} \cdot \alpha} \right|^2 d\xi.$$

Now we separate this integral into two parts, one with $\xi \in [-\frac{N}{2}, \frac{N}{2}]^n$ and the other with $\xi \notin [-\frac{N}{2}, \frac{N}{2}]^n$. For the first region, (5.8) in Lemma 5.10 with $\theta = \frac{\xi}{N}$

tells us that

$$\int_{\xi \in [-N/2, N/2]^n} \widehat{k}(\xi) \left| e^{-i\frac{\xi}{N} \cdot Nx} - \sum_{\alpha \in X_N} w_{\alpha, N}(x) e^{-i\frac{\xi}{N} \cdot \alpha} \right|^2 d\xi$$

$$\leq n^2 \left(1 + \frac{1}{2^N}\right)^{2n-2} \sum_{1 \leq j \leq n} \int_{\xi \in [-N/2, N/2]^n} \widehat{k}(\xi) \left(\frac{|\xi_j|}{N}\right)^{2N} d\xi.$$

For the second region, we apply (5.7) in Lemma 5.10 and obtain

$$\int_{\xi \notin [-N/2, N/2]^n} \widehat{k}(\xi) \left| e^{-i\frac{\xi}{N} \cdot Nx} - \sum_{\alpha \in X_N} w_{\alpha, N}(x) e^{-i\frac{\xi}{N} \cdot \alpha} \right|^2 d\xi$$

$$\leq (1 + (N2^N)^n)^2 \int_{\xi \notin [-N/2, N/2]^n} \widehat{k}(\xi) \, d\xi.$$

Combining the two cases above, we have

$$Q_N(x) \leq n^3 \left(1 + \frac{1}{2^N}\right)^{2n-2} \max_{1 \leq j \leq n} \left\{ (2\pi)^{-n} \int_{\xi \in [-N/2, N/2]^n} \widehat{k}(\xi) \left(\frac{|\xi_j|}{N}\right)^{2N} d\xi \right\}$$

$$+ \frac{(1 + (N2^N)^n)^2}{(2\pi)^n} \int_{\xi \notin [-N/2, N/2]^n} \widehat{k}(\xi) \, d\xi = \Upsilon_k(N).$$

Hence

$$\sup_{x \in [0,1]^n} \left\{ k(0) - 2 \sum_{\alpha \in X_N} w_{\alpha, N}(x) k \left(x - \frac{\alpha}{N}\right) \right.$$

$$\left. + \sum_{\alpha, \beta \in X_N} w_{\alpha, N}(x) k \left(\frac{\alpha - \beta}{N}\right) w_{\beta, N}(x) \right\} \leq \Upsilon_k(N). \qquad (5.14)$$

Since N satisfies (5.12), we have

$$\left\| f(x) - \sum_{\alpha \in X_N} f \left(\frac{\alpha}{N}\right) w_{\alpha, N}(x) \right\|_{\mathscr{C}(X)} \leq \frac{\eta}{2}.$$

Also, by the reproducing property, $|f(\frac{\alpha}{N})| = |\langle f, K_{\alpha/N} \rangle_K| \leq \|f\|_K$
$\sqrt{K(\alpha/N, \alpha/N)} \leq R\sqrt{k(0)}$. Hence

$$\left\| \left\{ f\left(\frac{\alpha}{N}\right) \right\} \right\|_{\ell^2(X_N)} \leq R\sqrt{k(0)}(N+1)^{n/2}.$$

Here $\ell^2(X_N)$ is the ℓ^2 space of sequences $\{x(\alpha)\}_{\alpha \in X_N}$ indexed by X_N.

Apply Theorem 5.3 to the ball of radius $r := R\sqrt{k(0)}(N+1)^{n/2}$ in the finite-dimensional space $\ell^2(X_N)$ and $\epsilon = \eta/(2(N2^N)^n)$. Then there are $\{c^l : l = 1, \ldots, [(2r/\epsilon + 1)^{\#X_N}]\} \subset \ell^2(X_N)$ such that for any $d \in \ell^2(X_N)$ with $\|d\|_{\ell^2(X_N)} \leq r$, we can find some l satisfying

$$\|d - c^l\|_{\ell^2(X_N)} \leq \epsilon.$$

This, together with Lemma 5.10, yields

$$\left\| \sum_{\alpha \in X_N} d_\alpha w_{\alpha,N}(x) - \sum_{\alpha \in X_N} c_\alpha^l w_{\alpha,N}(x) \right\|_{\mathscr{C}(X)} \leq \|d - c^l\|_{\ell^\infty(X_N)}$$

$$\left\| \sum_{\alpha \in X_N} |w_{\alpha,N}(x)| \right\|_{\mathscr{C}(X)} \leq (N2^N)^n \epsilon \leq \eta/2.$$

Here $\ell^\infty(X_N)$ is the ℓ^∞ space of sequences $\{x(\alpha)\}_{\alpha \in X_N}$ indexed by X_N that satisfies the relationship $\|c\|_{\ell^\infty(X_N)} \leq \|c\|_{\ell^2(X_N)}$ for all $c \in \ell^\infty(X_N)$.

Thus, with $d = \{f(\frac{\alpha}{N})\}$, we see that $\|f(x) - \sum_{\alpha \in X_N} c_\alpha^l w_{\alpha,N}(x)\|_{\mathscr{C}(X)}$ can be bounded by

$$\left\| f(x) - \sum_{\alpha \in X_N} f\left(\frac{\alpha}{N}\right) w_{\alpha,N}(x) \right\|_{\mathscr{C}(X)}$$

$$+ \left\| \sum_{\alpha \in X_N} d_\alpha w_{\alpha,N}(x) - \sum_{\alpha \in X_N} c_\alpha^l w_{\alpha,N}(x) \right\|_{\mathscr{C}(X)} \leq \eta.$$

We have covered $I_K(B_R)$ by balls with centers $\sum_{\alpha \in X_N} c_\alpha^l w_{\alpha,N}(x)$ and radius η. Therefore,

$$\mathcal{N}(I_K(B_R), \eta) \leq \left(\frac{2r}{\epsilon} + 1\right)^{\#X_N}.$$

That is,

$$\ln \mathcal{N}(I_K(B_R), \eta) \leq (N+1)^n \ln \left(\frac{2r}{\epsilon} + 1 \right)$$

$$\leq (N+1)^n \ln \left(8\sqrt{k(0)}(N+1)^{n/2}(N2^N)^n \frac{R}{\eta} \right).$$

The proof of Theorem 5.11 is complete. ∎

To see how to handle the function $\Upsilon_k(N)$ measuring the regularity of the kernel, and then to estimate the covering number, we turn to the example of Gaussian kernels.

Corollary 5.12 *Let $\sigma > 0$, $X = [0,1]^n$, and $K(x,y) = k(x-y)$ with*

$$k(x) = \exp \left\{ -\frac{\|x\|^2}{\sigma^2} \right\}, \quad x \in \mathbb{R}^n.$$

Then, for $0 < \eta < \frac{R}{2}$,

$$\ln \mathcal{N}(I_K(B_R), \eta) \leq \left(3 \ln \frac{R}{\eta} + \frac{54n}{\sigma^2} + 6 \right)^n \left((6n+1) \ln \frac{R}{\eta} + \frac{90n^2}{\sigma^2} + 11n + 3 \right) \tag{5.15}$$

holds. In particular, when $0 < \eta < R \exp\{-(90n^2/\sigma^2) - 11n - 3\}$, we have

$$\ln \mathcal{N}(I_K(B_R), \eta) \leq 4^n (6n+2) \left(\ln \frac{R}{\eta} \right)^{n+1}. \tag{5.16}$$

Proof. It is well known that

$$\widehat{k}(\xi) = (\sigma \sqrt{\pi})^n e^{-\sigma^2 \|\xi\|^2/4}. \tag{5.17}$$

Hence $\widehat{k}(\xi) > 0$ for any $\xi \in \mathbb{R}^n$.

Let us estimate the function Υ_k. For the first part, with $1 \leq j \leq n$, we have

$$(2\pi)^{-1} \int_{\xi_\ell \in [-N/2, N/2]} \sigma \sqrt{\pi} e^{-\sigma^2 \xi_\ell^2/4} \, d\xi_\ell < 1$$

when $\ell \neq j$. Hence

$$(2\pi)^{-n} \int_{\xi \in [-N/2, N/2]^n} (\sigma\sqrt{\pi})^n e^{-\sigma^2 \|\xi\|^2/4} \left(\frac{|\xi_j|}{N}\right)^{2N} d\xi$$

$$\leq \frac{\sigma\sqrt{\pi}}{2\pi} \int_{-N/2}^{N/2} e^{-\sigma^2 |\xi_j|^2/4} \left(\frac{|\xi_j|}{N}\right)^{2N} d\xi_j$$

$$\leq \frac{1}{\sqrt{\pi}} \left(\frac{2}{\sigma N}\right)^{2N} \Gamma\left(N + \frac{1}{2}\right).$$

If we apply Stirling's formula, this expression can be bounded by

$$\left(\frac{2}{\sigma N}\right)^{2N} \left(\frac{2N+1}{2e}\right)^{N+(1/2)} \frac{1}{\sqrt{2N+1}} e^{1/(6(2N+1))} \leq \left(\frac{2}{\sigma^2 eN}\right)^N.$$

As for the second term of Υ_k, we have

$$(2\pi)^{-n} \int_{\xi \notin [-N/2, N/2]^n} (\sigma\sqrt{\pi})^n e^{-\sigma^2 \|\xi\|^2/4} d\xi$$

$$\leq \frac{\sigma\sqrt{\pi}}{2\pi} \sum_{j=1}^{n} \int_{\xi_j \notin [-N/2, N/2]} e^{-\sigma^2 |\xi_j|^2/4} d\xi_j$$

$$\leq \frac{n\sigma\sqrt{\pi}}{\pi} \int_{N/2}^{+\infty} e^{-(\sigma^2/4)(t^2 - t/2)} e^{-(\sigma^2/4)(t/2)} dt$$

$$\leq \frac{n\sigma}{\sqrt{\pi}} e^{-\sigma^2 N(N-1)/16} \frac{8}{\sigma^2} e^{-\sigma^2 N/16}$$

$$= \frac{8n}{\sigma\sqrt{\pi}} e^{-(\sigma^2/16)N^2}.$$

If we combine these two estimates, the function Υ_k satisfies

$$\Upsilon_k(N) \leq n^3 \left(1 + \frac{1}{2^N}\right)^{2n-2} \left(\frac{2}{\sigma^2 eN}\right)^N + \frac{8n}{\sigma\sqrt{\pi}} (1 + (N2^N)^n)^2 e^{-(\sigma^2/16)N^2}.$$

Notice that when $N \geq n + 3$,

$$\left(1 + \frac{1}{2^N}\right)^{2n-2} \leq \left(1 + \frac{1}{2n}\right)^{2n-2} \leq e$$

and

$$(1 + (N2^N)^n)^2 \leq 2^{1-2n+4Nn}.$$

It follows that

$$\Upsilon_k(N) \le n^3 e \left(\frac{2}{\sigma^2 eN} \right)^N + \frac{n4^{2-n}}{\sigma\sqrt{\pi}} e^{-(\sigma^2/16)N^2 + 4nN \ln 2}.$$

Choose $N \ge 80n \ln 2/\sigma^2$. Then

$$\Upsilon_k(N) \le n^3 e \left(\frac{1}{16en \ln 2} \right)^N + \frac{4}{\sigma\sqrt{\pi}} 2^{-nN}$$

$$\le e \left(\frac{1}{16n} \right)^N + \frac{4}{\sigma\sqrt{\pi}} 2^{-nN} \tag{5.18}$$

$$\le \left(e + \frac{4}{\sigma\sqrt{\pi}} \right) \left(\max \left\{ \frac{1}{16n}, \frac{1}{2^n} \right\} \right)^N.$$

When $N \ge 2 \ln \frac{R}{\eta} + \frac{5}{2}$ and $N \ge \frac{3}{n} \ln \frac{R}{\eta} + \frac{5}{n} - \ln(\sigma\sqrt{\pi})/(n \ln 2)$, we know that each term in the estimates for Υ_k is bounded by $(\eta/(2R))^2/2$. Hence (5.12) holds.

Finally, we choose the smallest N satisfying

$$N \ge \frac{80n \ln 2}{\sigma^2} + 3 \ln \frac{R}{\eta} + 5.$$

Then, by checking the cases $\sigma \ge 1$ and $\sigma < 1$, we see that (5.12) is valid for any $0 < \eta < \frac{R}{2}$. By Theorem 5.11,

$$\ln \mathcal{N} (I_K(B_R), \eta) \le \left(3 \ln \frac{R}{\eta} + \frac{80n \ln 2}{\sigma^2} + 6 \right)^n \left(\left(\frac{5}{2} \ln 2 \right) nN + \ln \frac{R}{\eta} + \ln 8 \right)$$

$$\le \left(3 \ln \frac{R}{\eta} + \frac{54n}{\sigma^2} + 6 \right)^n \left((6n+1) \ln \frac{R}{\eta} + \frac{90n^2}{\sigma^2} + 11n + 3 \right).$$

This proves (5.15).

When $0 < \eta < Re^{(90n^2/\sigma^2) - 11n - 3}$, we have

$$\ln \mathcal{N} (I_K(B_R), \eta) \le 4^n (6n+2) \left(\ln \frac{R}{\eta} \right)^{n+1}.$$

This yields the last inequality in the statement. ∎

One can easily derive from the covering number estimates for $X = [0, 1]^n$ such estimates for an arbitrary X.

Proposition 5.13 *Let* $K(x, y) = k(x - y)$ *be a translation-invariant Mercer kernel on* $\mathbb{R}^n$ *and* $X \subseteq \mathbb{R}^n$. *Let* $\Delta > 0$ *and* K^Δ *be the Mercer kernel on* $X^\Delta = [0, 1]^n$ *given by*

$$K^\Delta(x, y) = k(\Delta(x - y)), \quad x, y \in [0, 1]^n.$$

Then

(i) *If* $X \subseteq x^* + [-\Delta/2, \Delta/2]^n$ *for some* $x^* \in X$, *then, for all* $\eta, R > 0$,

$$\mathcal{N}(I_K(B_R), \eta) \leq \mathcal{N}(I_{K^\Delta}(B_R), \eta).$$

(ii) *If* $X \supseteq x^* + [-\Delta/2, \Delta/2]^n$ *for some* $x^* \in X$, *then, for all* $\eta, R > 0$,

$$\mathcal{N}(I_K(B_R), \eta) \geq \mathcal{N}(I_{K^\Delta}(B_R), \eta).$$

Proof.

(i) Denote $t_0 = (\frac{1}{2}, \frac{1}{2}, \ldots, \frac{1}{2}) \in [0, 1]^n$. Let $g = \sum_{i=1}^m c_i K_{x_i} \in I_K(B_R)$. Then

$$\|g\|_K^2 = \sum_{i,j=1}^m c_i c_j k(x_i - x_j) = \sum_{i,j=1}^m c_i c_j k\left(\Delta\left(\frac{x_i - x^*}{\Delta} + t_0 - \frac{x_j - x^*}{\Delta} - t_0\right)\right)$$

$$= \sum_{i,j=1}^m c_i c_j K^\Delta\left(\frac{x_i - x^*}{\Delta} + t_0, \frac{x_j - x^*}{\Delta} + t_0\right) = \left\|\sum_{i=1}^m c_i K^\Delta_{\frac{x_i-x^*}{\Delta}+t_0}\right\|_{K^\Delta}^2,$$

the last line by the definition of K^Δ. Since $g \in I_K(B_R)$, we have

$$\sum_{i=1}^m c_i K^\Delta_{\frac{x_i-x^*}{\Delta}+t_0} \in I_{K^\Delta}(B_R).$$

If $\{f_1, \ldots, f_N\}$ is an η-net of $I_{K^\Delta}(B_R)$ on $X^\Delta = [0, 1]^n$ with $N = \mathcal{N}(I_{K^\Delta}(B_R), \eta)$, then there is some $j \in \{1, \ldots, N\}$ such that

$$\left\|\sum_{i=1}^m c_i K^\Delta_{\frac{x_i-x^*}{\Delta}+t_0} - f_j\right\|_{\mathscr{C}([0,1]^n)} \leq \eta.$$

This means that

$$\sup_{t \in [0,1]^n} \left|\sum_{i=1}^m c_i K^\Delta\left(t, \frac{x_i - x^*}{\Delta} + t_0\right) - f_j(t)\right| \leq \eta.$$

Take $t = ((x - x^*)/\Delta) + t_0$. When $x \in X \subseteq x^* + [-\Delta/2, \Delta/2]^n$, we have $t \in [0, 1]^n$. Hence

$$\sup_{x \in X} \left| \sum_{i=1}^{m} c_i K^{\Delta} \left(\frac{x - x^*}{\Delta} + t_0, \frac{x_i - x^*}{\Delta} + t_0 \right) - f_j \left(\frac{x - x^*}{\Delta} + t_0 \right) \right| \le \eta.$$

This is the same as

$$\sup_{x \in X} \left| \sum_{i=1}^{m} c_i k (x - x_i) - f_j \left(\frac{x - x^*}{\Delta} + t_0 \right) \right|$$

$$= \sup_{x \in X} \left| \sum_{i=1}^{m} c_i K_{x_i}(x) - f_j \left(\frac{x - x^*}{\Delta} + t_0 \right) \right| \le \eta.$$

This shows that if we define $f_j^*(x) := f_j(((x - x^*)/\Delta) + t_0)$, the set $\{f_1^*, \ldots, f_N^*\}$ is an η-net of the function set $\{\sum_{i=1}^{m} c_i K_{x_i} \in I_K(B_R)\}$ in $\mathscr{C}(X)$. Since this function set is dense in $I_K(B_R)$, we have

$$\mathcal{N}(I_K(B_R), \eta) \le N = \mathcal{N}(I_{K^{\Delta}}(B_R), \eta).$$

(ii) If $X \supseteq x^* + [-\Delta/2, \Delta/2]^n$ and $\{g_1, \ldots, g_N\}$ is an η-net of $I_K(B_R)$ with $N = \mathcal{N}(I_K(B_R), \eta)$, then, for each $g \in B_R$, we can find some $j \in \{1, \ldots, N\}$ such that $\|g - g_j\|_{\mathscr{C}(X)} \le \eta$.
Let $f = \sum_{i=1}^{m} c_i K_{t_i}^{\Delta} \in I_{K^{\Delta}}(B_R)$. Then, for any $t \in X^{\Delta}$,

$$f(t) = \sum_{i=1}^{m} c_i k(\Delta(t - t_i)) = \sum_{i=1}^{m} c_i k(x - x_i) = \sum_{i=1}^{m} c_i K_{x_i}(x),$$

where $x = x(t) = x^* + \Delta(t - t_0) \in X$ and $x_i = x^* + \Delta(t_i - t_0) \in X$. It follows from this expression that

$$\|f\|_{K^{\Delta}}^2 = \sum_{i,j=1}^{m} c_i c_j K^{\Delta}(t_i, t_j) = \sum_{i,j=1}^{m} c_i c_j k(\Delta(t_i - t_j))$$

$$= \sum_{i,j=1}^{m} c_i c_j K(x_i, x_j) = \|g\|_K^2 \le R,$$

where $g = \sum_{i=1}^{m} c_i K_{x_i}$. So, $g \in I_K(B_R)$ and we have $\|g - g_j\|_{\mathscr{C}(X)} = \sup_{x \in X} |g(x) - g_j(x)| \leq \eta$ for some $j \in \{1, \ldots, N\}$. But for $x = x(t) \in X$,

$$g(x) = \sum_{i=1}^{m} c_i k(x - x_i) = \sum_{i=1}^{m} c_i k(\Delta(t - t_i)) = f(t).$$

It follows that

$$\sup_{t \in [0,1]^n} \left| f(t) - g_j(x^* + \Delta(t - t_0)) \right| \leq \sup_{x \in X} \left| g(x) - g_j(x) \right| \leq \eta.$$

This shows that if we define $g_j^*(t) := g_j(x^* + \Delta(t - t_0))$, the set $\{g_1^*, \ldots, g_N^*\}$ is an η-net of $I_{K^\Delta}(B_R)$ in $\mathscr{C}([0, 1]^n)$. ∎

If we take $\Delta = 2\mathrm{Diam}(X)$ to be twice the diameter of X, then the condition $X \subseteq x^* + [-\Delta/2, \Delta/2]^n$ holds for any $x^* \in X$. If, moreover, $k(x) = \exp\{-\|x\|^2/\sigma^2\}$ then $K^\Delta(x, y) = \exp\{-\|x - y\|^2/(\sigma^2/\Delta^2)\}$. In this situation, Corollary 5.12 and Proposition 5.13 yield the following upper bound for the covering numbers of Gaussian kernels.

Corollary 5.14 *Let* $\sigma > 0$, $X \subset \mathbb{R}^n$ *with* $\mathrm{Diam}(X) < \infty$, *and* $K(x, y) = \exp\{-\|x - y\|^2/\sigma^2\}$. *Then, for any* $0 < \eta \leq \frac{R}{2}$, *we have*

$$\ln \mathcal{N}(I_K(B_R), \eta) \leq \left(3 \ln \frac{R}{\eta} + \frac{216n(\mathrm{Diam}(X))^2}{\sigma^2} + 6 \right)^n$$
$$\left((6n + 1) \ln \frac{R}{\eta} + \frac{360n^2(\mathrm{Diam}(X))^2}{\sigma^2} + 11n + 3 \right). \quad \blacksquare$$

We now note that the upper bound in Part (ii) of Theorem 5.1 follows from Corollary 5.14. We next apply Theorem 5.11 to kernels with exponentially decaying Fourier transforms.

Theorem 5.15 *Let k be as in Theorem 5.11, and assume that for some constants $C_0 > 0$ and $\lambda > n(6 + 2\ln 4)$,*

$$\widehat{k}(\xi) \leq C_0 e^{-\lambda \|\xi\|}, \quad \forall \xi \in \mathbb{R}^n.$$

Denote $\Lambda := \max\{1/e\lambda, 4^n/e^{\lambda/2}\}$. *Then for* $0 < \eta \leq 2R\sqrt{C_0} \Lambda^{(2n-1)/4}$,

$$\ln \mathcal{N}(I_K(B_R), \eta) \leq \left(\frac{4}{\ln(1/\Lambda)} \ln \frac{R}{\eta} + 1 + C_1 \right)^n \left\{ \left(\frac{4}{\ln(1/\Lambda)} + 1 \right) \ln \frac{R}{\eta} + C_2 \right\}$$

(5.19)

holds, where

$$C_1 := 1 + \frac{2\ln(32C_0)}{\ln(1/\Lambda)}, \quad C_2 := \ln\left(8\sqrt{\frac{C_0}{\lambda}}2^{n/2}(\Lambda^{-3/8}2^n)^{C_1}\right).$$

Proof. Let $N \in \mathbb{N}$ and $1 \leq j \leq n$. Since $|\xi_j|/N < 1$ for $\xi \in [-N/2, N/2]^n$, we have

$$\int_{\xi \in [-N/2,N/2]^n} \widehat{k}(\xi)\left(\frac{|\xi_j|}{N}\right)^{2N} d\xi \leq C_0 \int_{\xi \in [-N/2,N/2]^n} e^{-\lambda\|\xi\|}\left(\frac{|\xi_j|}{N}\right)^{2N} d\xi$$

$$\leq \frac{C_0}{N^N}N^{n-1}\int_{\xi_j \in [-N/2,N/2]} |\xi_j|^N e^{-\lambda|\xi_j|} d\xi_j$$

$$\leq \frac{2C_0}{\lambda^{N+1}N^N}N^{n-1}N!$$

$$\leq 2C_0\sqrt{2\pi}2^{(1/12)+1}\frac{N^{n-1/2}}{(e\lambda)^{N+1}},$$

the last inequality by Stirling's formula. Hence the first term of $\Upsilon_k(N)$ is at most

$$n^3\left(1+\frac{1}{2^N}\right)^{2n-2}(2\pi)^{-n}2C_0\sqrt{2\pi}2^{(1/12)+1}\frac{N^{n-(1/2)}}{(e\lambda)^{N+1}}$$

$$\leq 4C_0N^{n-(1/2)}\left(\frac{1}{e\lambda}\right)^{N+1}.$$

Here we have bounded the constant term as follows

$$n^3\left(1+\frac{1}{2^N}\right)^{2n-2}(2\pi)^{-n}2\sqrt{2\pi}2^{(1/12)+1} = n^3\left(\frac{(1+(1/2^N))^2}{2\pi}\right)^{n-1}\frac{2^{(1/12)+2}}{\sqrt{2\pi}}$$

$$\leq n^3\left(\frac{(1+(1/2))^2}{2\pi}\right)^{n-1}\frac{2^{(1/12)+2}}{\sqrt{2\pi}}$$

$$= n^3\left(\frac{9}{8\pi}\right)^{n-1}\frac{2^{(1/12)+2}}{\sqrt{2\pi}}$$

$$\leq 4, \quad \text{for all } n \in \mathbb{N}.$$

For the other term in $\Upsilon_k(N)$, we have

$$\int_{\xi \notin [-N/2,N/2]^n} \widehat{k}(\xi) \, d\xi \leq C_0 \int_{\|\xi\| \geq N/2} e^{-\lambda\|\xi\|}d\xi. \quad (5.20)$$

To estimate this integral, we recall spherical coordinates in $\mathbb{R}^n$:

$$\xi_1 = r \cos \theta_1$$
$$\xi_2 = r \sin \theta_1 \cos \theta_2$$
$$\xi_3 = r \sin \theta_1 \sin \theta_2 \cos \theta_3$$
$$\vdots$$
$$\xi_{n-1} = r \sin \theta_1 \sin \theta_2 \ldots \sin \theta_{n-2} \cos \theta_{n-1}$$
$$\xi_n = r \sin \theta_1 \sin \theta_2 \ldots \sin \theta_{n-2} \sin \theta_{n-1},$$

where $r \in (0, \infty)$, $\theta_1, \ldots, \theta_{n-2} \in [0, \phi)$, and $\theta_{n-1} \in [0, 2\pi)$. For a radial function $f(\|\xi\|)$ we have

$$\int_{r_1 \leq \|\xi\| \leq r_2} f(\|\xi\|) \, d\xi = w_{n-1} \int_{r_1}^{r_2} f(r) r^{n-1} \, dr,$$

where

$$w_{n-1} = \int_0^{2\pi} \int_0^{\pi} \ldots \int_0^{\pi} \sin^{n-2} \theta_1 \sin^{n-3} \theta_2 \ldots \sin \theta_{n-2} \, d\theta_1 \, d\theta_2 \ldots d\theta_{n-2} \, d\theta_{n-1}$$
$$= 2\pi \prod_{j=1}^{n-2} \int_0^{\pi} \sin^{n-j-1} \theta_j \, d\theta_j$$
$$= \frac{2\pi^{n/2}}{\Gamma(n/2)}.$$

Applying this with $f(\|\xi\|) = e^{-\lambda\|\xi\|}$, we find that

$$\int_{\|\xi\| \geq N/2} e^{-\lambda\|\xi\|} \, d\xi = \frac{2\pi^{n/2}}{\Gamma(n/2)} \int_{N/2}^{\infty} e^{-\lambda r} r^{n-1} \, dr.$$

By repeatedly integrating by parts, we see that

$$\int_{N/2}^{\infty} e^{-\lambda r} r^{n-1} \, dr = \frac{1}{\lambda} \left(\frac{N}{2}\right)^{n-1} e^{-\lambda N/2} + \frac{n-1}{\lambda} \int_{N/2}^{\infty} e^{-\lambda r} r^{n-2} \, dr$$
$$= \ldots = \sum_{j=1}^{n} \frac{1}{\lambda^j} \frac{(n-1)!}{(n-j)!} \left(\frac{N}{2}\right)^{n-j} e^{-\lambda N/2}.$$

Therefore, returning to (5.20), since $\lambda > 2n$,

$$\int_{\xi \notin [-N/2, N/2]^n} \widehat{k}(\xi) \, d\xi \leq C_0 \frac{2\pi^{n/2}}{\Gamma(n/2)} \sum_{j=1}^{n} \frac{1}{\lambda^j} \frac{(n-1)!}{(n-j)!} \left(\frac{N}{2}\right)^{n-j} e^{-\lambda N/2}$$

$$\leq C_0 \frac{2\pi^{n/2}}{\Gamma(n/2)} \sum_{j=1}^{n} \frac{1}{(2n)^j} \frac{(n-1)!}{(n-j)!} \left(\frac{N}{2}\right)^{n-j} e^{-\lambda N/2}$$

$$\leq C_0 \frac{2\pi^{n/2}}{\Gamma(n/2)} \sum_{j=1}^{n} \frac{(n-1)!}{(n-j)!j!} 2^{-n} N^{n-j} e^{-\lambda N/2}$$

$$\leq \frac{2C_0 \pi^{n/2}}{\Gamma(n/2)} 2 N^{n-1} e^{-\lambda N/2}.$$

It follows that the second term in $\Upsilon_k(N)$ is bounded by

$$\left(1 + \left(N 2^N\right)^n\right)^2 (2\pi)^{-n} \frac{2C_0 \pi^{n/2}}{\Gamma(n/2)} 2 N^{n-1} e^{-\lambda N/2} \leq 4 C_0 N^{3n} \left(\frac{4^n}{e^{\lambda/2}}\right)^N.$$

Combining the two bounds above, we have

$$\Upsilon_k(N) \leq 4 C_0 N^{n-1/2} \left(\frac{1}{e\lambda}\right)^{N+1} + 4 C_0 N^{3n} \left(\frac{4^n}{e^{\lambda/2}}\right)^N \leq 8 C_0 N^{3n} \Lambda^N.$$

Since $\lambda > n(6 + 2\ln 4)$, the definition of Λ yields

$$\Lambda \leq \max\left\{ \frac{1}{e(2n\ln 4 + n)}, \frac{4^n}{e^{n\ln 4 + 3n}} \right\} < e^{-3n}.$$

Since $xe^{-x} \leq e^{-1}$ for all $x \in (0, \infty)$, we have

$$N^{3n} \Lambda^{N/2} = \left(N \Lambda^{N/6n}\right)^{3n} = \left(N e^{-N/6n \ln(1/\Lambda)}\right)^{3n}$$

$$\leq \left(\frac{6n}{\log(1/\Lambda)} e^{-1}\right)^{3n} \leq \left(\frac{2}{e}\right)^{3n} < 1.$$

Then, for $N \geq 4n/\ln(1/\Lambda)$,

$$\Upsilon_k(N) \leq 8 C_0 \Lambda^{N/2}. \tag{5.21}$$

Thus, for $0 < \eta \leq 2R\sqrt{C_0} \Lambda^{(2n-1)/4}$, we may take $N \in \mathbb{N}$ such that $N \geq 4n/\ln(1/\Lambda)$ and $N > 2$ to obtain

$$8 C_0 \Lambda^{N/2} \leq \left(\frac{\eta}{2R}\right)^2 \leq 8 C_0 \Lambda^{(N-1)/2}. \tag{5.22}$$

Under this choice, (5.12) holds. Then, by Theorem 5.11,

$$\ln \mathcal{N}(I_K(B_R), \eta) \le (N+1)^n \ln\left(8\sqrt{k(0)}(N+1)^{n/2}(N2^N)^n \frac{R}{\eta}\right).$$

Now, by (5.22),

$$N \le 1 + \frac{2\ln(32C_0)}{\ln(1/\Lambda)} + \frac{4}{\ln(1/\Lambda)} \ln \frac{R}{\eta}.$$

Also, since

$$(N+1)^{n/2}(N2^N)^n \le 2^{n/2}(\Lambda^{-3/8}2^n)^N,$$

we have

$$\ln \mathcal{N}(I_K(B_R), \eta) \le \left(\frac{4}{\ln(1/\Lambda)} \ln \frac{R}{\eta} + 2 + \frac{2\ln(32C_0)}{\ln(1/\Lambda)}\right)^n \ln\Big(8\sqrt{k(0)}2^{n/2}$$

$$(\Lambda^{-3/8}2^n)^{1+2\ln(32C_0)/\ln(1/\Lambda)} (R/\eta)^{(4/\ln(1/\Lambda))+1}\Big).$$

Finally observe that

$$|k(0)| = |(2\pi)^{-n}\int \widehat{k}(\xi)\,d\xi| \le (2\pi)^{-n}\int C_0 e^{-\lambda\|\xi\|}\,d\xi \le \frac{C_0}{\lambda}.$$

Then (5.19) follows.   ∎

We can apply Theorem 5.15 to inverse multiquadric kernels.

Corollary 5.16 *Let* $c > n(6 + 2\ln 4)$, $\alpha > n/2$, *and*

$$k(x) = (c^2 + \|x\|^2)^{-\alpha}, \quad x \in \mathbb{R}^n.$$

Then there is a constant C_0 *depending only on* α *such that for* $0 < \eta \le 2R\sqrt{C_0}\Lambda^{(2n-1)/4}$,

$$\ln \mathcal{N}(I_K(B_R), \eta) \le \left(\frac{4}{\ln(1/\Lambda)} \ln \frac{R}{\eta} + 1 + C_1\right)^n \left\{\left(\frac{4}{\ln(1/\Lambda)} + 1\right) \ln \frac{R}{\eta} + C_2\right\}$$

holds, where $\Lambda = \max\{1/ec, 4^n/e^{c/2}\}$ *and* C_1, C_2 *are the constants defined in Theorem 5.15.*

Proof. For any $\epsilon > 0$, we know that there are positive constants $C_0 \geq 1$, depending only on α, and C_0^*, depending only on α and ϵ, such that

$$C_0^* e^{-(c+\epsilon)\|\xi\|} \leq \widehat{k}(\xi) \leq C_0 e^{-c\|\xi\|} \quad \forall \xi \in \mathbb{R}^n. \tag{5.23}$$

Then we can apply Theorem 5.15 with $\lambda = c$, and the desired estimate follows. ∎

5.4 Lower bounds for covering numbers

In this section we continue our discussion of the covering numbers of balls in RKHSs and provide some lower-bound estimates. This is done by bounding the related packing numbers.

Definition 5.17 Let S be a compact set in a metric space and $\eta > 0$. The *packing number* $\mathcal{M}(S, \eta)$ is the largest integer $m \in \mathbb{N}$ such that there exist m points $x_1, \ldots, x_m \in S$ being η-separated; that is, the distance between x_i and x_j is greater than η if $i \neq j$.

Covering and packing numbers are closely related.

Proposition 5.18 *For any* $\eta > 0$,

$$\mathcal{M}(S, 2\eta) \leq \mathcal{N}(S, \eta) \leq \mathcal{M}(S, \eta).$$

Proof. Let $k = \mathcal{M}(S, 2\eta)$ and $\{a_1, \ldots, a_k\}$ be a set of 2η-separated points in S. Then, by the triangle inequality, no closed ball of radius η can contain more than one a_i. This shows that $\mathcal{N}(S, \eta) \geq k$.

To prove the other inequality, let $k = \mathcal{M}(S, \eta)$ and $\{a_1, \ldots, a_k\}$ be a set of η-separated points in S. Then, the balls $B(a_i, \eta)$ cover S. Otherwise, there would exist a point a_{k+1} whose distance to $a_j, j = 1, \ldots, k$, was greater than η and one would have $\mathcal{M}(S, \eta) \geq k + 1$. ∎

The lower bounds for the packing numbers are presented in terms of the Gramian matrix

$$K[\mathbf{x}] = \left(K(x_i, x_j) \right)_{i,j=1}^{m}, \tag{5.24}$$

where $\mathbf{x} := \{x_1, \ldots, x_m\}$ is a set of points in X. Denote by $\|K[\mathbf{x}]^{-1}\|_2$ the norm of $K[\mathbf{x}]^{-1}$ (if it exists) as an operator on $\mathbb{R}^m$ with the 2-norm.

We use nodal functions in the RKHS $\mathcal{H}_K$ to provide lower bounds of covering numbers. They are used in the next chapter as well to construct interpolation schemes to estimate the approximation error.

Definition 5.19 Let $\mathbf{x} := \{x_1, \ldots, x_m\} \subseteq X$. We say that $\{u_i\}_{i=1}^m$ is a set of *nodal functions* associated with the nodes $x_1, \ldots, x_m$ if $u_i \in \mathrm{span}(K_{x_1}, \ldots, K_{x_m})$ and $u_i(x_j) = \delta_{ij}$.

The following result characterizes the existence of nodal functions.

Proposition 5.20 *Let K be a Mercer kernel on X and $\mathbf{x} := \{x_1, \ldots, x_m\} \subset X$. Then the following statements are equivalent:*

(i) *The nodal functions $\{u_i\}_{i=1}^m$ exist.*
(ii) *The functions $\{K_{x_i}\}_{i=1}^m$ are linearly independent.*
(iii) *The Gramian matrix $K[\mathbf{x}]$ is invertible.*
(iv) *There exists a set of functions $\{f_i\}_{i=1}^m \in \mathcal{H}_K$ such that $f_i(x_j) = \delta_{ij}$ for $i, j = 1, \ldots, m$.*

In this case, the nodal functions are uniquely given by

$$u_i(x) = \sum_{j=1}^m (K[\mathbf{x}]^{-1})_{i,j} K_{x_j}(x), \quad i = 1, \ldots, m. \tag{5.25}$$

Moreover, for each $x \in X$, the vector $(u_i(x))_{i=1}^m$ is the unique minimizer in $\mathbb{R}^m$ of the quadratic function Q given by

$$Q(w) = \sum_{i,j=1}^m w_i K(x_i, x_j) w_j - 2 \sum_{i=1}^m w_i K(x, x_i) + K(x, x), \quad w \in \mathbb{R}^m.$$

Proof.

(i) $\Rightarrow$ (ii). The nodal function property implies that the nodal functions $\{u_i\}$ are linearly independent. Hence (i) implies (ii), since the m-dimensional space $\mathrm{span}\{u_i\}_{i=1}^m$ is contained in $\mathrm{span}\{K_{x_i}\}_{i=1}^m$.

(ii) $\Rightarrow$ (iii). A solution $d = (d_1, \ldots, d_m) \in \mathbb{R}^m$ of the linear system

$$K[\mathbf{x}]d = 0$$

satisfies

$$\left\| \sum_{j=1}^m d_j K_{x_j} \right\|_K^2 = \sum_{i=1}^m d_i \left\{ \sum_{j=1}^m K(x_i, x_j) d_j \right\} = 0.$$

Then the linear independence of $\{K_{x_j}\}_{j=1}^m$ implies that the linear system has only the zero solution; that is, $K[\mathbf{x}]$ is invertible.

(iii) $\Rightarrow$ (iv). When $K[\mathbf{x}]$ is invertible, the functions $\{f_i\}_{i=1}^m$ given by $f_i = \sum_{j=1}^m (K[\mathbf{x}]^{-1})_{i,j} K_{x_j}$ satisfy

$$f_i(x_j) = \sum_{\ell=1}^m (K[\mathbf{x}]^{-1})_{i,\ell} K(x_\ell, x_j) = (K[\mathbf{x}]^{-1} K[\mathbf{x}])_{i,j} = \delta_{ij}.$$

These are the desired functions.

(iv) $\Rightarrow$ (i). Let $P_\mathbf{x}$ be the orthogonal projection from $\mathcal{H}_K$ onto $\text{span}\{K_{x_i}\}_{i=1}^m$. Then for $i, j = 1, \ldots, m$,

$$P_\mathbf{x}(f_i)(x_j) = \langle P_\mathbf{x}(f_i), K_{x_j} \rangle_K = \langle f_i, K_{x_j} \rangle_K = f_i(x_j) = \delta_{ij}.$$

So $\{u_i = P_\mathbf{x}(f_i)\}_{i=1}^m$ are the desired nodal functions.

The uniqueness of the nodal functions follows from the invertibility of the Gramian matrix $K[\mathbf{x}]$.

Since the quadratic form Q can be written as $Q(w) = w^T K[\mathbf{x}] w - 2b^T w + K(x, x)$ with the positive definite matrix $K[\mathbf{x}]$ and the vector $b = (K(x, x_i))_{i=1}^m$, we know that the minimizer w^* of Q in $\mathbb{R}^m$ is given by the linear system $K[\mathbf{x}] w^* = b$, which is exactly $(u_i(x))_{i=1}^m$. ∎

When the RKHS has finite dimension ℓ, then, for any $m \leq \ell$, we can find nodal functions $\{u_j\}_{j=1}^m$ associated with some subset $\mathbf{x} = \{x_1, \ldots, x_m\} \subseteq X$, whereas for $m > \ell$ no such nodal functions exist. When $\dim \mathcal{H}_K = \infty$, then, for any $m \in \mathbb{N}$, we can find a subset $\mathbf{x} = \{x_1, \ldots, x_m\} \subseteq X$ that possesses a set of nodal functions.

Theorem 5.21 *Let K be a Mercer kernel on X, $m \in \mathbb{N}$, and $\mathbf{x} = \{x_1, \ldots, x_m\} \subseteq X$ such that $K[\mathbf{x}]$ is invertible. Then*

$$\mathcal{M}(I_K(B_R), \eta) \geq 2^m - 1$$

for all $\eta > 0$ satisfying

$$\|K[\mathbf{x}]^{-1}\|_2 < \frac{1}{m} \left(\frac{R}{\eta}\right)^2.$$

Proof. By Proposition 5.20, the set of nodal functions $\{u_j(x)\}_{j=1}^m$ associated with $\mathbf{x}$ exists and can be expressed by

$$u_i(x) = \sum_{j=1}^m (K[\mathbf{x}]^{-1})_{ij} K_{x_j}(x), \quad i = 1, \ldots, m.$$

For each nonempty subset J of $\{1, \ldots, m\}$, we define the function $u_J(x) := \sum_{j \in J} \eta' u_j(x)$, where $\eta' > \eta$ satisfies $\|K[\mathbf{x}]^{-1}\|_2 < \frac{1}{m}(R/\eta')^2$. These $2^m - 1$ functions are η-separated in $\mathscr{C}(X)$.

For $J_1 \neq J_2$, there exists some $j_0 \in \{1, \ldots, m\}$ lying in one of the sets J_1, J_2, but not in the other. Hence

$$\|u_{J_1} - u_{J_2}\|_\infty \geq |u_{J_1}(x_{j_0}) - u_{J_2}(x_{j_0})| \geq \eta' > \eta.$$

What is left is to show that the functions u_J lie in B_R. To see this, take $\emptyset \neq J \subseteq \{1, \ldots, m\}$. Then

$$
\begin{aligned}
\|u_J\|_K^2 &= \left\langle \eta' \sum_{j \in J} \sum_{\ell=1}^m \left(K[\mathbf{x}]^{-1} \right)_{j\ell} K_{x_\ell}, \eta' \sum_{j' \in J} \sum_{s=1}^m \left(K[\mathbf{x}]^{-1} \right)_{j's} K_{x_s} \right\rangle_K \\
&= \eta'^2 \sum_{j,j' \in J} \sum_{\ell=1}^m \left(K[\mathbf{x}]^{-1} \right)_{j\ell} \sum_{s=1}^m \left(K[\mathbf{x}]^{-1} \right)_{j's} (K[\mathbf{x}])_{s\ell} \\
&= \eta'^2 \sum_{j,j' \in J} \left(K[\mathbf{x}]^{-1} \right)_{jj'} = \eta'^2 \sum_{i=1}^m \left((K[\mathbf{x}])^{-1}\mathbf{e} \right)_i \\
&\leq \eta'^2 \|(K[\mathbf{x}])^{-1}\mathbf{e}\|_{\ell^1(J)} \leq \eta'^2 \sqrt{m} \|(K[\mathbf{x}])^{-1}\mathbf{e}\|_{\ell^2(J)} \\
&\leq \eta'^2 \sqrt{m} \|\mathbf{e}\| \|(K[\mathbf{x}])^{-1}\|_2 = \eta'^2 m \|(K[\mathbf{x}])^{-1}\|_2,
\end{aligned}
$$

where $\mathbf{e}$ is the vector in $\ell^2(J)$ with all components 1. It follows that $\|u_J\|_K \leq \eta' \sqrt{m} \|K[\mathbf{x}]^{-1}\|_2^{1/2}$, and $u_J \in B_R$ since $\|K[\mathbf{x}]^{-1}\|_2 \leq \frac{1}{m}(R/\eta')^2$. ∎

Thus lower bounds for packing numbers and covering numbers can be obtained in terms of the norm of the inverse of the Gramian matrix. The latter can be estimated for convolution-type kernels, that is, kernels $K(x, y) = k(x-y)$ for some function k in $\mathbb{R}^n$, in terms of the Fourier transform $\hat{k}$ of k.

Proposition 5.22 *Suppose $K(x, y) = k(x-y)$ is a Mercer kernel on $X = [0, 1]^n$ and the Fourier transform of k is positive; that is,*

$$\hat{k}(\xi) > 0, \quad \forall \xi \in \mathbb{R}^n.$$

For $N \in \mathbb{N}$, if $X_N := \{0, 1, \ldots, N - 1\}^n$ and $\mathbf{x} = \{\alpha/N\}_{\alpha \in X_N}$, then

$$\|K[\mathbf{x}]^{-1}\|_2 \leq N^{-n} \left(\inf_{\xi \in [-N\pi, N\pi]^n} \widehat{k}(\xi) \right)^{-1}.$$

Proof. By the inverse Fourier transform,

$$k(x) = (2\pi)^{-n} \int_{\mathbb{R}^n} \widehat{k}(\xi) e^{ix \cdot \xi} \, d\xi,$$

we know that for any vector $c := (c_\alpha)_{\alpha \in X_N}$,

$$c^T K[\mathbf{x}] c = \sum_{\alpha, \beta} c_\alpha c_\beta (2\pi)^{-n} \int_{\mathbb{R}^n} \widehat{k}(\xi) e^{i((\alpha/N) - (\beta/N)) \cdot \xi} \, d\xi$$

$$= (2\pi)^{-n} \int_{\mathbb{R}^n} \widehat{k}(\xi) \left| \sum_\alpha c_\alpha e^{i\alpha \cdot \xi / N} \right|^2 d\xi$$

$$= (2\pi)^{-n} N^n \int_{\mathbb{R}^n} \widehat{k}(N\xi) \left| \sum_\alpha c_\alpha e^{i\alpha \cdot \xi} \right|^2 d\xi.$$

Bounding from below the integral over the subset $[-N\pi, N\pi]^n$, we see that

$$c^T K[\mathbf{x}] c \geq (2\pi)^{-n} N^n \left(\inf_{\eta \in [-N\pi, N\pi]^n} \widehat{k}(\eta) \right) \int_{[-\pi, \pi]^n} \left| \sum_\alpha c_\alpha e^{i\alpha \cdot \xi} \right|^2 d\xi$$

$$= \|c\|_{\ell^2(X_N)}^2 N^n \left(\inf_{\xi \in [-N\pi, N\pi]^n} \widehat{k}(\xi) \right).$$

It follows that the smallest eigenvalue of the matrix $K[\mathbf{x}]$ is at least

$$N^n \left(\inf_{\xi \in [-N\pi, N\pi]^n} \widehat{k}(\xi) \right),$$

from which the estimate for the norm of the inverse matrix follows. ∎

Combining Theorem 5.21 and Proposition 5.22, we obtain the following result.

Theorem 5.23 *Suppose $K(x, y) = k(x - y)$ is a Mercer kernel on $X = [0, 1]^n$ and the Fourier transform of k is positive. Then, for $N \in \mathbb{N}$,*

$$\ln \mathcal{N} \left(I_K(B_R), \frac{\eta}{2} \right) \geq \ln \mathcal{M}(I_K(B_R), \eta) \geq \ln 2\{N^n - 1\},$$

provided N satisfies

$$\inf_{\xi \in [-N\pi, N\pi]^n} \widehat{k}(\xi) \geq \left(\frac{\eta}{R}\right)^2.$$ ∎

As an example, we use Theorem 5.23 to give lower bounds for covering numbers of balls in RKHSs in the case of Gaussian kernels.

Corollary 5.24 *Let* $\sigma > 0, n \in \mathbb{N}$, *and*

$$k(x) = \exp\left\{-\frac{\|x\|^2}{\sigma^2}\right\}, \quad x \in \mathbb{R}^n.$$

Set $X = [0,1]^n$, *and let the kernel K be given by*

$$K(x,t) = k(x-t), \quad x,t \in [0,1]^n.$$

Then, for $0 < \eta \leq \frac{R}{2}(\sigma\sqrt{\pi}/2)^{n/2}e^{-n\sigma^2\pi^2/8}$,

$$\ln \mathcal{N}(I_K(B_R), \eta) \geq \ln 2\left(\frac{2}{\sigma\pi}\right)^n\left\{\frac{1}{n}\ln\left(\frac{R}{\eta}\right) + \ln(\sigma\sqrt{\pi})\right.$$
$$\left. -\left(\frac{2}{n}+1\right)\ln 2\right\}^{n/2} - \ln 2.$$

Proof. Since $\widehat{k}$ is positive, we may use Theorem 5.23.   ∎

5.5 On the smoothness of box spline kernels

The only result of this section, Proposition 5.25, shows a way to construct box spline kernels with a prespecified smoothness $r \in \mathbb{N}$.

Proposition 5.25 *Let* $B_0 = [b_1, \ldots, b_n]$ *be an invertible* $n \times n$ *matrix. Let* $B = [B_0 B_0 \ldots B_0]$ *be an s-fold copy of* B_0 *and* $k(x) = (M_B * M_B)(x)$ *be induced by the convolution of the box spline* M_B *with itself. Finally, let* $X \subseteq \mathbb{R}^n$, *and let* $K : X \times X \to \mathbb{R}$ *be defined by* $K(x,y) = k(x-y)$. *Then* $K \in \mathscr{C}^r(X \times X)$ *for all* $r < 2s - n$.

Proof. By Example 2.17, the Fourier transform $\widehat{k}$ of $k(x) = (M_B * M_B)(x)$ satisfies

$$\widehat{k}(\xi) = \left\{\prod_{j=1}^n \left(\frac{\sin((\xi \cdot b_j)/2)}{(\xi \cdot b_j)/2}\right)^2\right\}^s.$$

To get the smoothness of K we estimate the decay of $\widehat{k}$. First we observe that the function $t \mapsto (\sin t)/t$ satisfies, for all $t \notin (-1, 1)$,

$$\left| \frac{\sin t}{t} \right| \leq \frac{1}{|t|} \leq \frac{2}{1 + |t|}.$$

Also, when $t \in (-1, 1)$, $|(\sin t)/t| \leq 1$. Hence, for all $t \in \mathbb{R}$,

$$\left| \frac{\sin t}{t} \right|^2 \leq \left(\frac{2}{1 + |t|} \right)^2 \leq \frac{4}{1 + t^2}.$$

It follows that for all $\xi \in \mathbb{R}^n$,

$$\prod_{j=1}^{n} \left(\frac{\sin((\xi \cdot b_j)/2)}{(\xi \cdot b_j)/2} \right)^2 \leq \frac{4^n}{\prod_{j=1}^{n} \left(1 + \left| (\xi \cdot b_j)/2 \right|^2 \right)}.$$

If we denote $\eta = B_0^{\mathrm{T}} \xi$, then $\xi \cdot b_j = b_j^{\mathrm{T}} \xi = \eta_j$ and

$$\prod_{j=1}^{n} \left(1 + \left| \frac{\xi \cdot b_j}{2} \right|^2 \right) = \prod_{j=1}^{n} \left(1 + \frac{\eta_j^2}{4} \right) \geq 1 + \frac{1}{4} \sum_{j=1}^{n} \eta_j^2 = 1 + \frac{1}{4} \|\eta\|^2.$$

But $\|\eta\|^2 = \|B_0^{\mathrm{T}} \xi\|^2 \geq |\lambda_0|^2 \|\xi\|^2$, where λ_0 is the smallest (in modulus) eigenvalue of B_0. It follows that for all $\xi \in \mathbb{R}^n$,

$$\widehat{k}(\xi) \leq \left(\frac{4^n}{1 + \frac{1}{4} |\lambda_0|^2 \|\xi\|^2} \right)^s \leq \left(\frac{4^n}{\min\{1, |\lambda_0|^2/4\}} \right)^s (1 + \|\xi\|^2)^{-s}.$$

Therefore, for any $p < 2s - \frac{n}{2}$,

$$\int_{\mathbb{R}^n} (1 + \|\xi\|^2)^p |\widehat{k}(\xi)|^2 \, d\xi \leq \left(\frac{4^n}{\min\{1, |\lambda_0|^2/4\}} \right)^{2s}$$
$$\int_{\mathbb{R}^n} \left(\frac{1}{1 + \|\xi\|^2} \right)^{2s-p} d\xi < \infty$$

and, thus, $k \in H^p(\mathbb{R}^n)$. By the Sobolev embedding theorem, $K \in \mathscr{C}^r(X \times X)$ for all $r < 2s - n$. ∎

5.6 References and additional remarks

Properties of function spaces on bounded domains X are discussed in [120]. In particular, one can find conditions on X (such as having a minimally smooth boundary) for the extension of function classes on a bounded domain X to the corresponding classes on $\mathbb{R}^n$.

Estimating covering numbers for various function spaces is a standard theme in the fields of function spaces [47] and approximation theory [78, 100]. The upper and lower bounds (5.2) for generalized Lipschitz spaces and, more generally, Triebel–Lizorkin spaces can be found in [47].

The upper bounds for covering numbers of balls of RKHSs associated with Sobolev smooth kernels described in Section 5.2 (Theorem 5.8) and the lower bounds given in Section 5.4 (Theorem 5.21) can be found in [156]. The bounds for analytic translation invariant kernels discussed in Section 5.3 are taken from [155].

The bounds (5.23) for the Fourier transform of the inverse multiquadrics can be found in [82] and [105], where properties of nodal functions and Proposition 5.20 can also be found.

For estimates of smoothness of general box splines sharper than those in Proposition 5.25, see [41].

6

Logarithmic decay of the approximation error

In Chapter 4 we characterized the regression functions and kernels for which the approximation error has a decay of order $\mathcal{O}(R^{-\theta})$. This characterization was in terms of the integral operator L_K and interpolation spaces. In this chapter we continue this discussion.

We first show, in Theorem 6.2, that for a $\mathscr{C}^\infty$ kernel K (and under a mild condition on ρ_X) the approximation error can decay as $\mathcal{O}(R^{-\theta})$ only if f_ρ is $\mathscr{C}^\infty$ as well. Since the latter is too strong a requirement on f_ρ, we now focus on regression functions and kernels for which a logarithmic decay in the approximation error holds. Our main result, Theorem 6.7, is very general and allows for several applications. The result, which will be proved in Section 6.4, shows some such consequences for our two main examples of analytic kernels.

Theorem 6.1 *Let X be a compact subset of $\mathbb{R}^n$ with piecewise smooth boundary and $f_\rho \in H^s(X)$ with $s > 0$. Let $\sigma, c > 0$.*

(i) *(Gaussian) For $K(x,t) = e^{-|x-t|^2/\sigma^2}$ we have*

$$\inf_{\|g\|_K \leq R} \|f_\rho - g\|_{\mathscr{L}^2(X)} \leq C(\ln R)^{-s/8}, \quad R \geq 1,$$

where C is a constant independent of R. When $s > \frac{n}{2}$,

$$\inf_{\|g\|_K \leq R} \|f_\rho - g\|_{\mathscr{C}(X)} \leq C(\ln R)^{(n/16)-(s/8)}, \quad R \geq 1.$$

(ii) *(Inverse multiquadrics) For $K(x,t) = (c^2 + |x-t|^2)^{-\alpha}$ with $\alpha > 0$ we have*

$$\inf_{\|g\|_K \leq R} \|f_\rho - g\|_{\mathscr{L}^2(X)} \leq C(\ln R)^{-s/2}, \quad R \geq 1,$$

109

where C is a constant independent of R. When $s > \frac{n}{2}$,

$$\inf_{\|g\|_K \leq R} \|f_\rho - g\|_{\mathscr{C}(X)} \leq C(\ln R)^{(n/4)-(s/2)}, \quad R \geq 1.$$

The quantity $\inf_{\|g\|_K \leq R} \|f_\rho - g\|_{\mathscr{L}^2(X)}$ is not the approximation error $(-\sigma_\rho^2)$ $\mathcal{A}(f_\rho, R)$ unless ρ_X is the Lebesgue measure μ. It is, however, possible to obtain bounds on $\mathcal{A}(f_\rho, R)$ using Theorem 6.1. If $s > \frac{n}{2}$, one can use the bound in $\mathscr{C}(X)$ for bounding $\inf_{\|g\|_K \leq R} \|f_\rho - g\|_{\mathscr{L}_{\rho_X}^2(X)}$ for an arbitrary ρ_X. This is so since $\sup\{\|f\|_{\mathscr{L}_{\rho_X}^2(X)} : \rho_X \text{ is a probability measure on } X\} = \|f\|_{\mathscr{C}(X)}$. In the general case,

$$\inf_{\|g\|_K \leq R} \|f_\rho - g\|_{\mathscr{L}_{\rho_X}^2(X)} \leq \mathcal{D}_{\mu\rho} \inf_{\|g\|_K \leq R} \|f_\rho - g\|_{\mathscr{L}^2(X)},$$

where $\mathcal{D}_{\mu\rho}$ denotes the operator norm of the identity

$$\mathscr{L}^2(X) \xrightarrow{\text{Id}} \mathscr{L}_{\rho_X}^2(X).$$

We call $\mathcal{D}_{\mu\rho}$ the *distortion* of ρ (with respect to μ). It measures how much ρ_X distorts the ambient measure μ. It is often reasonable to suppose that the distortion $\mathcal{D}_{\mu\rho}$ is finite.

Since ρ is not known, neither, in general is $\mathcal{D}_{\mu\rho}$. In some cases, however, the context may provide some information about $\mathcal{D}_{\mu\rho}$. An important case is the one in which, despite ρ not being known, we do know ρ_X. In this case $\mathcal{D}_{\mu\rho}$ may be derived.

In Theorem 6.1 we assume Sobolev regularity only for the approximated function f_ρ. To have better approximation orders, more information about ρ should be used: for instance, analyticity of f_ρ or degeneracy of the marginal distribution ρ_X.

6.1 Polynomial decay of the approximation error for $\mathscr{C}^\infty$ kernels

In this section we use Corollary 4.17, Theorem 5.5, and the embedding relation (5.1) to prove that a $\mathscr{C}^\infty$ kernel cannot yield a polynomial decay in the approximation error unless f_ρ is $\mathscr{C}^\infty$ itself, assuming a mild condition on the measure ρ_X.

We say that a measure ν *dominates* the Lebesgue measure on X when $d\nu(x) \geq C_0 dx$ for some constant $C_0 > 0$.

Theorem 6.2 *Assume* $X \subseteq \mathbb{R}^n$ *has piecewise smooth boundary and* K *is a* $\mathscr{C}^\infty$ *Mercer kernel on* X. *Assume as well that* ρ_X *dominates the Lebesgue measure on* X. *If for some* $\theta > 0$

$$\mathcal{A}(f_\rho, R) := \inf_{\|g\|_K \le R} \|f - g\|_{\mathscr{L}^2_{\rho_X}(X)} = \mathcal{O}(R^{-\theta}),$$

then f_ρ *is* $\mathscr{C}^\infty$ *on* X.

Proof. Since ρ_X dominates the Lebesgue measure μ we have that ρ_X is nondegenerate. Hence, $\mathcal{H}_K^+ = \mathcal{H}_K$ by Remark 4.18. By Corollary 4.17, our decay assumption implies that $f_\rho \in (\mathscr{L}^2_{\rho_X}(X), \mathcal{H}_K)_{\theta/(2+\theta)}$. We show that for all $s > 0$, $f_\rho \in \mathrm{Lip}^*(s, \mathscr{L}^2_\mu(X))$. To do so, we take $r \in \mathbb{N}, r \ge 2s(2+\theta)/\theta > s$, and $t \in \mathbb{R}^n$. Let $g \in \mathcal{H}_K$ and $x \in X_{r,t}$. Then

$$\Delta^r_t f_\rho(x) = \Delta^r_t (f_\rho - g)(x) + \Delta^r_t g(x) = \sum_{j=0}^r \binom{r}{j} (-1)^{r-j} (f_\rho - g)(x + jt) + \Delta^r_t g(x).$$

Let $\ell = 2s(2+\theta)/\theta$. Using the triangle inequality and the definition of $\| \ \|_{\mathrm{Lip}^*(\ell/2, \mathscr{C}(X))}$, it follows that

$$\left\{ \int_{X_{r,t}} |\Delta^r_t f_\rho(x)|^2 \, dx \right\}^{1/2} \le \sum_{j=0}^r \binom{r}{j} \|f_\rho - g\|_{\mathscr{L}^2_\mu(X)} + \|\Delta^r_t g\|_{\mathscr{L}^2_\mu(X)}$$

$$\le 2^r \|f_\rho - g\|_{\mathscr{L}^2_\mu(X)} + \sqrt{\mu(X)} \|g\|_{\mathrm{Lip}^*(\ell/2, \mathscr{C}(X))} \|t\|^{\ell/2}.$$

Since K is $\mathscr{C}^\infty$, $g \in \mathcal{H}_K$, and $r > \frac{\ell}{2}$, we can apply Theorem 5.5 to deduce that

$$\|g\|_{\mathrm{Lip}^*(\ell/2, \mathscr{C}(X))} \le \sqrt{2^{r+1} \|K\|_{\mathrm{Lip}^*(\ell)}} \|g\|_K.$$

Also, $d\rho_X(x) \ge C_0 dx$ implies that $\|f_\rho - g\|_{\mathscr{L}^2_\mu(X)} \le (1/\sqrt{C_0}) \|f_\rho - g\|_{\mathscr{L}^2_{\rho_X}(X)}$ and $\mu(X) \le 1/C_0$. By taking the infimum over $g \in \mathcal{H}_K$, we see that

$$\left\{ \int_{X_{r,t}} |\Delta^r_t f_\rho(x)|^2 \, dx \right\}^{1/2} \le \frac{1}{\sqrt{C_0}} \inf_{g \in \mathcal{H}_K} \left\{ 2^r \|f_\rho - g\|_{\mathscr{L}^2_{\rho_X}(X)} \right.$$

$$\left. + \sqrt{2^{r+1} \|K\|_{\mathrm{Lip}^*(\ell)}} \|g\|_K \|t\|^{\ell/2} \right\}$$

$$\leq \frac{1}{\sqrt{C_0}} \left(2^r + \sqrt{2^{r+1} \| K \|_{\mathsf{Lip}^*(\ell)}} \right)$$

$$\inf_{g \in \mathcal{H}_K} \left\{ \| f_\rho - g \|_{\mathscr{L}^2_{\rho_X}(X)} + \| t \|^{\ell/2} \| g \|_K \right\}.$$

Since $f_\rho \in (\mathscr{L}^2_{\rho_X}(X), \mathcal{H}_K)_{\theta/(2+\theta)}$, by the definition of the interpolation space in terms of the $\mathbb{K}$-functional, we have

$$\inf_{g \in \mathcal{H}_K} \left\{ \| f_\rho - g \|_{\mathscr{L}^2_{\rho_X}(X)} + \| t \|^{\ell/2} \| g \|_K \right\} = \mathbb{K}\left(f_\rho, \| t \|^{\ell/2} \right)$$

$$\leq C_0' \| t \|^{\theta\ell/2(2+\theta)} = C_0' \| t \|^s,$$

where C_0' may be taken as the norm of f_ρ in the interpolation space. It follows that

$$| f_\rho |_{\mathsf{Lip}^*(s, \mathscr{L}^2_\mu(X))} = \sup_{t \in \mathbb{R}^n} \| t \|^{-s} \left\{ \int_{X_{r,t}} | \Delta^r_t f_\rho(x) |^2 \, dx \right\}^{1/2}$$

$$\leq \frac{1}{\sqrt{C_0}} \left(2^r + \sqrt{2^{r+1} \| K \|_{\mathsf{Lip}^*(\ell)}} \right) C_0' < \infty.$$

Therefore, $f_\rho \in \mathsf{Lip}^*(s, \mathscr{L}^2_\mu(X))$. By (5.1) this implies $f_\rho \in \mathscr{C}^d(X)$ for any integer $d < s - \frac{n}{2}$. But s can be arbitrarily large, from which it follows that $f_\rho \in \mathscr{C}^\infty(X)$.   ∎

6.2 Measuring the regularity of the kernel

The approximation error depends not only on the regularity of the approximated function but also on the regularity of the Mercer kernel. We next measure the regularity of a Mercer kernel K on a finite set of points $\mathbf{x} = \{x_1, \ldots, x_m\} \subset X$. To this end, we introduce the following function:

$$\epsilon_K(\mathbf{x}) := \sup_{x \in X} \left\{ \inf_{w \in \mathbb{R}^m} \left\{ K(x,x) - 2 \sum_{i=1}^m w_i K(x, x_i) + \sum_{i,j=1}^m w_i K(x_i, x_j) w_j \right\}^{1/2} \right\}. \tag{6.1}$$

We show that by choosing w_i appropriately, one has $\epsilon_K(\mathbf{x}) \to 0$ as $\mathbf{x}$ becomes dense in X. It is the order of decay of $\epsilon_K(\mathbf{x})$ with respect to the density of $\mathbf{x}$ in X that now measures the regularity of functions in $\mathcal{H}_K$. The faster the decay, the more regular the functions.

As an example to see how $\epsilon_K(\mathbf{x})$ measures the regularity of functions in $\mathcal{H}_K$, suppose that for some $0 < s \le 1$, the kernel K is $\mathsf{Lip}(s)$; that is,

$$|K(x,y) - K(x,t)| \le C(d(y,t))^s, \quad \forall x, y, t \in X,$$

where C is a constant independent of x, y, t. Define the number

$$d_{\mathbf{x}} := \max_{x \in X} \min_{i \le m} d(x, x_i)$$

to measure the density of $\mathbf{x}$ in X. Let $x \in X$. Choose $x_\ell \in \mathbf{x}$ such that $d(x, x_\ell) \le d_{\mathbf{x}}$. Set the coefficients $\{w_j\}_{j=1}^m$ as $w_\ell = 1$, and $w_j = 0$ if $j \ne \ell$. Then

$$K(x,x) - 2\sum_{i=1}^m w_i K(x, x_i) + \sum_{i,j=1}^m w_i K(x_i, x_j) w_j = K(x,x) - 2K(x, x_\ell) + K(x_\ell, x_\ell).$$

The $\mathsf{Lip}(s)$ regularity and the symmetry of K yield

$$K(x,x) - 2K(x, x_\ell) + K(x_\ell, x_\ell) \le 2C(d(x, x_\ell))^s \le 2C d_{\mathbf{x}}^s.$$

Hence

$$\epsilon_K(\mathbf{x}) \le 2C d_{\mathbf{x}}^s.$$

In particular, if $X = [0,1]$ and $\mathbf{x} = \{j/N\}_{j=0}^N$, then $d_{\mathbf{x}} \le \frac{1}{2N}$, and therefore $\epsilon_K(\mathbf{x}) \le 2^{1-s} C N^{-s}$. We obtain a polynomial decay with exponent s.

When K is $\mathscr{C}^s$, the function $\epsilon_K(\mathbf{x})$ decays as $\mathcal{O}(d_{\mathbf{x}}^s)$. For analytic kernels, $\epsilon_K(\mathbf{x})$ often decays exponentially. In this section we derive these decaying rates for the function $\epsilon_K(\mathbf{x})$.

Recall the Lagrange interpolation polynomials $\{w_{l,s-1}(t)\}_{l=0}^{s-1}$ on $[0,1]$ with interpolating points $\{0, 1/s - 1, \ldots, 1\}$ defined by (5.3) with s replaced by $s - 1$. For any polynomial p of degree at most $s - 1$,

$$\sum_{l=0}^{s-1} w_{l,s-1}(t) p(l/(s-1)) = p(t).$$

This, together with the Taylor expansion of f at t,

$$f(y) = \sum_{j=0}^{s-1} \frac{f^{(j)}(t)}{j!}(y - t)^j + R_s(f)(y, t),$$

implies that

$$\left| \sum_{l=0}^{s-1} w_{l,s-1}(t)(f(l/(s-1)) - f(t)) \right| = \left| \sum_{l=0}^{s-1} w_{l,s-1}(t)R_s(f)(l/(s-1),t) \right|.$$

Here $R_s(f)$ is the linear operator representing the remainder of the Taylor expansion and satisfies

$$|R_s(f)(y,t)| = \left| \frac{1}{(s-1)!} \int_t^y (y-u)^{s-1} f^{(s)}(u)du \right|$$

$$\leq \frac{|y-t|^s}{s!} \|f^{(s)}\|_\infty \leq \frac{1}{s!} \|f^{(s)}\|_\infty.$$

Using Lemma 5.9, we now obtain

$$\left| \sum_{l=0}^{s-1} w_{l,s-1}(t)(f(l/(s-1)) - f(t)) \right| \leq \frac{(s-1)2^{s-1}}{s!} \|f^{(s)}\|_\infty. \tag{6.2}$$

Recall also the multivariate Lagrange interpolation polynomials

$$w_{\alpha,s-1}(x) = \prod_{j=1}^n w_{\alpha_j,s-1}(x_j),$$

$$x = (x_1,\ldots,x_n), \alpha = (\alpha_1,\ldots,\alpha_n) \in \{0,\ldots,s-1\}^n$$

defined by (5.6).

Now we can estimate $\epsilon_K(\mathbf{x})$ for $\mathscr{C}^s$ kernels as follows.

Theorem 6.3 *Let $X = [0,1]^n$, $s \in \mathbb{N}$, and K be a Mercer kernel on X such that for each $\alpha \in \mathbb{N}^n$ with $|\alpha| \leq s$,*

$$\frac{\partial^\alpha}{\partial y^\alpha} K(x,y) = \frac{\partial^{|\alpha|}}{\partial y_1^{\alpha_1} \cdots \partial y_n^{\alpha_n}} K(x,y) \in \mathscr{C}([0,1]^{2n}).$$

Then, for $N \geq s$ and $\mathbf{x} = \{\alpha/N\}_{\alpha \in \{0,1,\ldots,N-1\}^n}$, we have

$$\epsilon_K(\mathbf{x}) \leq \frac{2^{2sn+1} s^{s+2n}}{s!} \left(\sum_{i=1}^n \left\| \frac{\partial^s K}{\partial y_i^s} \right\|_{\mathscr{C}(X \times X)} \right) N^{-s}.$$

Proof. Let $x \in X$. Then $x = \frac{\beta}{N} + \frac{s-1}{N}t$ for some $\beta \in \{0, 1, \ldots, N - s + 1\}^n$ and $t \in [0, 1]^n$. Choose the coefficients in the definition of $\epsilon_K(\mathbf{x})$ to be

$$
w_\alpha = \begin{cases} w_{\gamma, s-1}(t) & \text{if } \alpha = \beta + \gamma, \ \gamma \in \{0, \ldots, s-1\}^n \\ 0 & \text{otherwise.} \end{cases}
$$

Then we can see that the expression in the definition of ϵ_K is

$$
K(x, x) - 2 \sum_{\gamma \in \{0, \ldots, s-1\}^n} w_{\gamma, s-1}(t) K\left(x, \frac{\beta + \gamma}{N}\right)
$$

$$
+ \sum_{\gamma, \eta \in \{0, \ldots, s-1\}^n} w_{\gamma, s-1}(t) K\left(\frac{\beta + \gamma}{N}, \frac{\beta + \eta}{N}\right) w_{\eta, s-1}(t).
$$

But $\sum_{\gamma \in \{0, \ldots, s-1\}^n} w_{\gamma, s-1}(t) \equiv 1$. So the above expression equals

$$
\sum_{\gamma \in \{0, \ldots, s-1\}^n} w_{\gamma, s-1}(t) \left\{ K(x, x) - K\left(x, \frac{\beta + \gamma}{N}\right) \right\} + \sum_{\gamma \in \{0, \ldots, s-1\}^n} w_{\gamma, s-1}(t)
$$

$$
\left\{ \sum_{\eta \in \{0, \ldots, s-1\}^n} w_{\eta, s-1}(t) \left(K\left(\frac{\beta + \gamma}{N}, \frac{\beta + \eta}{N}\right) - K\left(\frac{\beta + \gamma}{N}, x\right) \right) \right\}.
$$

Using Equation (6.2) for the univariate function $g(z) = f\big(\gamma_1/(s-1), \ldots, (\gamma_{i-1})/(s-1), z, t_{i+1}, \ldots, t_n\big)$ with $z \in [0, 1]$ and all the other variables fixed, we get

$$
\left| \sum_{\gamma_i = 0}^{s-1} w_{\gamma_i, s-1}(t_i) \left(g\left(\frac{\gamma_i}{s-1}\right) - g(t_i) \right) \right| \leq \frac{(s-1)2^{s-1}}{s!} \left\| \frac{\partial^s f}{\partial t_i^s} \right\|_{\mathscr{C}([0,1]^n)}.
$$

Using Lemma 5.9 for $\gamma_j, j \neq i$, we conclude that for a function f on $[0, 1]^n$ and for $i = 1, \ldots, n$,

$$
\left| \sum_{\gamma \in \{0, \ldots, s-1\}^n} w_{\gamma, s-1}(t) \left(f\left(\frac{\gamma_1}{s-1}, \ldots, \frac{\gamma_{i-1}}{s-1}, \frac{\gamma_i}{s-1}, t_{i+1}, \ldots, t_n\right) \right. \right.
$$

$$
\left. \left. -f\left(\frac{\gamma_1}{s-1}, \ldots, \frac{\gamma_{i-1}}{s-1}, t_i, t_{i+1}, \ldots, t_n\right) \right) \right| \leq ((s-1)2^{s-1})^{n-1}
$$

$$
\frac{(s-1)2^{s-1}}{s!} \left\| \frac{\partial^s f}{\partial t_i^s} \right\|_{\mathscr{C}([0,1]^n)}.
$$

Replacing $\gamma_i/(s-1)$ by t_i each time for one $i \in \{1,\ldots,n\}$, we obtain

$$\left| \sum_{\gamma \in \{0,\ldots,s-1\}^n} w_{\gamma,s-1}(t) \left(f\left(\frac{\gamma}{s-1}\right) - f(t) \right) \right|$$

$$\leq \frac{((s-1)2^{s-1})^n}{s!} \sum_{i=1}^n \left\| \frac{\partial^s f}{\partial t_i^s} \right\|_{\mathscr{C}([0,1]^n)}.$$

Applying this estimate to the functions $f(t) = K\left(x, \frac{\beta+(s-1)t}{N}\right)$ and $K\left(\frac{\beta+\gamma}{N}, \frac{\beta+(s-1)t}{N}\right)$, we find that the expression for ϵ_K can be bounded by

$$\left(1 + ((s-1)2^{s-1})^n\right) \frac{((s-1)2^{s-1})^n}{s!} \left(\frac{s}{N}\right)^s \sum_{i=1}^n \left\| \frac{\partial^s K}{\partial y_i^s} \right\|_{\mathscr{C}(X \times X)}.$$

This bound is valid for each $x \in X$. Therefore, we obtain the required estimate for $\epsilon_K(\mathbf{x})$ by taking the supremum for $x \in X$. ∎

The behavior of the quantity $\epsilon_K(\mathbf{x})$ is better if the kernel is of convolution type, that is, if $K(x,y) = k(x-y)$ for an analytic function k.

Theorem 6.4 *Let $X = [0,1]^n$ and $K(x,y) = k(x-y)$ be a Mercer kernel on X with*

$$\widehat{k}(\xi) \leq C_0 e^{-\lambda|\xi|}, \quad \forall \xi \in \mathbb{R}^n$$

for some constants $C_0 > 0$ and $\lambda > 4 + 2n \ln 4$. Then, for $\mathbf{x} = \{\frac{\alpha}{N}\}_{\alpha \in \{0,1,\ldots,N-1\}^n}$ with $N \geq 4n/\ln\min\{e\lambda, 4^{-n}e^{\lambda/2}\}$, we have

$$\epsilon_K(\mathbf{x}) \leq 4C_0 \left(\max\left\{ \frac{1}{e\lambda}, \frac{4^n}{e^{\lambda/2}} \right\} \right)^{N/2}.$$

Proof. Let $X_N := \{0,\ldots,N-1\}^n$. For a fixed $x \in X$, choose the coefficients w_i in (6.1) to be $w_{\alpha,N}(x)$. Then the expression of the definition of $\epsilon_K(\mathbf{x})$ becomes $Q_N(x)$ given by (5.13). It follows that $\epsilon_K(\mathbf{x}) \leq \sup_{x \in X} Q_N(x)$. Hence by (5.14),

$$\epsilon_K(\mathbf{x}) \leq \Upsilon_k(N).$$

But the assumption of the kernel here verifies the condition in Theorem 5.15. Thus, we can apply the estimate (5.21) for $\Upsilon_k(N)$ to draw our conclusion here. ∎

6.3 Estimating the approximation error in RKHSs

Recall the nodal functions $\{u_i = u_{i,\mathbf{x}}\}_{i=1}^m$ associated with a finite subset $\mathbf{x}$ of X, given by (5.25). We use them in the RKHS $\mathcal{H}_K$ on a compact metric space (X, d) to construct an interpolation scheme. This scheme is defined as follows:

$$I_{\mathbf{x}}(f)(x) = \sum_{i=1}^m f(x_i) u_i(x), \quad x \in X, f \in \mathscr{C}(X). \tag{6.3}$$

It satisfies $I_{\mathbf{x}}(f)(x_i) = f(x_i)$ for $i = 1, \dots, m$.

The error of the interpolation scheme for functions in $\mathcal{H}_K$ can be estimated as follows.

Proposition 6.5 *Let K be a Mercer kernel and $\mathbf{x} = \{x_1, \dots, x_m\} \subset X$ such that $K[\mathbf{x}]$ is invertible. Define the interpolation scheme $I_{\mathbf{x}}$ by (6.3). Then, for $f \in \mathcal{H}_K$,*

$$\|I_{\mathbf{x}}(f) - f\|_{\mathscr{C}(X)} \leq \epsilon_K(\mathbf{x}) \|f\|_K$$

and

$$\|I_{\mathbf{x}}(f)\|_K \leq \|f\|_K.$$

Proof. For $x \in X$

$$I_{\mathbf{x}}(f)(x) - f(x) = \sum_{i=1}^m f(x_i) u_i(x) - f(x) = \sum_{i=1}^m u_i(x) \langle K_{x_i}, f \rangle_K - \langle K_x, f \rangle_K$$

$$= \left\langle \sum_{i=1}^m u_i(x) K_{x_i} - K_x, f \right\rangle_K,$$

the second equality by the reproducing property of $\mathcal{H}_K$ applied to f. By the Cauchy–Schwarz inequality in $\mathcal{H}_K$,

$$|I_{\mathbf{x}}(f)(x) - f(x)| \leq \left\| \sum_{i=1}^m u_i(x) K_{x_i} - K_x \right\|_K \|f\|_K.$$

Since $\langle K_s, K_t \rangle_K = K(s, t)$, we have

$$\left\| \sum_{i=1}^m u_i(x) K_{x_i} - K_x \right\|_K^2 = K(x, x) - 2 \sum_{i=1}^m u_i(x) K(x, x_i) + \sum_{i,j=1}^m u_i(x) K(x_i, x_j) u_j(x).$$

By Proposition 5.20, the quadratic function

$$Q(w) = K(x, x) - 2 \sum_{i=1}^m w_i K(x, x_i) + \sum_{i,j=1}^m w_i K(x_i, x_j) w_j$$

is minimized over $\mathbb{R}^m$ at $(u_i(x))_{i=1}^m$. Therefore,

$$\left\| \sum_{i=1}^m u_i(x) K_{x_i} - K_x \right\|_K \le \epsilon_K(\mathbf{x}).$$

It follows that

$$|I_{\mathbf{x}}(f)(x) - f(x)| \le \epsilon_K(\mathbf{x}) \|f\|_K.$$

This proves the estimate for $\|I_{\mathbf{x}}(f) - f\|_{\mathscr{C}(X)}$.

To prove the other inequality, note that, since $I_{\mathbf{x}}(f) \in \mathcal{H}_K$ and $I_{\mathbf{x}}(f)(x_i) = f(x_i)$, for $i = 1, \ldots, m$,

$$0 = I_{\mathbf{x}}(f)(x_i) - f(x_i) = \langle K_{x_i}, I_{\mathbf{x}}(f) - f \rangle_K.$$

This means that $I_{\mathbf{x}}(f) - f$ is orthogonal to $\mathrm{span}\{K_{x_i}\}_{i=1}^m$. Hence $I_{\mathbf{x}}(f)$ is the orthogonal projection of f onto $\mathrm{span}\{K_{x_i}\}_{i=1}^m$ and therefore $\|I_{\mathbf{x}}(f)\|_K \le \|f\|_K$. ∎

Proposition 6.5 bounds the interpolation error $\|I_{\mathbf{x}}(f) - f\|_{\mathscr{C}(X)}$ in terms of the regularity of K and the density of $\mathbf{x}$ in X measured by $\epsilon_K(\mathbf{x})$, when the approximated function f lies in the RKHS (i.e., $f \in \mathcal{H}_K$).

In the remainder of this section, we deal with the interpolation error when the approximated function is from a larger function space (e.g., a Sobolev space), not necessarily from $\mathcal{H}_K$. To this end, we need to know how large $\mathcal{H}_K$ is compared with the space where the approximated function lies. For a convolution-type kernel, that is, a kernel $K(x, y) = k(x - y)$ with $\widehat{k}(\xi) > 0$, this depends on how slowly $\widehat{k}$ decays. So, in addition to the function ϵ_K measuring the smoothness of K, we use the function $\Lambda_k : \mathbb{R}_+ \to \mathbb{R}_+$ defined by

$$\Lambda_k(r) := \left(\inf_{[-r\pi, r\pi]^n} \widehat{k}(\xi) \right)^{-1/2},$$

measuring the speed of decay of $\widehat{k}$. Note that Λ_k is a nondecreasing function. We also use the function $\Upsilon_k(N)$, which measures the regularity of K, introduced in Section 5.3.

Lemma 6.6 *Let $k \in \mathscr{L}^2(\mathbb{R}^n)$ be an even function with $\widehat{k}(\xi) > 0$, and K be the kernel on $X = [0, 1]^n$ given by $K(x, y) = k(x - y)$. For $f \in \mathscr{L}^2(\mathbb{R}^n)$ and $M \le N \in \mathbb{N}$, we define $f_M \in \mathscr{L}^2(\mathbb{R}^n)$ by*

$$\widehat{f_M}(\xi) = \begin{cases} \widehat{f}(\xi) & \text{if } \xi \in [-M\pi, M\pi]^n \\ 0 & \text{otherwise.} \end{cases}$$

Then, for $\mathbf{x} = \{0, \frac{1}{N}, \ldots, \frac{N-1}{N}\}^n$, we have

(i) $\|I_{\mathbf{x}}(f_M)\|_K \leq \|f\|_{\mathscr{L}^2} \Lambda_k(N)$.

(ii) $\|f_M - I_{\mathbf{x}}(f_M)\|_{\mathscr{C}(X)} \leq \|f\|_{\mathscr{L}^2} \Lambda_k(M) \Upsilon_k(N)$.

(iii) $\|f - f_M\|_{\mathscr{L}^2(X)}^2 \leq (2\pi)^{-n} \int_{\xi \notin [-M\pi, M\pi]^n} |\widehat{f}(\xi)|^2 \, d\xi \to 0 \quad (as \ M \to \infty)$.

Proof.

(i) For $i, j \in X_N := \{0, 1, \ldots, N-1\}^n$ and $x_i = i/N \in \mathbf{x}$, expression (5.25) for the nodal function u_i associated with $\mathbf{x}$ gives

$$\langle u_i, u_j \rangle_K = \sum_{s,t \in X_N} (K[\mathbf{x}]^{-1})_{is} (K[\mathbf{x}]^{-1})_{jt} \langle K_{x_s}, K_{x_t} \rangle_K$$

$$= \sum_{s,t \in X_N} (K[\mathbf{x}]^{-1})_{is} (K[\mathbf{x}]^{-1})_{jt} (K[\mathbf{x}])_{ts} = (K[\mathbf{x}]^{-1})_{ij}.$$

Then, for $g \in \mathscr{C}(X)$, we have

$$\|I_{\mathbf{x}}(g)\|_K^2 = \left\| \sum_{i \in X_N} g(x_i) u_i(x) \right\|_K^2$$

$$= \sum_{i,j \in X_N} g(x_i) g(x_j) \langle u_i, u_j \rangle_K = (g_{|\mathbf{x}})^T K[\mathbf{x}]^{-1} (g_{|\mathbf{x}}),$$

where $g_{|\mathbf{x}}$ is the vector $(g(x_i))_{i \in X_N} \in \mathbb{R}^{N^n}$. It follows that

$$\|I_{\mathbf{x}}(g)\|_K^2 \leq \|K[\mathbf{x}]^{-1}\|_2 \|g_{|\mathbf{x}}\|_{\ell^2(X_N)}^2 = \|K[\mathbf{x}]^{-1}\|_2 \sum_{i \in X_N} |g(x_i)|^2.$$

We now apply this analysis to the function f_M satisfying $\widehat{f_M}(\xi) = 0$ for $\xi \in [-N\pi, N\pi]^n \setminus [-M\pi, M\pi]^n$ to obtain

$$\sum_{j \in X_N} |f_M(x_j)|^2 = \sum_{j \in X_N} \left| (2\pi)^{-n} \int_{\xi \in [-M\pi, M\pi]^n} \widehat{f_M}(\xi) e^{i(j/N) \cdot \xi} \, d\xi \right|^2$$

$$\leq \sum_{j \in X_N} \left| (2\pi)^{-n} \int_{\xi \in [-\pi, \pi]^n} \widehat{f_M}(N\xi) e^{ij \cdot \xi} N^n \, d\xi \right|^2$$

$$\leq (2\pi)^{-n} \int_{\xi \in [-\pi, \pi]^n} \left| \widehat{f_M}(N\xi) N^n \right|^2 \, d\xi \leq N^n \|f\|_{\mathscr{L}^2}^2.$$

Then

$$\|I_{\mathbf{x}}(f_M)\|_K^2 \leq \|K[\mathbf{x}]^{-1}\|_2 N^n \|f\|_{\mathscr{L}^2}^2.$$

But, by Proposition 5.22, $\|K[\mathbf{x}]^{-1}\|_2 \leq N^{-n} \big(\Lambda_k(N)\big)^2$. Therefore,

$$\|I_\mathbf{x}(f_M)\|_K \leq \|f\|_{\mathscr{L}^2} \Lambda_k(N).$$

This proves the statement in (i).

(ii) Let $x \in X$. Then

$$f_M(x) - I_\mathbf{x}(f_M)(x) = (2\pi)^{-n} \int_{\xi \in [-M\pi, M\pi]^n} \widehat{f}(\xi) \left\{ e^{ix\cdot\xi} - \sum_{j \in X_N} u_j(x) e^{ix_j\cdot\xi} \right\} d\xi.$$

By the Cauchy–Schwarz inequality,

$$|f_M(x) - I_\mathbf{x}(f_M)(x)| \leq \left\{ (2\pi)^{-n} \int_{\xi \in [-M\pi, M\pi]^n} \frac{|\widehat{f}(\xi)|^2}{\widehat{k}(\xi)} d\xi \right\}^{1/2}$$

$$\left\{ (2\pi)^{-n} \int_{\mathbb{R}^n} \widehat{k}(\xi) \left| e^{ix\cdot\xi} - \sum_{j \in X_N} u_j(x) e^{ix_j\cdot\xi} \right|^2 d\xi \right\}^{1/2}.$$

The first term on the right is bounded by $\|f\|_{\mathscr{L}^2} \Lambda_k(M)$, since $\widehat{k}(\xi) \geq \Lambda_k^{-2}(M)$ for $\xi \in [-M\pi, M\pi]^n$. The second term is

$$\left(k(0) - 2 \sum_{j \in X_N} u_j(x) k(x - x_j) + \sum_{i,j \in X_N} u_i(x) k(x_i - x_j) u_j(x) \right)^{1/2},$$

which can be bounded by $\Upsilon_k(N)$ according to (5.14) with $\{0, 1, \ldots, N\}^n$ replaced by $\{0, 1, \ldots, N-1\}^n$. Therefore,

$$\|f_M - I_\mathbf{x}(f_M)\|_{\mathscr{C}(X)} \leq \|f\|_{\mathscr{L}^2} \Lambda_k(M) \Upsilon_k(N).$$

(iii) By Plancherel's formula (Theorem 2.3),

$$\|f - f_M\|_{\mathscr{L}^2(\mathbb{R}^n)}^2 = (2\pi)^{-n} \int_{\xi \notin [-M\pi, M\pi]^n} |\widehat{f}(\xi)|^2 d\xi.$$

Thus, all the statements hold true.   ∎

Lemma 6.6 provides quantitative estimates for the interpolation error:

$$\|f - I_\mathbf{x}(f_M)\|_{\mathscr{L}^2(X)} \leq \left\{ (2\pi)^{-n} \int_{\xi \notin [-M\pi, M\pi]^n} |\widehat{f}(\xi)|^2 d\xi \right\}^{1/2}$$

$$+ \|f\|_{\mathscr{L}^2} \Lambda_k(M) \Upsilon_k(N)$$

with

$$\|I_{\mathbf{x}}(f_M)\|_K \leq \|f\|_{\mathscr{L}^2} \Lambda_k(N).$$

Choose $N = N(M) \geq M$ such that $\Lambda_k(M)\Upsilon_k(N) \to 0$ as $M \to +\infty$. We then have $\|f - I_{\mathbf{x}}(f_M)\|_{\mathscr{L}^2(X)} \to 0$. Also, the RKHS norm of $I_{\mathbf{x}}(f_M)$ is asymptotically controlled by $\Lambda_k(N)$.

We can now state the main estimates for the approximation error for balls in the RKHS $\mathcal{H}_K$ on $X = [0,1]^n$. Denote by Λ_k^{-1} the inverse function of the nondecreasing function Λ_k,

$$\Lambda_k^{-1}(R) := \max\{r > 0 : \Lambda_k(r) \leq R\}, \quad \text{for } R > 1/\widehat{k}(0).$$

Theorem 6.7 *Let* $X = [0,1]^n$, *$s > 0$, and $f \in H^s(\mathbb{R}^n)$. Then, for $R > \|f\|_2$,*

$$\inf_{\|g\|_K \leq R} \|f - g\|_{\mathscr{L}^2(X)} \leq \inf_{0 < M \leq N_R} \left\{ \Lambda_k(M)\|f\|_2 \Upsilon_k(N_R) + \|f\|_s(\pi M)^{-s} \right\},$$

where $N_R = \lfloor \Lambda_k^{-1}(R/\|f\|_2) \rfloor$, the integer part of $\Lambda_k^{-1}(R/\|f\|_2)$. If $s > \frac{n}{2}$, then

$$\inf_{\|g\|_K \leq R} \|f - g\|_{\mathscr{C}(X)} \leq \inf_{0 < M \leq N_R} \left\{ \Lambda_k(M)\|f\|_2 \Upsilon_k(N_R) + \frac{\|f\|_s}{\sqrt{s - n/2}} M^{(n/2)-s} \right\}.$$

Proof. Take N to be N_R.

Let $M \in (0, N]$. Set the function f_M as in Lemma 6.6. Then, by Lemma 6.6,

$$\|I_{\mathbf{x}}(f_M)\|_K \leq \|f\|_2 \Lambda_k(N) \leq R$$

and

$$\|f - I_{\mathbf{x}}(f_M)\|_{\mathscr{L}^2(X)} \leq \|f_M - I_{\mathbf{x}}(f_M)\|_{\mathscr{C}(X)} + \|f - f_M\|_{\mathscr{L}^2(X)}$$
$$\leq \Lambda_k(M)\|f\|_2 \Upsilon_k(N) + \|f\|_s(\pi M)^{-s}.$$

If $s > \frac{n}{2}$, then

$$\|f - f_M\|_{\mathscr{C}(X)} \leq (2\pi)^{-n} \int_{\xi \notin [-M\pi, M\pi]^n} |\widehat{f}(\xi)| d\xi \leq \frac{\|f\|_s}{\sqrt{s - n/2}} M^{\frac{n}{2}-s}.$$

Hence the second statement of the theorem also holds. ∎

Corollary 6.8 *Let* $X = [0,1]^n$, $s > 0$, *and* $f \in H^s(\mathbb{R}^n)$. *If for some $\alpha_1, \alpha_2, C_1, C_2 > 0$, one has*

$$\widehat{k}(\xi) \geq C_1(1 + |\xi|)^{-\alpha_1}, \quad \forall \xi \in \mathbb{R}^n$$

and

$$\Upsilon_k(N) \leq C_2 N^{-\alpha_2}, \qquad \forall N \in \mathbb{N},$$

then, for $R > (1 + 1/C_1)(1 + 1/\widehat{k}(0))\|f\|_2,$

$$\inf_{\|g\|_K \leq R} \|f - g\|_{\mathscr{L}^2(X)} \leq \left\{ C_3 \|f\|_2 + C_1^{-2s/\alpha_1} \|f\|_s \right\} \left(\frac{R}{\|f\|_2} \right)^{-\gamma},$$

where $C_3 := 2^{\alpha_2} C_2 C_1^{-2\alpha_2/\alpha_1} \left(2^{\alpha_1}/C_1 + \left(\sqrt{n}\pi\right)^{\alpha_1} C_1 \right)^{1/2}$ *and*

$$\gamma = \begin{cases} \frac{4\alpha_2 s}{\alpha_1(\alpha_1 + 2s)} & \text{if } \alpha_1 + 2s \geq 2\alpha_2 \\ \frac{2s}{\alpha_1}, & \text{if } \alpha_1 + 2s < 2\alpha_2. \end{cases}$$

If, in addition, $s > n/2$*, then*

$$\inf_{\|g\|_K \leq R} \|f - g\|_{\mathscr{C}(X)} \leq \left\{ C_3 \|f\|_2 + \frac{C_1^{(n-2s)/\alpha_1} 2^{s-n/2}}{\sqrt{s - n/2}} \|f\|_s \right\} \left(\frac{R}{\|f\|_2} \right)^{-\gamma'},$$

where

$$\gamma' = \begin{cases} \frac{\alpha_2(4s - 2n)}{\alpha_1(\alpha_1 + 2s - n)} & \text{if } \alpha_1 + 2s - n \geq 2\alpha_2, \\ \frac{2s - n}{\alpha_1}, & \text{if } \alpha_1 + 2s - n < 2\alpha_2. \end{cases}$$

Proof.   By the assumption on the lower bound of $\widehat{k}$, we find

$$\Lambda_k(r) \leq \frac{1}{\sqrt{C_1}} (1 + \sqrt{n}\pi r)^{\alpha_1/2}.$$

It follows that for $R/\|f\|_2 \geq \max\{1/C_1, 1/\widehat{k}(0)\}$,

$$\Lambda_k^{-1}(R/\|f\|_2) \geq C_1^{2/\alpha_1} (R/\|f\|_2)^{2/\alpha_1}$$

and

$$N_R \geq \tfrac{1}{2} C_1^{2/\alpha_1} (R/\|f\|_2)^{2/\alpha_1}.$$

Also,

$$\Upsilon_k(N) \leq C_2 \left([\Lambda_k^{-1}(R/\|f\|_2)] \right)^{-\alpha_2} \leq 2^{\alpha_2} C_2 (C_1 R/\|f\|_2)^{-2\alpha_2/\alpha_1}.$$

Then, by Theorem 6.7,

$$\inf_{\|g\|_K \leq R} \|f - g\|_{\mathscr{L}^2(X)} \leq \inf_{0 < M \leq N_R} \left\{ \left(\frac{(1 + \sqrt{n}\pi M)^{\alpha_1}}{C_1} \right)^{1/2} \right.$$
$$\left. 2^{\alpha_2} C_2 \left(\frac{C_1 R}{\|f\|_2} \right)^{-2\alpha_2/\alpha_1} \|f\|_2 + \|f\|_s (\pi M)^{-s} \right\}.$$

Take $M = \frac{1}{2} C_1^{2/\alpha_1} (R/\|f\|_2)^{\gamma/s}$ with γ as in the statement. Then, $M \leq N_R$, and we can see that

$$\inf_{\|g\|_K \leq R} \|f - g\|_{\mathscr{L}^2(X)} \leq \left\{ C_3 \|f\|_2 + C_1^{-2s/\alpha_1} \|f\|_s \right\} \left(\frac{R}{\|f\|_2} \right)^{-\gamma}.$$

This proves the first statement of the corollary. The second statement can be proved in the same way. ∎

Corollary 6.9 *Let* $X = [0,1]^n$, $s > 0$, *and* $f \in H^s(\mathbb{R}^n)$. *If for some* $\alpha_1, \alpha_2, \delta_1, \delta_2, C_1, C_2 > 0$, *one has*

$$\widehat{k}(\xi) \geq C_1 \exp\left\{ -\delta_1 |\xi|^{\alpha_1} \right\}, \qquad \forall \xi \in \mathbb{R}^n$$

and

$$\Upsilon_k(N) \leq C_2 \exp\left\{ -\delta_2 N^{\alpha_2} \right\}, \qquad \forall N \in \mathbb{N},$$

then, for $R > (1 + A/C_1)\|f\|_2$,

$$\inf_{\|g\|_K \leq R} \|f - g\|_{\mathscr{L}^2(X)} \leq \left\{ \frac{C_2 B}{\sqrt{C_1}} \|f\|_2 + \delta_1^{s/\alpha_1} 2^s n^{s/2} \|f\|_s \right\} \left(\ln R + \ln \frac{C_1}{\|f\|_2} \right)^{-\gamma s},$$

where A, B *are constants depending only on* $\alpha_1, \alpha_2, \delta_1, \delta_2$, *and*

$$\gamma = \begin{cases} \frac{1}{\alpha_1} & \text{if } \alpha_1 < \alpha_2 \\ \frac{\alpha_2}{2\alpha_1^2} & \text{if } \alpha_1 \geq \alpha_2. \end{cases}$$

If, in addition, $s > \frac{n}{2}$, *then, for* $R > (1 + A/C_1)\|f\|_2$,

$$\inf_{\|g\|_K \leq R} \|f - g\|_{\mathscr{C}(X)} \leq B' \left\{ \frac{C_2 B}{\sqrt{C_1}} \|f\|_2 + \|f\|_s \right\} \left(\ln R + \ln \frac{C_1}{\|f\|_2} \right)^{\gamma(\frac{n}{2} - s)},$$

where B' *is a constant depending only on* $\alpha_1, \alpha_2, \delta_1, \delta_2, n$, *and* s.

Proof. By the assumption on the lower bound of $\widehat{k}$, we find

$$\Lambda_k(r) \le \frac{1}{\sqrt{C_1}} \exp\left\{\frac{\delta_1}{2}(\sqrt{n}\pi r)^{\alpha_1}\right\}.$$

It follows that for $R/\|f\|_2 \ge \max\{1/\sqrt{C_1}, 1/\widehat{k}(0)\}$,

$$\Lambda_k^{-1}(R/\|f\|_2) \ge \frac{1}{\sqrt{n}\pi}\left(\frac{2}{\delta_1}\ln\frac{R\sqrt{C_1}}{\|f\|_2}\right)^{1/\alpha_1},$$

and its integer part N_R satisfies

$$N_R \ge \widetilde{R} := \frac{1}{2\sqrt{n}\pi}\left(\frac{2}{\delta_1}\ln\frac{R\sqrt{C_1}}{\|f\|_2}\right)^{1/\alpha_1}.$$

Also, for $R > \|f\|_2\exp\{(2/\delta_1)_1(2\sqrt{n}\pi)^{\alpha_1}\}/\sqrt{C_1}$,

$$\Upsilon_k(N_R) \le C_2\exp\left\{-\delta_2[\Lambda_k^{-1}(R/\|f\|_2)]^{\alpha_2}\right\} \le C_2\exp\left\{-\delta_2\widetilde{R}^{\alpha_2}\right\}.$$

Then, by Theorem 6.7, for $R > (1 + (1 + \exp\{(2/\delta_1)(2\sqrt{n}\pi)^{\alpha_1}\}/\sqrt{C_1}))\|f\|_2$,

$$\inf_{\|g\|_K \le R}\|f - g\|_{\mathscr{L}^2(X)} \le \inf_{0 < M \le \widetilde{R}}\left\{\frac{C_2\|f\|_2}{\sqrt{C_1}}\exp\left\{\frac{\delta_1}{2}(\sqrt{n}\pi M)^{\alpha_1} - \delta_2\widetilde{R}^{\alpha_2}\right\}\right.$$
$$\left. + \|f\|_s(\pi M)^{-s}\right\}.$$

Take

$$M = \frac{\delta_1^{-1/\alpha_1}2^{1/\alpha_1-1}}{\sqrt{n}\pi}\left(\ln\frac{R\sqrt{C_1}}{\|f\|_2}\right)^{\gamma},$$

where γ is given in our statement. For $R > \|f\|_2/\sqrt{C_1}, M \le \widetilde{R} \le \Lambda_k^{-1}(R/\|f\|_2)$
holds. Therefore, if $R > \exp\left\{(\delta_2 2^{-\alpha_2}\delta_1^{-\alpha_2/\alpha_1}(\sqrt{n}\pi)^{-\alpha_2})^{\alpha_1/(\gamma\alpha_1^2-\alpha_2)}\right\}\|f\|_2/\sqrt{C_1}$,
we have

$$\inf_{\|g\|_K \le R}\|f - g\|_{\mathscr{L}^2(X)} \le \frac{C_2\|f\|_2}{\sqrt{C_1}}\exp\left\{-\delta_2 2^{-\alpha_2-1}\delta_1^{-\alpha_2/\alpha_1}(\sqrt{n}\pi)^{-\alpha_2}\right.$$
$$\left.\left(\ln\frac{R\sqrt{C_1}}{\|f\|_2}\right)^{\alpha_2/\alpha_1}\right\} + \delta_1^{s/\alpha_1}2^s n^{s/2}\|f\|_s\left(\ln\frac{R\sqrt{C_1}}{\|f\|_2}\right)^{-\gamma s}$$
$$\le \left\{\frac{C_2\|f\|_2}{\sqrt{C_1}}C_3 + \delta_1^{s/\alpha_1}2^s n^{s/2}\|f\|_s\right\}\left(\ln R + \ln\frac{\sqrt{C_1}}{\|f\|_2}\right)^{-\gamma s}$$

where $C_3 := \sup_{x \geq 1} x^{-\gamma s} \exp\{-\delta x^{\alpha_2/\alpha_1}\}$ and $\delta = \delta_2^{2-\alpha_2-1} \delta_1^{-\alpha_2/\alpha_1} (\sqrt{n}\pi)^{-\alpha_2}$. This proves the first statement of the corollary. The second statement follows using the same argument. ∎

6.4 Proof of Theorem 6.1

Now we can apply Corollary 6.9 to verify Theorem 6.1. We may assume that $X \subseteq [0, 1]^n$. Since X has piecewise smooth boundary, every function $f \in H^s$ can be extended to a function $F \in H^s(\mathbb{R}^n)$ such that $\|f\|_{H^s(\mathbb{R}^n)} \leq C_X \|f\|_{H^s(X)}$, where the constant C_X depends only on X, not on $f \in H^s(X)$.

(i) From (5.17) and (5.18), we know that the condition of Corollary 6.9 is satisfied with

$$C_1 = (\sigma\sqrt{\pi})^n, \delta_1 = \frac{\sigma^2}{4}, \alpha_1 = 2$$

and

$$C_2 > 0, \alpha_2 = 1, \delta_2 = \ln\min\{16n, 2^n\}.$$

Then the bounds given in Corollary 6.9 hold. Since $\alpha_1 > \alpha_2$, we have $\gamma = \frac{1}{8}$ and the first statement of Theorem 6.1 follows with bounds depending on C_X.

(ii) By (5.23), we see that the condition of Corollary 6.9 is valid. Moreover,

$$C_1 > 0, \quad \delta_1 = c + \varepsilon, \quad \alpha_1 = 1$$

and

$$C_2 > 0, \quad \alpha_2 = 1, \quad \delta_2 = \frac{1}{2} \ln\min\left\{ec, \frac{e^{c/2}}{4^n}\right\}.$$

Then $\alpha_1 = \alpha_2$ and the bounds of Corollary 6.9 hold with $\gamma = \frac{1}{2}$. This yields the second statement of Theorem 6.1.

6.5 References and additional remarks

Logarithmic decays of the approximation error can be characterized by general interpolation spaces [16], as done for polynomial decays in Chapter 4. However, sharp bounds for the decay of the $\mathbb{K}$-functional are hard to obtain. For example, it is unknown how far the power index $\frac{s}{8}$ in Theorem 6.1 can be improved.

The function $\epsilon_K(\mathbf{x})$ defined by (6.1) is called the *power function* in the literature on radial basis functions. It was introduced by Madych and

Nelson [82], and extensively used by Wu and Schaback [147], and it plays an important role in error estimates for scattered data interpolation using radial basis functions. In that literature (e.g., [147, 66]), the interpolation scheme (6.3) is essential. What is different in learning theory is the presence of an RKHS $\mathcal{H}_K$, not necessarily a Sobolev space.

   Theorem 6.1 was proved in [113], and the approach in Section 6.3 was presented in [157].

7

On the bias–variance problem

Let K be a Mercer kernel, and $\mathcal{H}_K$ its induced RKHS. Assume that

(i) $K \in \mathcal{C}^s(X \times X)$, and
(ii) the regression function f_ρ satisfies $f_\rho \in \text{Range}(L_K^{\theta/(4+2\theta)})$, for some $\theta > 0$.

Fix a sample size m and a confidence $1 - \delta$, with $0 < \delta < 1$. To each $R > 0$ we associate a hypothesis space $\mathcal{H} = \mathcal{H}_{K,R}$, and we can consider $f_\mathcal{H}$ and, for $\mathbf{z} \in Z^m$, $f_\mathbf{z}$. The *bias–variance problem* consists of finding the value of R that minimizes a natural bound for the error $\mathcal{E}(f_\mathbf{z})$ (with confidence $1 - \delta$). This value of R determines a particular hypothesis space in the family of such spaces parameterized by R, or, to use a terminology common in the learning literature, it selects a *model*.

Theorem 7.1 *Let K be a Mercer kernel on $X \subset \mathbb{R}^n$ satisfying conditions (i) and (ii) above.*

(i) *We exhibit, for each $m \in \mathbb{N}$ and $\delta \in [0, 1)$, a function*

$$E_{m,\delta} = E : \mathbb{R}^+ \to \mathbb{R}$$

such that for all $R > 0$ and randomly chosen $\mathbf{z} \in Z^m$,

$$\int_X (f_\mathbf{z} - f_\rho)^2 \, d\rho_X \leq E(R)$$

with confidence $1 - \delta$.
(ii) *There is a unique minimizer R_* of $E(R)$.*
(iii) *When $m \to \infty$, we have $R_* \to \infty$ and $E(R_*) \to 0$.*

The proof of Theorem 7.1 relies on the main results of Chapters 3, 4, and 5. We show in Section 7.3 that R_* and $E(R_*)$ have the asymptotic expressions $R_* = \mathcal{O}\left(m^{1/((2+\theta)(1+2n/s))}\right)$ and $E(R_*) = \mathcal{O}\left(m^{-\theta/((2+\theta)(1+2n/s))}\right)$.

127

It follows from the proof of Theorem 7.1 that R_* may be easily computed from m, δ, $\|I_K\|$, M_ρ, $\|f_\rho\|_\infty$, $\|g\|_{\mathscr{L}^2_{\rho_X}}$, and θ. Here $g \in \mathscr{L}^2_{\rho_X}(X)$ is such that $L_K^{\theta/(4+2\theta)}(g) = f_\rho$. Note that this requires substantial information about ρ and, in particular, about f_ρ. The next chapter provides an alternative approach to the one considered thus far whose corresponding bias–variance problem can be solved without information on ρ.

7.1 A useful lemma

The following lemma will be used here and in Chapter 8.

Lemma 7.2 Let $c_1, c_2, \ldots, c_\ell > 0$ and $s > q_1 > q_2 > \ldots > q_{\ell-1} > 0$. Then the equation

$$x^s - c_1 x^{q_1} - c_2 x^{q_2} - \cdots - c_{\ell-1} x^{q_{\ell-1}} - c_\ell = 0$$

has a unique positive solution x^*. In addition,

$$x^* \le \max \left\{ (\ell c_1)^{1/(s-q_1)}, (\ell c_2)^{1/(s-q_2)}, \ldots, (\ell c_{\ell-1})^{1/(s-q_{\ell-1})}, (\ell c_\ell)^{1/s} \right\}.$$

Proof. We prove the first assertion by induction on ℓ. If $\ell = 1$, then the equation is $x^s - c_1 = 0$, which has a unique positive solution.

For $\ell > 1$ let $\varphi(x) = x^s - c_1 x^{q_1} - c_2 x^{q_2} - \cdots - c_{\ell-1} x^{q_{\ell-1}} - c_\ell$. Then, taking the derivative with respect to x,

$$\varphi'(x) = sx^{s-1} - q_1 c_1 x^{q_1-1} - \cdots - c_{\ell-1} q_{\ell-1} x^{q_{\ell-1}-1}$$

$$= sx^{q_{\ell-1}-1} \left(x^{s-q_{\ell-1}} - \frac{q_1 c_1}{s} x^{q_1-q_{\ell-1}} - \cdots - \frac{c_{\ell-1} q_{\ell-1}}{s} \right)$$

$$=: sx^{q_{\ell-1}-1} \psi(x).$$

By induction, hypothesis φ' has a unique positive zero that is the unique positive zero $\bar{x}$ of ψ. Since $\psi(0) < 0$ and $\lim_{x \to +\infty} \psi(x) = +\infty$, we deduce that $\psi(x) < 0$ for $x \in [0, \bar{x})$ and $\psi(x) > 0$ for $x \in (\bar{x}, +\infty)$. This implies that $\varphi'(x) < 0$ for $x \in (0, \bar{x})$ and $\varphi'(x) > 0$ for $x \in (\bar{x}, +\infty)$. Therefore, φ is strictly decreasing on $[0, \bar{x})$ and strictly increasing on $(\bar{x}, +\infty)$. But $\varphi(0) < 0$ and, hence $\varphi(\bar{x}) < 0$. Since φ is strictly increasing on $(\bar{x}, +\infty)$ and $\lim_{x \to +\infty} \varphi(x) = +\infty$, we conclude that φ has a unique zero x^* on $(\bar{x}, +\infty)$ which is its unique positive zero. The shape of φ is as in Figure 7.1. This proves the first statement.

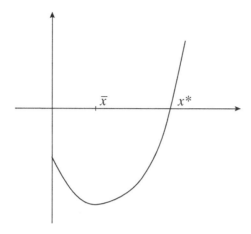

Figure 7.1

To prove the second statement, let $x > \max\left\{(\ell c_i)^{1/(s-q_i)} \mid i = 1, \dots, \ell\right\}$, where we set $q_\ell = 0$. Then, for $i = 1, \dots, \ell$, $c_i < \frac{1}{\ell}x^{s-q_i}$. It follows that

$$\sum_{i=1}^{\ell} c_i x^{q_i} < \sum_{i=1}^{\ell} \frac{1}{\ell} x^{s-q_i} x^{q_i} = x^s;$$

that is, $\varphi(x) > 0$.   ∎

Remark 7.3 Note that given $c_1, c_2, \dots, c_\ell$ and $s, q_1, q_2, \dots, q_{\ell-1}$, one can efficiently compute (a good approximation of) x^* using algorithms such as Newton's method.

7.2 Proof of Theorem 7.1

We first describe the natural bound we plan to minimize. Recall that $\mathcal{E}(f_{\mathbf{z}})$ equals the sum $\mathcal{E}_{\mathcal{H}}(f_{\mathbf{z}}) + \mathcal{E}(f_{\mathcal{H}})$ of the sample and approximation errors, or, equivalently,

$$\int_X (f_{\mathbf{z}} - f_\rho)^2 = \mathcal{E}_{\mathcal{H}}(f_{\mathbf{z}}) + \inf_{\|f\|_K \le R} \|f - f_\rho\|^2_{\mathscr{L}^2_{\rho_X}}.$$

We first want to bound the sample error. Let

$$M = M(R) = \|I_K\|R + M_\rho + \|f_\rho\|_\infty. \tag{7.1}$$

Then, for all $f \in \mathcal{H}_{K,R}$, $|f(x) - y| \leq M$ almost everywhere since

$$|f(x) - y| \leq |f(x)| + |y| \leq |f(x)| + |y - f_\rho(x)| + |f_\rho(x)|$$
$$\leq \|I_K\|R + M_\rho + \|f_\rho\|_\infty.$$

The sample error ε satisfies, with confidence $1 - \delta$, by Theorem 3.3,

$$\mathcal{N}\left(\mathcal{H}_{K,R}, \frac{\varepsilon}{12M}\right) e^{-m\varepsilon/300M^2} \geq \delta$$

and therefore, by Theorem 5.1(i) with $\eta = \frac{\varepsilon}{12M}$ (which applies due to assumption (i)),

$$\exp\left\{\overline{C}\left(\frac{12MR}{\varepsilon}\right)^{2n/s}\right\} \exp\left\{-\frac{m\varepsilon}{300M^2}\right\} \geq \delta$$

(with $\overline{C} = C(\mathrm{Diam}(X))^n \|K\|_{\mathscr{C}^s(X \times X)}^{n/s}$ and C depending on X and s but independent of R, ε, and M) or

$$\frac{m\varepsilon}{300M^2} - \ln\left(\frac{1}{\delta}\right) - \overline{C}\left(\frac{12M^2}{\|I_K\|\varepsilon}\right)^{2n/s} \leq 0,$$

where we have also used that $R\|I_K\| \leq M$. Write $\nu = \varepsilon/M^2$. Multiplying by $\nu^{2n/s}$, the inequality above takes the form

$$c_0\nu^{d+1} - c_1\nu^d - c_2 \leq 0, \tag{7.2}$$

where $d = \frac{2n}{s}$, $c_0 = \frac{m}{300}$, $c_1 = \ln\left(\frac{1}{\delta}\right)$, and $c_2 = \overline{C}\left(12/\|I_K\|\right)^d$.

If we take the equality in (7.2), we obtain an equation that, by Lemma 7.2, has exactly one positive solution for ν. Let $\nu^*(m, \delta)$ be this solution. Then $\varepsilon(R) = M^2\nu^*(m, \delta)$ is the best bound we can obtain from Theorem 3.3 for the sample error.

Now consider the approximation error. Owing to assumption (ii), Theorem 4.1 applies to yield

$$\mathcal{A}(f_\rho, R) \leq 2^{2+\theta}\|g\|_{\mathscr{L}^2_{\rho_X}}^{2+\theta} R^{-\theta} =: \mathcal{A}(R),$$

where $g \in \mathscr{L}^2_{\rho_X}(X)$ is such that $L_K^{\theta/(4+2\theta)}(g) = f_\rho$ and

$$\mathcal{A}(f_\rho, R) = \inf_{f \in \mathcal{H}_{K,R}} \mathcal{E}(f) - \mathcal{E}(f_\rho) = \inf_{\|f\|_K \leq R} \|f - f_\rho\|_{\mathscr{L}^2_{\rho_X}}^2.$$

We can therefore take $E(R) = \mathcal{A}(R) + \varepsilon(R)$ and Part (i) is proved.

We now proceed with Part (ii). For a point $R > 0$ to be a minimum of $\mathcal{A}(R) + \varepsilon(R)$, it is necessary that $\mathcal{A}'(R) + \varepsilon'(R) = 0$. Taking derivatives and noting that by (7.1), $M'(R) = \|I_K\|$, we get

$$\mathcal{A}'(R) = -2^{2+\theta}\|g\|^{2+\theta}_{\mathscr{L}^2_{\rho_X}}\theta R^{-\theta-1} \quad \text{and} \quad \varepsilon'(R) = 2M\|I_K\|v^*(m,\delta).$$

Therefore, writing $Q = \frac{1}{R}$, it is necessary that

$$\left(2^{2+\theta}\|g\|^{2+\theta}\theta\right)Q^{\theta+2} - \left(2(M_\rho + \|f_\rho\|_\infty)\|I_K\|v^*(m,\delta)\right)Q$$
$$- 2\|I_K\|^2 v^*(m,\delta) = 0. \tag{7.3}$$

By Lemma 7.2, it follows that there is a unique positive solution Q_* of (7.3) and, thus, a unique positive solution R_* of $\mathcal{A}'(R) + \varepsilon'(R) = 0$. This solution is the only minimum of E since $E(R) \to \infty$ when $R \to 0$ and when $R \to \infty$.

We finally prove Part (iii). Note that by Lemma 7.2, the solution of the equation induced by (7.2) satisfies

$$v^*(m,\delta) \le \max\left\{\frac{600\ln(1/\delta)}{m}, \left(\frac{600\overline{C}}{m}\left(\frac{12}{\|I_K\|}\right)^d\right)^{1/(d+1)}\right\}.$$

Therefore, $v^*(m,\delta) \to 0$ when $m \to \infty$. Also, since $1/R_*$ is a root of (7.3), Lemma 7.2 applies again to yield

$$\frac{1}{R_*} \le \max\left\{\left(\frac{4(M_\rho + \|f_\rho\|_\infty)\|I_K\|v^*(m,\delta)}{2^{2+\theta}\|g\|^{2+\theta}\theta}\right)^{1/(\theta+1)},\right.$$
$$\left.\times \left(\frac{4\|I_K\|^2 v^*(m,\delta)}{2^{2+\theta}\|g\|^{2+\theta}\theta}\right)^{1/(\theta+2)}\right\},$$

from which it follows that $R_* \to \infty$ when $m \to \infty$. Note that this implies that

$$\lim_{m\to\infty}\mathcal{A}(R_*) \le \lim_{m\to\infty}2^{2+\theta}\|g\|^{2+\theta}R_*^{-\theta} = 0.$$

Finally, since Q_* is a solution of equation (7.3),

$$\left(2^{2+\theta}\|g\|^{2+\theta}\theta\right)Q_*^\theta - \left[2(M_\rho + \|f_\rho\|_\infty)\|I_K\|Q_* - 2\|I_K\|^2\right]$$
$$\times \left(\frac{v^*(m,\delta)}{Q_*^2}\right) = 0,$$

and therefore $v^*(m,\delta)R_*^2 = v^*(m,\delta)/Q_*^2 \to 0$ when $m \to \infty$, and, by (7.1),

$$\lim_{m\to\infty} \varepsilon(R_*) = \lim_{m\to\infty} M^2 v^*(m,\delta)$$

$$= \lim_{m\to\infty} (\|I_K\| R_* + M_\rho + \|f_\rho\|_\infty)^2 v^*(m,\delta) = 0.$$

This finishes the proof of the theorem. ∎

7.3 A concrete example of bias–variance

Let $R \geq 1$ in the proof of Theorem 7.1. Then $M \leq (\|I_K\| + M_\rho + \|f_\rho\|_\infty)R$ and we may take

$$\varepsilon(R) = (\|I_K\| + M_\rho + \|f_\rho\|_\infty)^2 R^2 v^*(m,\delta)$$

as an upper bound for the sample error with confidence $1 - \delta$. Hence, under conditions (i) and (ii), we may choose

$$E(R) = (\|I_K\| + M_\rho + \|f_\rho\|_\infty)^2 R^2 v^*(m,\delta) + 2^{2+\theta} \|g\|^{2+\theta} R^{-\theta}.$$

With this choice,

$$R_* = (v^*(m,\delta))^{-1/(2+\theta)} 2\|g\| (\|I_K\| + M_\rho + \|f_\rho\|_\infty)^{-2/(2+\theta)} \left(\frac{\theta}{2}\right)^{1/(2+\theta)}$$

tends to infinity as m does so and

$$E(R_*) = \left[\left(\frac{\theta}{2}\right)^{2/(2+\theta)} + \left(\frac{2}{\theta}\right)^{\theta/(2+\theta)}\right]$$

$$\times 4\|g\|^2 (\|I_K\| + M_\rho + \|f_\rho\|_\infty)^{2\theta/(2+\theta)} (v^*(m,\delta))^{\theta/(2+\theta)}$$

$$= \mathcal{O}\left(m^{-\theta/((2+\theta)(1+2n/s))}\right) \to 0 \text{ as } m \to \infty.$$

Example 7.4 Let K be the spline kernel on $X = [-1,1]$ given in Example 4.19. If ρ_X is the Lebesgue measure and $\|f_\rho(x+t) - f_\rho(x)\|_{\mathscr{L}^2([-1,1-t])} = \mathcal{O}(t^\theta)$ for some $\theta > 0$, then $\mathcal{A}(f_\rho, R) = \mathcal{O}(R^{-\theta})$. Take $s = \frac{1}{2}$ and $n = 1$. Then we have

$$E(R_*) = \mathcal{O}\left(m^{-\theta/(5(2+\theta))}\right).$$

When θ is sufficiently large, $\|f_\mathbf{z} - f_\rho\|^2_{\mathscr{L}^2_{\rho_X}} = \mathcal{O}\left(m^{-(1/5)+\varepsilon}\right)$ for an arbitrarily small ε.

7.4 References and additional remarks

In this chapter we have considered a form of the bias–variance problem that optimizes the parameter R, fixing all the others. One can consider other forms of the bias–variance problem by optimizing other parameters. For instance, in Example 2.24, one can consider the degree of smoothness of the kernel K. The smoother K is, the smaller $\mathcal{H}_K$ is. Therefore, the sample error decreases and the approximation error increases with a parameter reflecting this smoothness.

We have already discussed the bias–variance problem in Section 1.5. Further ideas on this problem can be found in Chapter 9 of 18 and in [95].

Bounds for the roots of real and complex polynomials such as those in Lemma 7.2 are a standard theme in algebra going back to Gauss. A reference for several such bounds is [91]. Theorem 7.1 was originally proved in [39].

8

Least squares regularization

We now abandon the setting of a compact hypothesis space adopted thus far and change the perspective slightly. We will consider as a hypothesis space an RKHS $\mathcal{H}_K$ but we will add a penalization term in the error to avoid overfitting, as in the setting of compact hypothesis spaces.

In what follows, we consider as a hypothesis space $\mathcal{H} = \mathcal{H}_K$ – that is, $\mathcal{H}$ is a whole linear space – and the *regularized error* $\mathcal{E}_\gamma$ defined by

$$\mathcal{E}_\gamma(f) = \int_Z (f(x) - y)^2 \, d\rho + \gamma \|f\|_K^2$$

for a fixed $\gamma > 0$. For a sample $\mathbf{z}$, the *regularized empirical error* $\mathcal{E}_{\mathbf{z},\gamma}$ is defined by

$$\mathcal{E}_{\mathbf{z},\gamma}(f) = \frac{1}{m} \sum_{i=1}^{m} (y_i - f(x_i))^2 + \gamma \|f\|_K^2.$$

One can consider a target function f_γ minimizing $\mathcal{E}_\gamma(f)$ over $\mathcal{H}_K$ and an empirical target $f_{\mathbf{z},\gamma}$ minimizing $\mathcal{E}_{\mathbf{z},\gamma}$ over $\mathcal{H}_K$. We prove in Section 8.2 the existence and uniqueness of these target and empirical target functions. One advantage of this new approach, which becomes apparent from the results in this section, is that the empirical target function can be given an explicit form, readily computable, in terms of the sample $\mathbf{z}$, the parameter γ, and the kernel K.

Our discussion of Sections 1.4 and 1.5 remains valid in this context and the following questions concerning $f_{\mathbf{z},\gamma}$ require an answer: *Given* $\gamma > 0$, *how large is the excess generalization error* $\mathcal{E}(f_{\mathbf{z},\gamma}) - \mathcal{E}(f_\rho)$? *Which value of* γ *minimizes the excess generalization error?* The main result of this chapter provides some answer to these questions.

Theorem 8.1 *Assume that K satisfies $\log \mathcal{N}(B_1, \eta) \leq C_0(1/\eta)^{s^*}$ for some $s^* > 0$, and ρ satisfies $f_\rho \in \mathrm{Range}(L_K^{\theta/2})$ for some $0 < \theta \leq 1$. Take $\gamma_* = m^{-\zeta}$*

with $\zeta < 1/(1 + s^*)$. Then, for every $0 < \delta < 1$ and $m \geq m_\delta$, with confidence $1 - \delta$,

$$\int_X \big(f_{\mathbf{z},\gamma_*}(x) - f_\rho(x)\big)^2 \, d \, \rho_X \leq C_0' \log(2/\delta) m^{-\theta\zeta}$$

holds.

Here C_0' is a constant depending only on $s^*, \zeta, \mathbf{C}_K, \mathbf{M}, C_0$, and $\|L_K^{-\theta/2} f_\rho\|$, and m_δ depends also on δ. We may take

$$m_\delta := \max \left\{ \big(108/C_0\big)^{1/s^*} \big(\log(2/\delta)\big)^{1+1/s^*}, \big(1/(2c)\big)^{2/(\zeta - 1/(1+s^*))} \right\},$$

where $c = (2\mathbf{C}_K + 5)\big(108C_0\big)^{1/(1+s^*)}$.

At the end of this chapter, in Section 8.6, we show that the regularization approach just introduced and the minimization in compact hypothesis spaces considered thus far are closely related.

The parameter γ is said to be the *regularization parameter*. The whole approach outlined above is called a *regularization scheme*.

Note that γ_* can be computed from knowledge of m and s^* only. No information on f_ρ is required. The next example shows a simple situation where Theorem 8.1 applies and yields bounds on the generalization error from a simple assumption on f_ρ.

Example 8.2 Let K be the spline kernel on $X = [-1, 1]$ given in Example 4.19. If ρ_X is the Lebesgue measure and $\|f_\rho(x + t) - f_\rho(x)\|_{\mathscr{L}^2([-1, 1-|t|])} = \mathcal{O}(t^\theta)$ for some $0 < \theta \leq 1$, then, by the conclusion of Example 4.19 and Theorem 4.1, $f_\rho \in \mathrm{Range}\big(L_K^{(\theta-\varepsilon)/2}\big)$ for any $\varepsilon > 0$. Theorem 5.8 also tells us that $\log \mathcal{N}(B_1, \eta) \leq C_0 (1/\eta)^2$. So, we may take $s^* = 2$. Choose $\gamma_* = m^{-\zeta}$ with $\zeta = \frac{1-2\varepsilon}{3} < \frac{1}{3}$. Then Theorem 8.1 yields

$$\mathcal{E}(f_{\mathbf{z},\gamma_*}) - \mathcal{E}(f_\rho) = \|f_{\mathbf{z},\gamma_*} - f_\rho\|_{\mathscr{L}_{\rho_X}^2}^2 = \mathcal{O}\left(\log \frac{2}{\delta} m^{-(\theta/3)+\varepsilon}\right)$$

with confidence $1 - \delta$.

8.1 Bounds for the regularized error

Let X, K, f_γ, and $f_{\mathbf{z},\gamma} = f_{\mathbf{z}}$ be as above. Assume, for the time being, that f_γ and $f_{\mathbf{z},\gamma}$ exist.

Theorem 8.3 *Let $f_\gamma \in \mathcal{H}_K$ and $f_\mathbf{z}$ be as above. Then $\mathcal{E}(f_\mathbf{z}) - \mathcal{E}(f_\rho) \leq \mathcal{E}(f_\mathbf{z}) - \mathcal{E}(f_\rho) + \gamma \|f_\mathbf{z}\|_K^2$, which can be bounded by*

$$\left\{ \mathcal{E}(f_\gamma) - \mathcal{E}(f_\rho) + \gamma \|f_\gamma\|_K^2 \right\} + \left\{ \mathcal{E}(f_\mathbf{z}) - \mathcal{E}_\mathbf{z}(f_\mathbf{z}) + \mathcal{E}_\mathbf{z}(f_\gamma) - \mathcal{E}(f_\gamma) \right\}. \quad (8.1)$$

Proof. Write $\mathcal{E}(f_\mathbf{z}) - \mathcal{E}(f_\rho) + \gamma \|f_\mathbf{z}\|_K^2$ as

$$\{\mathcal{E}(f_\mathbf{z}) - \mathcal{E}_\mathbf{z}(f_\mathbf{z})\} + \left\{ \left(\mathcal{E}_\mathbf{z}(f_\mathbf{z}) + \gamma \|f_\mathbf{z}\|_K^2 \right) - \left(\mathcal{E}_\mathbf{z}(f_\gamma) + \gamma \|f_\gamma\|_K^2 \right) \right\}$$
$$+ \left\{ \mathcal{E}_\mathbf{z}(f_\gamma) - \mathcal{E}(f_\gamma) \right\} + \left\{ \mathcal{E}(f_\gamma) - \mathcal{E}(f_\rho) + \gamma \|f_\gamma\|_K^2 \right\}.$$

The definition of $f_\mathbf{z}$ implies that the second term is at most zero. Hence $\mathcal{E}(f_\mathbf{z}) - \mathcal{E}(f_\rho) + \gamma \|f_\mathbf{z}\|_K^2$ is bounded by (8.1). ∎

The first term in (8.1) is the regularized error of f_γ. We denote it by $\mathcal{D}(\gamma)$; that is,

$$\mathcal{D}(\gamma) := \mathcal{E}(f_\gamma) - \mathcal{E}(f_\rho) + \gamma \|f_\gamma\|_K^2 = \inf_{f \in \mathcal{H}_K} \left\{ \mathcal{E}(f) - \mathcal{E}(f_\rho) + \gamma \|f\|_K^2 \right\}. \quad (8.2)$$

Note that by Proposition 1.8,

$$\mathcal{D}(\gamma) = \|f_\gamma - f_\rho\|_\rho^2 + \gamma \|f_\gamma\|_K^2 \geq \|f_\gamma - f_\rho\|_\rho^2.$$

We call the second term in (8.1) the *sample error* (this use of the expression differs slightly from the one in Section 1.4).

In this section we give bounds for the regularized error. The bounds (Proposition 8.5 below) easily follow from the next general result.

Theorem 8.4 *Let H be a Hilbert space and A a self-adjoint, positive compact operator on H. Let $s > 0$ and $\gamma > 0$. Then*

(i) *For all $a \in H$, the minimizer $\widehat{b}$ of the optimization problem $\min_{b \in H} \left(\|b - a\|^2 + \gamma \|A^{-s}b\|^2 \right)$ exists and is given by*

$$\widehat{b} = \left(A^{2s} + \gamma \mathrm{Id} \right)^{-1} A^{2s} a.$$

(ii) *For $0 < \theta \leq 2s$,*

$$\|\widehat{b} - a\| \leq \gamma^{\theta/(2s)} \|A^{-\theta} a\|,$$

where we define $\|A^{-\theta} a\| = \infty$ if $a \notin \mathrm{Range}(A^\theta)$.

(iii) *For* $0 < \theta \le s$,

$$\min_{b \in H} \left(\|b - a\|^2 + \gamma \|A^{-s}b\|^2 \right) \le \gamma^{\theta/s} \|A^{-\theta}a\|^2.$$

(iv) *For* $s < \theta \le 3s$,

$$\|A^{-s}(\widehat{b} - a)\| \le \gamma^{\theta/(2s) - 1/2} \|A^{-\theta}a\|.$$

Proof. First note that replacing A by A^s and $\theta/(2s)$ by θ we can reduce the problem to the case $s = 1$ where $A^{-\theta}a = \left[(A^s)^2\right]^{-\theta/(2s)} a$.

(i) Consider

$$\varphi(b) = \|b - a\|^2 + \gamma \|A^{-1}b\|^2.$$

If a point $\widehat{b}$ minimizes φ, then it must be a zero of the derivative $D\varphi$ whose value at $b \in \text{Range}(A)$ satisfies $\varphi(b + \varepsilon f) - \varphi(b) = \langle D\varphi(b), \varepsilon f \rangle + o(\varepsilon)$ for $f \in \text{Range}(A)$. But $\varphi(b + \varepsilon f) - \varphi(b) = 2\langle b - a, \varepsilon f \rangle + 2\gamma \langle A^{-2}b, \varepsilon f \rangle + \varepsilon^2 \|f\|^2 + \varepsilon^2 \gamma \|A^{-1}f\|^2$. So $\widehat{b}$ satisfies $(\text{Id} + \gamma A^{-2})\widehat{b} = a$, which implies $\widehat{b} = (\text{Id} + \gamma A^{-2})^{-1}a = (A^2 + \gamma \text{Id})^{-1}A^2 a$. Note that the operator $\text{Id} + \gamma A^{-2}$ is invertible since it is the sum of the identity and a positive (but maybe unbounded) operator.

We use the method from Chapter 4 to prove the remaining statements. If $\lambda_1 \ge \lambda_2 \ge \ldots$ denote the eigenvalues of A^2 corresponding to normalized eigenvectors $\{\phi_k\}$, then

$$\widehat{b} = \sum_{k \ge 1} \frac{\lambda_k}{\lambda_k + \gamma} a_k \phi_k,$$

where $a = \sum_{k \ge 1} a_k \phi_k$. It follows that

$$\widehat{b} - a = \sum_{k \ge 1} \frac{-\gamma}{\lambda_k + \gamma} a_k \phi_k.$$

Assume $\|A^{-\theta}a\| = \left\{ \sum_k a_k^2 / \lambda_k^\theta \right\}^{1/2} < \infty$.

(ii) For $0 < \theta \le 2$, we have

$$\|\widehat{b} - a\|^2 = \sum_{k \ge 1} \left(\frac{-\gamma}{\lambda_k + \gamma} \right)^2 a_k^2 = \gamma^\theta \sum_{k \ge 1} \left(\frac{\gamma}{\lambda_k + \gamma} \right)^{2 - \theta} \left(\frac{\lambda_k}{\lambda_k + \gamma} \right)^\theta \frac{a_k^2}{\lambda_k^\theta}$$

$$\le \gamma^\theta \|A^{-\theta}a\|^2.$$

(iii) For $0 < \theta \leq 1$, $A^{-1}\widehat{b} = \sum_{k\geq 1}\left((\sqrt{\lambda_k})/(\lambda_k + \gamma)\right)a_k\phi_k$. Hence

$$\|\widehat{b} - a\|^2 + \gamma\|A^{-1}\widehat{b}\|^2 = \sum_{k\geq 1}\left(\frac{\gamma}{\lambda_k + \gamma}\right)^2 a_k^2 + \gamma\sum_{k\geq 1}\frac{\lambda_k}{(\lambda_k + \gamma)^2}a_k^2$$

$$= \sum_{k\geq 1}\frac{\gamma}{\lambda_k + \gamma}a_k^2,$$

which is bounded by

$$\gamma^\theta\sum_{k\geq 1}\left(\frac{\gamma}{\lambda_k + \gamma}\right)^{1-\theta}\left(\frac{\lambda_k}{\lambda_k + \gamma}\right)^\theta\frac{a_k^2}{\lambda_k^\theta} \leq \gamma^\theta\|A^{-\theta}a\|^2.$$

(iv) When $1 < \theta \leq 3$, we find that

$$A^{-1}(\widehat{b} - a) = \sum_{k\geq 1}\frac{-\gamma}{(\lambda_k + \gamma)\sqrt{\lambda_k}}a_k\phi_k.$$

It follows that

$$\|A^{-1}(\widehat{b} - a)\|^2 = \sum_{k\geq 1}\frac{\gamma^2}{(\lambda_k + \gamma)^2\lambda_k}a_k^2$$

$$= \gamma^{\theta-1}\sum_{k\geq 1}\left(\frac{\gamma}{\lambda_k + \gamma}\right)^{3-\theta}\left(\frac{\lambda_k}{\lambda_k + \gamma}\right)^{\theta-1}\frac{a_k^2}{\lambda_k^\theta} \leq \gamma^{\theta-1}\|A^{-\theta}a\|^2.$$

Thus all the estimates have been verified.   ∎

Bounds for the regularized error $\mathcal{D}(\gamma)$ follow from Theorem 8.4.

Proposition 8.5 *Let $X \subset \mathbb{R}^n$ be a compact domain and K a Mercer kernel such that for some $0 < \theta \leq 2 f_\rho \in \text{Range}(L_K^{\theta/2})$. Then*

(i) $\|f_\gamma - f_\rho\|_\rho^2 = \mathcal{E}(f_\gamma) - \mathcal{E}(f_\rho) \leq \gamma^\theta\|L_K^{-\theta/2}f_\rho\|^2$.

(ii) *For* $0 < \theta \leq 1$,

$$\mathcal{D}(\gamma) = \mathcal{E}(f_\gamma) - \mathcal{E}(f_\rho) + \gamma\|f_\gamma\|_K^2 \leq \gamma^\theta\|L_K^{-\theta/2}f_\rho\|_\rho^2.$$

(iii) *For* $1 < \theta \leq 3$,

$$\|f_\gamma - f_\rho\|_K \leq \gamma^{(\theta-1)/2}\|L_K^{-\theta/2}f_\rho\|_\rho.$$

Proof. Apply Theorem 8.4 with $H = \mathscr{L}_\rho^2(X)$, $s = 1$, $A = L_K^{1/2}$, and $a = f_\rho$, and use that $\|L_K^{-1/2}f\| = \|A^{-1}f\| = \|f\|_K$. We know that f_γ is the minimizer of

$$\min_{f \in \mathscr{L}_\rho^2(X)} (\|f - f_\rho\|^2 + \gamma \|f\|_K^2) = \min_{f \in \mathcal{H}_K} (\|f - f_\rho\|^2 + \gamma \|f\|_K^2)$$

since $\|f\|_K = \infty$ for $f \notin \mathcal{H}_K$. Our conclusion follows from Theorem 8.4 and Proposition 1.8. ∎

8.2 On the existence of target functions

Let X, K, f_γ, and $f_{\mathbf{z},\gamma} = f_\mathbf{z}$ be as above. Since the hypothesis space $\mathcal{H}_K$ is not compact, the existence of f_γ and $f_{\mathbf{z},\gamma}$ is not obvious. The goal of this section is to prove that both f_γ and $f_{\mathbf{z},\gamma}$ exist and are unique. In addition, we show that $f_{\mathbf{z},\gamma}$ is easily computable from γ, the sample $\mathbf{z}$, and the kernel K on the compact metric space X.

Proposition 8.6 *Let $\nu = \rho_X$ in the definition of the integral operator L_K. For all $\gamma > 0$ the function $f_\gamma = (L_K + \gamma \mathrm{Id})^{-1}L_K f_\rho$ is the unique minimizer of $\mathcal{E}_\gamma$ over $\mathcal{H}_K$.*

Proof. Apply Theorem 8.4 with $H = \mathscr{L}_\nu^2(X)$, $s = 1$, $A = L_K^{1/2}$, and $a = f_\rho$. Since, for all $f \in \mathcal{H}_K$, $\|f\|_K = \|L_K^{-1/2}f\|_{\mathscr{L}_\nu^2(X)}$, we have

$$\|b - a\|^2 + \gamma \|A^{-s}b\|^2 = \|b - f_\rho\|_{\mathscr{L}_{\rho_X}(X)}^2 + \gamma \|b\|_K^2 = \mathcal{E}_\gamma(b) - \sigma_\rho^2.$$

Thus, the minimizer f_γ is $\widehat{b}$ in Theorem 8.4, and the proposition follows. ∎

Proposition 8.7 *Let $\mathbf{z} \in Z^m$ and $\gamma > 0$. The empirical target function can be expressed as*

$$f_\mathbf{z}(x) = \sum_{i=1}^m a_i K(x, x_i),$$

where $a = (a_1, \ldots, a_m)$ is the unique solution of the well-posed linear system in $\mathbb{R}^m$

$$(\gamma m \, \mathrm{Id} + K[\mathbf{x}])a = \mathbf{y}.$$

Here, we recall that $K[\mathbf{x}]$ is the $m \times m$ matrix whose (i,j) entry is $K(x_i, x_j)$, $\mathbf{x} = (x_1, \ldots, x_m) \in X^m$, and $\mathbf{y} = (y_1, \ldots, y_m) \in Y^m$ such that $\mathbf{z} = ((x_1, y_1), \ldots, (x_m, y_m))$.

Proof. Let $H(f) = \frac{1}{m} \sum_{i=1}^{m} (y_i - f(x_i))^2 + \gamma \|f\|_K^2$. Take ν to be a Borel, nondegenerate measure on X and L_K to be the corresponding integral operator. Let $\{\phi_k\}_{k \geq 1}$ be an orthonormal basis of $\mathcal{L}_\nu^2(X)$ consisting of eigenfunctions of L_K, and let $\{\lambda_k\}_{k \geq 1}$ be their corresponding eigenvalues. By Theorem 4.12, we can then write, for any $f \in \mathcal{H}_K$, $f = \sum_{\lambda_k > 0} c_k \phi_k$ with $\|f\|_K^2 = \sum_{\lambda_k > 0} c_k^2 / \lambda_k$.

For every k with $\lambda_k > 0$, $\partial H / \partial c_k = \frac{-2}{m} \sum_{i=1}^{m} (y_i - f(x_i)) \phi_k(x_i) + 2\gamma (c_k / \lambda_k)$. If f is a minimum of H, then, for each k with $\lambda_k > 0$, we must have $\partial H / \partial c_k = 0$ or, solving for c_k,

$$c_k = \lambda_k \sum_{i=1}^{m} a_i \phi_k(x_i),$$

where $a_i = (y_i - f(x_i))/\gamma m$. Thus,

$$f(x) = \sum_{\lambda_k > 0} c_k \phi_k(x) = \sum_{\lambda_k > 0} \lambda_k \sum_{i=1}^{m} a_i \phi_k(x_i) \phi_k(x)$$

$$= \sum_{i=1}^{m} a_i \sum_{\lambda_k > 0} \lambda_k \phi_k(x_i) \phi_k(x) = \sum_{i=1}^{m} a_i K(x_i, x),$$

where we have applied Theorem 4.10 in the last equality. Replacing $f(x_i)$ in the definition of a_i above we obtain

$$a_i = \frac{y_i - \sum_{j=1}^{m} a_j K(x_j, x_i)}{\gamma m}.$$

Multiplying both sides by γm and writing the result in matrix form we obtain $(\gamma m \, \mathrm{Id} + K[\mathbf{x}]) a = \mathbf{y}$, and this system is well posed since $K[\mathbf{x}]$ is positive semidefinite and the result of adding a positive semidefinite matrix and the identity is positive definite. ∎

8.3 A first estimate for the excess generalization error

In this section we bound the confidence for the sample error to be small enough. The main result is Theorem 8.10.

In what follows we assume that

$$\mathbf{M} = \inf \left\{ \bar{M} \geq 0 \mid \{(x, y) \in Z \mid |y| \geq \bar{M}\} \text{ has measure zero} \right\}$$

is finite. Note that

$$|y| \leq \mathbf{M} \quad \text{and} \quad |f_\rho(x)| \leq \mathbf{M}$$

almost surely.

For $R > 0$ let $B_R = \{f \in \mathcal{H}_K : \|f\|_K \leq R\}$. Recall that for each $f \in \mathcal{H}_K$, $\|f\|_\infty \leq \mathbf{C}_K \|f\|_K$, where $\mathbf{C}_K = \sup_{x \in X} \sqrt{K(x,x)}$.

For the sample error estimates, we require the confidence $\mathcal{N}(B_1, \eta) \exp\{-\frac{m\eta}{54}\}$ to be at most δ. So we define the following quantity to realize this confidence.

Definition 8.8 Let $g = g_{K,m} : \mathbb{R}_+ \to \mathbb{R}$ be the function given by

$$g(\eta) = \log \mathcal{N}(B_1, \eta) - \frac{m\eta}{54}.$$

The function g is strictly decreasing in $(0, +\infty)$ with $g(0) = +\infty$ and $g(+\infty) = -\infty$. Also, $g(1) = -\frac{m}{54}$. Moreover, $\lim_{\eta \to \varepsilon_+} \mathcal{N}(B_1, \eta) = \mathcal{N}(B_1, \varepsilon)$ for all $\varepsilon > 0$. Therefore, for $0 < \delta < 1$, the inequality

$$g(\eta) \leq \log \delta \tag{8.3}$$

has a unique minimal solution $v^*(m, \delta)$. Moreover,

$$\lim_{m \to \infty} v^*(m, \delta) = 0.$$

More quantitatively, when K is $\mathscr{C}^s$ on $X \subset \mathbb{R}^n$, $\log \mathcal{N}(B_1, \eta) \leq C_0 (1/\eta)^{s^*}$ with $s^* = \frac{2n}{s}$ (cf. Theorem 5.1(i)). In this case the following decay holds.

Lemma 8.9 *If the Mercer kernel K satisfies $\log \mathcal{N}(B_1, \eta) \leq C_0(1/\eta)^{s^*}$ for some $s^* > 0$, then*

$$v^*(m, \delta) \leq \max\left\{ \frac{108 \log(1/\delta)}{m}, \left(\frac{108 C_0}{m} \right)^{1/(1+s^*)} \right\}.$$

Proof. Observe that $g(\eta) \leq h(\eta) := C_0(1/\eta)^{s^*} - \frac{m\eta}{54}$. Since h is also strictly decreasing and continuous on $(0, +\infty)$, we can take Δ to be the unique positive solution of the equation $h(t) = \log \delta$. We know that $v^*(m, \delta) \leq \Delta$. The equation $h(t) = \log \delta$ can be expressed as

$$t^{1+s^*} - \frac{54 \log(1/\delta)}{m} t^{s^*} - \frac{54 C_0}{m} = 0.$$

Then Lemma 7.2 with $d = 2$ yields $\Delta \leq \max\{108 \log(1/\delta)/m, \left(108 C_0/m\right)^{1/(1+s^*)}\}$. This verifies the bound for $v^*(m, \delta)$. ∎

Theorem 8.10 *For all* $\gamma \in (0, 1]$ *and* $0 < \delta < 1$, *with confidence* $1 - \delta$,

$$
\mathcal{E}(f_{\mathbf{z}}) - \mathcal{E}(f_\rho) \leq \frac{2(\mathbf{C}_K + 3)^2 \mathbf{M}^2 v^*(m, \delta/2)}{\gamma} + \left(\frac{8\mathbf{C}_K^2 \log(2/\delta)}{m\gamma} + 6\mathbf{M} + 4 \right)
$$

$$
\times \mathcal{D}(\gamma) + \frac{(48\mathbf{M}^2 + 6\mathbf{M}) \log(2/\delta)}{m}.
$$

holds.

Theorem 8.10 will follow from some lemmas and propositions given in the remainder of this section. Before proceeding with these results, however, we note that from Theorem 8.10, a convergence property for the regularized scheme follows.

Corollary 8.11 *Let* $0 < \delta < 1$ *be arbitrary. Take* $\gamma = \gamma(m)$ *to satisfy* $\gamma(m) \to 0$, $\underline{\lim}_{m \to \infty} m\gamma(m) \geq 1$, *and* $\gamma(m)/(v^*(m, \delta/2)) \to +\infty$. *If* $\mathcal{D}(\gamma) \to 0$, *then, for any* $\varepsilon > 0$, *there is some* $M_{\delta, \varepsilon} \in \mathbb{N}$ *such that with confidence* $1 - \delta$,

$$
\mathcal{E}(f_{\mathbf{z}}) - \mathcal{E}(f_\rho) \leq \varepsilon, \quad \forall m \geq M_{\delta, \varepsilon}
$$

holds. ∎

As an example, for $\mathscr{C}^s$ kernels on $X \subset \mathbb{R}^n$, the decay of $v^*(m, \delta)$ shown in Lemma 8.9 with $s^* = \frac{2n}{s}$ yields the following convergence rate.

Corollary 8.12 *Assume that* K *satisfies* $\log \mathcal{N}(B_1, \eta) \leq C_0(1/\eta)^{s^*}$ *for some* $s^* > 0$, *and* ρ *satisfies* $f_\rho \in \text{Range}(L_K^{\theta/2})$ *for some* $0 < \theta \leq 1$. *Then, for all* $\gamma \in (0, 1]$ *and all* $0 < \delta < 1$, *with confidence* $1 - \delta$,

$$
\int_X \left(f_{\mathbf{z}}(x) - f_\rho(x) \right)^2 d\rho_X \leq C_1 \left\{ \frac{\log(\frac{2}{\delta})}{m\gamma} + \frac{1}{m^{1/(1+s^*)}\gamma} + \gamma^\theta + \right.
$$

$$
\left. \log\left(\frac{2}{\delta}\right)\left(\frac{\gamma^\theta}{m\gamma} + \frac{1}{m}\right) \right\}
$$

holds, where C_1 *is a constant depending only on* $s, \mathbf{C}_K, \mathbf{M}, C_0$, *and* $\|L_K^{-\theta/2} f_\rho\|$. *If* $\gamma = m^{-1/((1+\theta)(1+s^*))}$, *then the convergence rate is* $\int_X \left(f_{\mathbf{z}}(x) - f_\rho(x) \right)^2 d\rho_X \leq 6C_1 \log(2/\delta) m^{-\theta/((1+\theta)(1+s^*))}$.

Proof. The proof is an easy consequence of Theorem 8.10, Lemma 8.9, and Proposition 8.5. ∎

For $\mathscr{C}^\infty$ kernels, s^* can be arbitrarily small. Then the decay rate exhibited in Corollary 8.12 is $m^{-(1/2)+\varepsilon}$ for any $\varepsilon > 0$, achieved with $\theta = 1$. We improve

Theorem 8.10 in the next section, where more satisfactory bounds (with decay rate $m^{-1+\varepsilon}$) are presented. The basic ideas of the proof are included in this section.

To move toward the proof of Theorem 8.10, we write the sample error as

$$\mathcal{E}(f_{\mathbf{z}}) - \mathcal{E}_{\mathbf{z}}(f_{\mathbf{z}}) + \mathcal{E}_{\mathbf{z}}(f_\gamma) - \mathcal{E}(f_\gamma) = \left\{ \mathbf{E}(\xi_1) - \frac{1}{m} \sum_{i=1}^{m} \xi_1(z_i) \right\}$$

$$+ \left\{ \frac{1}{m} \sum_{i=1}^{m} \xi_2(z_i) - \mathbf{E}(\xi_2) \right\}, \qquad (8.4)$$

where

$$\xi_1 := \big(f_{\mathbf{z}}(x) - y\big)^2 - \big(f_\rho(x) - y\big)^2 \quad \text{and} \quad \xi_2 := \big(f_\gamma(x) - y\big)^2 - \big(f_\rho(x) - y\big)^2.$$

The second term on the right-hand side of (8.4) is about the random variable ξ_2 on Z. Since its mean $\mathbf{E}(\xi_2) = \mathcal{E}(f_\gamma) - \mathcal{E}(f_\rho)$ is nonnegative, we may apply the Bernstein inequality to estimate this term. To do so, however, we need bounds for $\|f_\gamma\|_\infty$.

Lemma 8.13 *For all $\gamma > 0$,*

$$\|f_\gamma\|_K \le \sqrt{\mathcal{D}(\gamma)/\gamma} \quad \text{and} \quad \|f_\gamma\|_\infty \le C_K \sqrt{\mathcal{D}(\gamma)/\gamma}.$$

Proof. Since f_γ is a minimizer of (8.2), we know that

$$\gamma \|f_\gamma\|_K^2 \le \mathcal{E}(f_\gamma) - \mathcal{E}(f_\rho) + \gamma \|f_\gamma\|_K^2 = \mathcal{D}(\gamma).$$

Thus, the first inequality holds. The second follows from $\|f_\gamma\|_\infty \le C_K \|f_\gamma\|_K$. ∎

Proposition 8.14 *For every $0 < \delta < 1$, with confidence at least $1 - \delta$,*

$$\frac{1}{m} \sum_{i=1}^{m} \xi_2(z_i) - \mathbf{E}(\xi_2) \le \left(\frac{4 C_K^2 \log(1/\delta)}{m\gamma} + 3\mathbf{M} + 1 \right) \mathcal{D}(\gamma)$$

$$+ \frac{\big(24\mathbf{M}^2 + 3\mathbf{M}\big) \log(1/\delta)}{m}$$

holds.

Proof. From the definition of ξ_2, it follows that

$$\xi_2 = \big(f_\gamma(x) - f_\rho(x)\big)\big\{\big(f_\gamma(x) - y\big) + \big(f_\rho(x) - y\big)\big\}.$$

Almost everywhere, since $|f_\rho(x)| \le \mathbf{M}$, we have

$$|\xi_2| \le (\|f_\gamma\|_\infty + \mathbf{M})(\|f_\gamma\|_\infty + 3\mathbf{M}) \le c := (\|f_\gamma\|_\infty + 3\mathbf{M})^2.$$

Hence $|\xi_2(z) - \mathbf{E}(\xi_2)| \le B := 2c$. Moreover, we have

$$\mathbf{E}(\xi_2^2) = \mathbf{E}\left((f_\gamma(x) - f_\rho(x))^2\{(f_\gamma(x) - y) + (f_\rho(x) - y)\}^2\right)$$

$$\le \|f_\gamma - f_\rho\|^2 (\|f_\gamma\|_\infty + 3\mathbf{M})^2,$$

which implies that $\sigma^2(\xi_2) \le \mathbf{E}(\xi_2^2) \le c\mathcal{D}(\gamma)$. Now we apply the one-side Bernstein inequality in Corollary 3.6 to ξ_2. It asserts that for any $t > 0$,

$$\frac{1}{m} \sum_{i=1}^m \xi_2(z_i) - \mathbf{E}(\xi_2) \le t$$

with confidence at least

$$1 - \exp\left\{-\frac{mt^2}{2(\sigma^2(\xi_2) + \frac{1}{3}Bt)}\right\} \ge 1 - \exp\left\{-\frac{mt^2}{2c(\mathcal{D}(\gamma) + \frac{2}{3}t)}\right\}.$$

Choose t^* to be the unique positive solution of the quadratic equation

$$-\frac{mt^2}{2c(\mathcal{D}(\gamma) + \frac{2}{3}t)} = \log \delta.$$

Then, with confidence $1 - \delta$,

$$\frac{1}{m} \sum_{i=1}^m \xi_2(z_i) - \mathbf{E}(\xi_2) \le t^*$$

holds. But

$$t^* = \left(\frac{2c}{3}\log(1/\delta) + \sqrt{\left(\frac{2c}{3}\log(1/\delta)\right)^2 + 2cm\log(1/\delta)\mathcal{D}(\gamma)}\right) \Big/ m$$

$$\le \frac{4c\log(1/\delta)}{3m} + \sqrt{2c\log(1/\delta)\mathcal{D}(\gamma)/m}.$$

By Lemma 8.13, $c \le 2\mathbf{C}_K^2 \mathcal{D}(\gamma)/\gamma + 18\mathbf{M}^2$. It follows that

$$\sqrt{2c\log(1/\delta)\mathcal{D}(\gamma)/m} \le \sqrt{\log(1/\delta)}\left(\frac{2\mathbf{C}_K\mathcal{D}(\gamma)}{\sqrt{m\gamma}} + 6\mathbf{M}\sqrt{\mathcal{D}(\gamma)/m}\right)$$

and, therefore, that

$$t^* \leq \frac{8C_K^2 \log(1/\delta)}{3m\gamma} \mathcal{D}(\gamma) + \frac{72M^2 \log(1/\delta)}{3m} + 2C_K \frac{\sqrt{\log(1/\delta)}}{\sqrt{m\gamma}} \mathcal{D}(\gamma)$$
$$+ \frac{3M \log(1/\delta)}{m} + 3M\mathcal{D}(\gamma).$$

This implies the desired estimate. ∎

The first term on the right-hand side of (8.4) is more difficult to deal with because ξ_1 involves the sample $\mathbf{z}$ through $f_{\mathbf{z}}$. We use a result from Chapter 3, Lemma 3.19, to bound this term by means of a covering number. For $R > 0$, define $\mathcal{F}_R$ to be the set of functions from Z to $\mathbb{R}$

$$\mathcal{F}_R := \left\{ (f(x) - y)^2 - (f_\rho(x) - y)^2 : f \in B_R \right\}. \tag{8.5}$$

Proposition 8.15 *For all $\varepsilon > 0$ and $R \geq M$,*

$$\Prob_{\mathbf{z} \in Z^m} \left\{ \sup_{f \in B_R} \frac{\mathcal{E}(f) - \mathcal{E}(f_\rho) - (\mathcal{E}_{\mathbf{z}}(f) - \mathcal{E}_{\mathbf{z}}(f_\rho))}{\sqrt{\mathcal{E}(f) - \mathcal{E}(f_\rho) + \varepsilon}} \leq \sqrt{\varepsilon} \right\}$$
$$\geq 1 - \mathcal{N} \left(B_1, \frac{\varepsilon}{(C_K + 3)^2 R^2} \right) \exp \left\{ - \frac{m\varepsilon}{54(C_K + 3)^2 R^2} \right\}.$$

Proof. Consider the set $\mathcal{F}_R$. Each function $g \in \mathcal{F}_R$ has the form $g(z) = (f(x) - y)^2 - (f_\rho(x) - y)^2$ with $f \in B_R$. Hence $\mathbf{E}(g) = \mathcal{E}(f) - \mathcal{E}(f_\rho) = \|f - f_\rho\|^2 \geq 0$, $\mathbf{E}_{\mathbf{z}}(g) = \mathcal{E}_{\mathbf{z}}(f) - \mathcal{E}_{\mathbf{z}}(f_\rho)$, and

$$g(z) = (f(x) - f_\rho(x))\{(f(x) - y) + (f_\rho(x) - y)\}.$$

Since $\|f\|_\infty \leq C_K \|f\|_K \leq C_K R$ and $|f_\rho(x)| \leq M$ almost everywhere, we find that

$$|g(z)| \leq (C_K R + M)(C_K R + 3M) \leq c := (C_K R + 3M)^2.$$

So we have $|g(z) - \mathbf{E}(g)| \leq B := 2c$ almost everywhere.

In addition,

$$\mathbf{E}(g^2) = \mathbf{E} \left[(f(x) - f_\rho(x))^2 \{(f(x) - y) + (f_\rho(x) - y)\}^2 \right]$$
$$\leq (C_K R + 3M)^2 \|f - f_\rho\|^2.$$

Thus, $\mathbf{E}(g^2) \leq c\mathbf{E}(g)$ for each $g \in \mathcal{F}_R$.

Applying Lemma 3.19 with $\alpha = \frac{1}{4}$ to the function set $\mathcal{F}_R$, we deduce that

$$\sup_{f \in B_R} \frac{\mathcal{E}(f) - \mathcal{E}(f_\rho) - \left(\mathcal{E}_\mathbf{z}(f) - \mathcal{E}_\mathbf{z}(f_\rho)\right)}{\sqrt{\mathcal{E}(f) - \mathcal{E}(f_\rho) + \varepsilon}} = \sup_{g \in \mathcal{F}_R} \frac{\mathbf{E}(g) - \frac{1}{m}\sum_{i=1}^{m} g(z_i)}{\sqrt{\mathbf{E}(g) + \varepsilon}} \leq \sqrt{\varepsilon}$$

with confidence at least

$$1 - \mathcal{N}\left(\mathcal{F}_R, \varepsilon/4\right) \exp\left\{-\frac{m\varepsilon/16}{2c + \frac{2}{3}B}\right\} \geq 1 - \mathcal{N}\left(\mathcal{F}_R, \varepsilon/4\right) \exp\left\{-\frac{m\varepsilon}{54(\mathbf{C}_K + 3)^2 R^2}\right\}.$$

Here we have used the expressions for c, $B = 2c$, and the restriction $R \geq \mathbf{M}$.

What is left is to bound the covering number $\mathcal{N}\left(\mathcal{F}_R, \varepsilon/4\right)$. To do so, we note that

$$\left|\left(f_1(x) - y\right)^2 - \left(f_2(x) - y\right)^2\right| \leq \|f_1 - f_2\|_\infty \left|\left(f_1(x) - y\right) + \left(f_2(x) - y\right)\right|.$$

But $|y| \leq \mathbf{M}$ almost surely, and $\|f\|_\infty \leq \mathbf{C}_K \|f\|_K \leq \mathbf{C}_K R$ for each $f \in B_R$. Therefore, almost surely,

$$\left|\left(f_1(x) - y\right)^2 - \left(f_2(x) - y\right)^2\right| \leq 2\left(\mathbf{M} + \mathbf{C}_K R\right)\|f_1 - f_2\|_\infty, \quad \forall f_1, f_2 \in B_R.$$

Since an $\eta/(2(\mathbf{M}R + \mathbf{C}_K R^2))$-covering of B_1 yields an $\eta/(2(\mathbf{M} + \mathbf{C}_K R))$-covering of B_R, and vice versa, we see that for any $\eta > 0$, an $\eta/(2(\mathbf{M}R + \mathbf{C}_K R^2))$-covering of B_1 provides an η-covering of $\mathcal{F}_R$. That is,

$$\mathcal{N}(\mathcal{F}_R, \eta) \leq \mathcal{N}\left(B_1, \frac{\eta}{2(\mathbf{M}R + \mathbf{C}_K R^2)}\right), \quad \forall \eta > 0. \tag{8.6}$$

But $R \geq \mathbf{M}$ and $2(1 + \mathbf{C}_K) \leq (\mathbf{C}_K + 3)^2$. So our desired estimate follows. ∎

Now we can derive the error bounds. For $R > 0$, denote

$$\mathcal{W}(R) := \{\mathbf{z} \in Z^m : \|f_\mathbf{z}\|_K \leq R\}.$$

Proposition 8.16 *For all $0 < \delta < 1$ and $R \geq \mathbf{M}$, there is a set $V_R \subset Z^m$ with $\rho(V_R) \leq \delta$ such that for all $\mathbf{z} \in \mathcal{W}(R) \setminus V_R$, the regularized error $\mathcal{E}_\gamma(f_\mathbf{z}) = \mathcal{E}(f_\mathbf{z}) - \mathcal{E}(f_\rho) + \gamma \|f_\mathbf{z}\|_K^2$ is bounded by*

$$2(\mathbf{C}_K + 3)^2 R^2 v^*(m, \delta/2) + \left(\frac{8\mathbf{C}_K^2 \log(2/\delta)}{m\gamma} + 6\mathbf{M} + 4\right)\mathcal{D}(\gamma)$$

$$+ \frac{(48\mathbf{M}^2 + 6\mathbf{M})\log(2/\delta)}{m}.$$

Proof. Note that $\sqrt{\mathcal{E}(f) - \mathcal{E}(f_\rho) + \varepsilon} \sqrt{\varepsilon} \leq \frac{1}{2}\big(\mathcal{E}(f) - \mathcal{E}(f_\rho)\big) + \varepsilon$. Using the quantity $v^*(m, \delta)$, Proposition 8.15 with $\varepsilon = (\mathbf{C}_K + 3)^2 R^2 v^*(m, \frac{\delta}{2})$ tells us that there is a set $V_R' \subset Z^m$ of measure at most $\frac{\delta}{2}$ such that

$$\mathcal{E}(f) - \mathcal{E}(f_\rho) - \big(\mathcal{E}_{\mathbf{z}}(f) - \mathcal{E}_{\mathbf{z}}(f_\rho)\big) \leq \frac{1}{2}\big(\mathcal{E}(f) - \mathcal{E}(f_\rho)\big) + (\mathbf{C}_K + 3)^2$$

$$R^2 v^*(m, \delta/2), \quad \forall f \in B_R, \mathbf{z} \in Z^m \setminus V_R'.$$

In particular, when $\mathbf{z} \in \mathcal{W}(R) \setminus V_R', f_{\mathbf{z}} \in B_R$ and

$$\mathbf{E}(\xi_1) - \frac{1}{m}\sum_{i=1}^m \xi_1(z_i) = \mathcal{E}(f_{\mathbf{z}}) - \mathcal{E}(f_\rho) - \big(\mathcal{E}_{\mathbf{z}}(f_{\mathbf{z}}) - \mathcal{E}_{\mathbf{z}}(f_\rho)\big)$$

$$\leq \frac{1}{2}\big(\mathcal{E}(f_{\mathbf{z}}) - \mathcal{E}(f_\rho)\big) + (\mathbf{C}_K + 3)^2 R^2 v^*(m, \delta/2).$$

Now apply Proposition 8.14 with δ replaced by $\frac{\delta}{2}$. We can find another set $V_R'' \subset Z^m$ of measure at most $\frac{\delta}{2}$ such that for all $\mathbf{z} \in Z^m \setminus V_R''$,

$$\frac{1}{m}\sum_{i=1}^m \xi_2(z_i) - \mathbf{E}(\xi_2) \leq \left(\frac{4\mathbf{C}_K^2 \log(2/\delta)}{m\gamma} + 3\mathbf{M} + 1\right)\mathcal{D}(\gamma)$$

$$+ \frac{(24\mathbf{M}^2 + 3\mathbf{M})\log(2/\delta)}{m}.$$

Combining these two bounds with (8.4), we see that for all $\mathbf{z} \in \mathcal{W}(R) \setminus (V_R' \cup V_R'')$,

$$\mathcal{E}(f_{\mathbf{z}}) - \mathcal{E}_{\mathbf{z}}(f_{\mathbf{z}}) + \mathcal{E}_{\mathbf{z}}(f_\gamma) - \mathcal{E}(f_\gamma) \leq \frac{1}{2}\big(\mathcal{E}(f_{\mathbf{z}}) - \mathcal{E}(f_\rho)\big) + (\mathbf{C}_K + 3)^2 R^2 v^*(m, \delta/2)$$

$$+ \left(\frac{4\mathbf{C}_K^2 \log(2/\delta)}{m\gamma} + 3\mathbf{M} + 1\right)\mathcal{D}(\gamma)$$

$$+ \frac{(24\mathbf{M}^2 + 3\mathbf{M})\log(2/\delta)}{m}.$$

This inequality, together with Theorem 8.3, tells us that for all $\mathbf{z} \in \mathcal{W}(R) \setminus (V_R' \cup V_R'')$,

$$\mathcal{E}(f_{\mathbf{z}}) - \mathcal{E}(f_\rho) + \gamma \|f_{\mathbf{z}}\|_K^2 \leq \mathcal{D}(\gamma) + \frac{1}{2}\big(\mathcal{E}(f_{\mathbf{z}}) - \mathcal{E}(f_\rho)\big) + (\mathbf{C}_K + 3)^2$$

$$R^2 v^*(m, \delta/2) + \left(\frac{4\mathbf{C}_K^2 \log(2/\delta)}{m\gamma} + 3\mathbf{M} + 1\right)\mathcal{D}(\gamma)$$

$$+ \frac{(24\mathbf{M}^2 + 3\mathbf{M})\log(2/\delta)}{m}.$$

This gives the desired bound with $V_R = V_R' \cup V_R''$.   ∎

To prove Theorem 8.10, we still need an R satisfying $\mathcal{W}(R) = Z^m$.

Lemma 8.17 *For all $\gamma > 0$ and almost all $\mathbf{z} \in Z^m$,*

$$\|f_\mathbf{z}\|_K \leq \frac{\mathbf{M}}{\sqrt{\gamma}}.$$

Proof. Since $f_\mathbf{z}$ minimizes $\mathcal{E}_{\mathbf{z},\gamma}$, we have

$$\gamma \|f_\mathbf{z}\|_K^2 \leq \mathcal{E}_{\mathbf{z},\gamma}(f_\mathbf{z}) \leq \mathcal{E}_{\mathbf{z},\gamma}(0) = \frac{1}{m} \sum_{i=1}^m (y_i - 0)^2 \leq \mathbf{M}^2,$$

the last almost surely. Therefore, $\|f_\mathbf{z}\|_K \leq \mathbf{M}/\sqrt{\gamma}$ for almost all $\mathbf{z} \in Z^m$.   ∎

Lemma 8.17 says that $\mathcal{W}(\mathbf{M}/\sqrt{\gamma}) = Z^m$ up to a set of measure zero (we ignore this null set later). Take $R := \mathbf{M}/\sqrt{\gamma} \geq \mathbf{M}$. Theorem 8.10 follows from Proposition 8.16.

8.4 Proof of Theorem 8.1

In this section we improve the excess generalization error estimate of Theorem 8.10. The method in the previous section was rough because we used the bound $\|f_\mathbf{z}\|_K \leq \mathbf{M}/\sqrt{\gamma}$ shown in Lemma 8.17. This is much worse than the bound for f_γ given in Lemma 8.13, namely, $\|f_\gamma\|_K \leq \sqrt{\mathcal{D}(\gamma)}/\sqrt{\gamma}$. Yet we expect the minimizer $f_\mathbf{z}$ of $\mathcal{E}_{\mathbf{z},\gamma}$ to be a good approximation of the minimizer f_γ of $\mathcal{E}_\gamma$. In particular, we expect $\|f_\mathbf{z}\|_K$ also to be bounded by, essentially, $\sqrt{\mathcal{D}(\gamma)}/\sqrt{\gamma}$. We prove that this is the case with high probability by applying Proposition 8.16 iteratively. As a consequence, we obtain better bounds for the excess generalization error.

Lemma 8.18 *For all $0 < \delta < 1$ and $R \geq \mathbf{M}$, there is a set $V_R \subset Z^m$ with $\rho(V_R) \leq \delta$ such that*

$$\mathcal{W}(R) \subseteq \mathcal{W}(a_m R + b_m) \cup V_R,$$

where $a_m := (2\mathbf{C}_K + 5)\sqrt{v^*(m, \delta/2)/\gamma}$ and

$$b_m := \left(\frac{2\mathbf{C}_K \sqrt{2 \log(2/\delta)}}{\sqrt{m\gamma}} + \sqrt{6\mathbf{M} + 4} \right) \sqrt{\frac{\mathcal{D}(\gamma)}{\gamma}} + \frac{(7\mathbf{M} + 1)\sqrt{2 \log(2/\delta)}}{\sqrt{m\gamma}}.$$

Proof. By Proposition 8.16, there is a set $V_R \subset Z^m$ with $\rho(V_R) \leq \delta$ such that for all $\mathbf{z} \in \mathcal{W}(R) \setminus V_R$,

$$\gamma \|f_{\mathbf{z}}\|_K^2 \leq 2(\mathbf{C}_K + 3)^2 R^2 v^*(m, \delta/2) + \left(\frac{8\mathbf{C}_K^2 \log(2/\delta)}{m\gamma} + 6\mathbf{M} + 4 \right) \mathcal{D}(\gamma)$$
$$+ \frac{(48\mathbf{M}^2 + 6\mathbf{M}) \log(2/\delta)}{m}.$$

This implies that

$$\|f_{\mathbf{z}}\|_K \leq a_m R + b_m, \quad \forall \mathbf{z} \in \mathcal{W}(R) \setminus V_R,$$

with a_m and b_m as given in our statement. In other words, $\mathcal{W}(R) \setminus V_R \subseteq \mathcal{W}(a_m R + b_m)$. ∎

Lemma 8.19 *Assume that K satisfies $\log \mathcal{N}(B_1, \eta) \leq C_0(1/\eta)^{s^*}$ for some $s^* > 0$. Take $\gamma = m^{-\zeta}$ with $\zeta < 1/(1 + s^*)$. For all $0 < \delta < 1$ and $m \geq m_\delta$, with confidence $1 - 3\delta/\big(1/(1 + s^*) - \zeta\big)$*

$$\|f_{\mathbf{z}}\|_K \leq C_2 \sqrt{\log(2/\delta)} \big(\sqrt{\mathcal{D}(\gamma)/\gamma} + 1 \big)$$

holds. Here $C_2 > 0$ is a constant that depends only on $s^*, \zeta, \mathbf{C}_K, C_0$, and $\mathbf{M}$, and $m_\delta \in \mathbb{N}$ depends also on δ.

Proof. By Lemma 8.9, when $m \geq \big(108/C_0\big)^{1/s^*} \big(\log(2/\delta)\big)^{1+1/s^*}$,

$$v^*(m, \delta/2) \leq \big(108 C_0/m\big)^{1/(1+s^*)} \tag{8.7}$$

holds. It follows that

$$a_m \leq (2\mathbf{C}_K + 5)\big(108 C_0\big)^{1/(2+2s^*)} m^{\zeta/2 - 1/(2+2s^*)}. \tag{8.8}$$

Denote $c := (2\mathbf{C}_K + 5)\big(108 C_0\big)^{1/(2+2s^*)}$. When $m \geq \big(1/(2c)\big)^{2/(\zeta - 1/(1+s^*))}$, we have $a_m \leq \frac{1}{2}$.

Choose $m_\delta := \max\left\{ \big(108/C_0\big)^{1/s^*} \big(\log(2/\delta)\big)^{1+1/s^*}, \big(1/(2c)\big)^{2/(\zeta - 1/(1+s^*))} \right\}$.

Define a sequence $\{R^{(j)}\}_{j \in \mathbb{N}}$ by $R^{(0)} = \mathbf{M}/\sqrt{\gamma}$ and, for $j \geq 1$,

$$R^{(j)} = a_m R^{(j-1)} + b_m.$$

Then Lemma 8.17 proves $\mathcal{W}(R^{(0)}) = Z^m$, and Lemma 8.18 asserts that for each $j \geq 1$, $\mathcal{W}(R^{(j-1)}) \subseteq \mathcal{W}(R^{(j)}) \cup V_{R^{(j-1)}}$ with $\rho(V_{R^{(j-1)}}) \leq \delta$. Apply this inclusion for $j = 1, 2, \ldots, J$, with J satisfying $2/\big(1/(1 + s^*) - \zeta\big) \leq J \leq 3/\big(1/(1 + s^*) - \zeta\big)$. We see that

$$Z^m = \mathcal{W}(R^{(0)}) \subseteq \mathcal{W}(R^{(1)}) \cup V_{R^{(0)}} \subseteq \cdots \subseteq \mathcal{W}(R^{(J)}) \cup \left(\bigcup_{j=0}^{J-1} V_{R^{(j)}} \right).$$

It follows that the measure of the set $\mathcal{W}(R^{(J)})$ is at least $1 - J\delta \geq 1 - 3\delta/\big(1/(1 + s^*) - \zeta\big)$. By the definition of the sequence, we have

$$R^{(J)} = a_m^J R^{(0)} + b_m \sum_{j=0}^{J-1} a_m^j \leq \mathbf{M}c^J m^{J\left(\zeta/2 - 1/(2+2s^*)\right)+\zeta/2} + b_m \leq \mathbf{M}c^J + b_m.$$

Here we have used (8.8) and $a_m \leq \frac{1}{2}$ in the first inequality, and then the restriction $J \geq 2/(1/(1 + s^*) - \zeta) > \zeta/(1/(1 + s^*) - \zeta)$ in the second inequality. Note that $c^J \leq \big((2\mathbf{C}_K + 5)(108C_0 + 1)\big)^{3/\left(1/(1+s^*)-\zeta\right)}$. Since $\gamma = m^{-\zeta}$, b_m can be bounded as

$$b_m \leq \sqrt{2\log(2/\delta)}\left((2\mathbf{C}_K + \sqrt{6\mathbf{M} + 4})\sqrt{\mathcal{D}(\gamma)/\gamma} + 7\mathbf{M} + 1 \right).$$

Thus, $R^{(J)} \leq C_2\sqrt{\log(2/\delta)}(\sqrt{\mathcal{D}(\gamma)/\gamma} + 1)$ with C_2 depending only on $s^*, \zeta, \mathbf{C}_K, C_0$, and $\mathbf{M}$. ∎

Proof of Theorem 8.1 Applying Proposition 8.16 with $R := C_2\sqrt{\log(2/\delta)}$ $(\sqrt{\mathcal{D}(\gamma)/\gamma} + 1)$, and using that $\gamma_* = m^{-\zeta}$ and $\mathcal{D}(\gamma_*) \leq \gamma_*^\theta \|L_K^{-\theta/2} f_\rho\|^2$, we deduce from (8.7) that for $m \geq m_\delta$ and all $\mathbf{z} \in \mathcal{W}(R) \setminus V_R$,

$$\mathcal{E}(f_\mathbf{z}) - \mathcal{E}(f_\rho) \leq 2(\mathbf{C}_K + 3)^2 C_2^2 \log(2/\delta) 2 m^{\zeta(1-\theta)}(108C_0)^{1/(1+s^*)} m^{-1/(1+s^*)}$$
$$+ C_2' \log(2/\delta) m^{-\theta\zeta}$$
$$\leq C_3 \log(2/\delta) m^{-\theta\zeta}$$

holds. Here C_2', C_3 are positive constants depending only on $s^*, \zeta, \mathbf{C}_K, C_0, \mathbf{M}$, and $\|L_K^{-\theta/2} f_\rho\|$.

By Lemma 8.19, the set $\mathcal{W}(R)$ has measure at least $1 - 3\delta/\big(1/(1 + s^*) - \zeta\big)$ when $m \geq m_\delta$. Replacing $3\delta/\big(1/(1 + s^*) - \zeta\big) + \delta$ (the last δ from the bound on V_R) by δ and letting C_0' be the resulting C_3, our conclusion follows. ∎

When $s^* \to 0$ and $\theta \to 1$, we see that the convergence rate $\theta \zeta$ can be arbitrarily close to 1.

8.5 Reminders V

We use the following result on Lagrange multipliers.

Proposition 8.20 *Let U be a Hilbert space and F, H be real-valued $\mathscr{C}^1$ functions on U. Let $c \in U$ be a solution of the problem*

$$\min \ F(f)$$
$$s.t. \ H(f) \le 0.$$

Then, there exist real numbers μ, λ, not both zero, such that

$$\mu DF(c) + \lambda DH(c) = 0. \tag{8.9}$$

Here D means derivative. Furthermore, if $H(c) < 0$, then $\lambda = 0$. Finally, if either $H(c) < 0$ or $DH(c) \ne 0$, then $\mu \ne 0$ and $\frac{\lambda}{\mu} \ge 0$. ∎

If $\mu \ne 0$ above, we can take $\mu = 1$ and we call the resulting λ the *Lagrange multiplier* of the problem at c.

8.6 Compactness and regularization

Let $\mathbf{z} = (z_1, \ldots, z_m)$ with $z_i = (x_i, y_i) \in X \times Y$ for $i = 1, \ldots, m$. We also write $\mathbf{x} = (x_1, \ldots, x_m)$ and $\mathbf{y} = (y_1, \ldots, y_m)$. Assume that $y \ne 0$ and $K[\mathbf{x}]$ is invertible. Let $a^* = K[\mathbf{x}]^{-1}\mathbf{y}, f^* = \sum_{i=1}^{m} a_i^* K_{x_i}$, and $R_0^2 = \|f^*\|_K^2 = \mathbf{y}K[\mathbf{x}]^{-1}\mathbf{y}$. Let $E_{\mathbf{z}}(\gamma)$ and $\overline{E_{\mathbf{z}}}(R)$ be the problems

$$\min \ \frac{1}{m} \sum_{i=1}^{m} (f(x_i) - y_i)^2 + \gamma \|f\|_K^2$$

$$s.t. \ f \in \mathcal{H}_K$$

and

$$\min \quad \frac{1}{m}\sum_{i=1}^{m}(f(x_i) - y_i)^2$$

$$\text{s.t.} \quad f \in B(\mathcal{H}_K, R),$$

respectively, where $R, \gamma > 0$.

In Proposition 8.7 and Corollary 1.14 we have seen that the minimizers $f_{z,\gamma}$ and $f_{z,R}$ of $E_z(\gamma)$ and $\overline{E_z}(R)$, respectively, exist. A natural question is, What is the relationship between the problems $E_z(\gamma)$ and $\overline{E_z}(R)$ and their minimizers? The main result in this section answers this question.

Theorem 8.21 *There exists a decreasing global homeomorphism*

$$\Lambda_z : (0, +\infty) \to (0, R_0)$$

satisfying

(i) *for all $\gamma > 0$, $f_{z,\gamma}$ is the minimizer of $\overline{E_z}(\Lambda_z(\gamma))$, and*
(ii) *for all $R \in (0, R_0)$, $f_{z,R}$ is the minimizer of $E_z(\Lambda_z^{-1}(R))$.*

To prove Theorem 8.21, we use Proposition 8.20 for the problems $\overline{E_z}(R)$ with $U = \mathcal{H}_K$, $F(f) = \frac{1}{m}\sum_{i=1}^{m}(f(x_i) - y_i)^2$, and $H(f) = \|f\|_K^2 - R^2$. Note that for $x \in X$, the mapping

$$\mathcal{H}_K \to \mathbb{R}$$

$$f \mapsto f(x) = \langle f, K_x\rangle_K$$

is a bounded linear functional and therefore $\mathscr{C}^1$ with derivative K_x. It follows that F is $\mathscr{C}^1$ as well and $DF(f) = \frac{2}{m}\sum_{i=1}^{m}(f(x_i) - y_i)K_{x_i}$. Also, H is $\mathscr{C}^1$ and $DH(f) = 2f$. Define

$$\Lambda_z : (0, +\infty) \to (0, R_0)$$

$$\gamma \mapsto \|f_{z,\gamma}\|_K.$$

Also, for each $R \in (0, R_0)$, choose one minimizer $f_{z,R}$ of $\overline{E_z}(R)$ and let

$$\Gamma_z : (0, R_0) \to (0, +\infty)$$

$$R \mapsto \text{the Lagrange multiplier of } \overline{E_z}(R) \text{ at } f_{z,R}.$$

Lemma 8.22 *The function* Γ_z *is well defined.*

Proof. We apply Proposition 8.20 to the problem $\overline{E}_z(R)$ and claim that $f_{z,R}$ is not the zero function. Otherwise, $H(f_{z,R}) = H(0) < 0$, which implies $DF(f_{z,R}) = DF(0) = -\frac{2}{m}\sum_{i=1}^{m} y_i K_{x_i} = 0$ by (8.9), contradicting the invertibility of $K[\mathbf{x}]$.

Since $f_{z,R}$ is not the zero function, $DH(f_{z,R}) = 2f_{z,R} \neq 0$. Also, $DF(f_{z,R}) = \frac{2}{m}\sum_{i=1}^{m}(f_{z,R}(x_i) - y_i)K_{x_i} \neq 0$, since $K[\mathbf{x}]$ is invertible. By Proposition 8.20, $\mu \neq 0$ and $\lambda \neq 0$. Taking $\mu = 1$, we conclude that the Lagrange multiplier λ is positive. ∎

Proposition 8.23
(i) *For all* $\gamma > 0$, $f_{z,\gamma}$ *is the minimizer of* $\overline{E}_z(\Lambda_z(\gamma))$.
(ii) *Let* $R \in (0, R_0)$. *Then* $f_{z,R}$ *is the minimizer of* $E_z(\Gamma_z(R))$.

Proof. Assume that

$$\frac{1}{m}\sum_{i=1}^{m}(f(x_i) - y_i)^2 < \frac{1}{m}\sum_{i=1}^{m}(f_{z,\gamma}(x_i) - y_i)^2$$

for some $f \in B(\mathcal{H}_K, \Lambda_z(\gamma))$. Then

$$\frac{1}{m}\sum_{i=1}^{m}(f(x_i) - y_i)^2 + \gamma\|f\|_K^2 < \frac{1}{m}\sum_{i=1}^{m}(f_{z,\gamma}(x_i) - y_i)^2 + \gamma\Lambda_z(\gamma)^2$$

$$= \frac{1}{m}\sum_{i=1}^{m}(f_{z,\gamma}(x_i) - y_i)^2 + \gamma\|f_{z,\gamma}\|_K^2,$$

contradicting the requirement that $f_{z,\gamma}$ minimizes the objective function of $E_z(\gamma)$. This proves (i).

For Part (ii) note that the proof of Lemma 8.22 yields $\mu = 1$ and $\lambda = \Gamma_z(R) > 0$. Since $f_{z,R}$ minimizes $\overline{E}_z(R)$, by Proposition 8.20, $f_{z,R}$ satisfies

$$D(F + \lambda H)(f_{z,R}) = D(F + \lambda\|f\|_K^2)(f_{z,R}) = 0;$$

that is, the derivative of the objective function of $E_z(\lambda)$ vanishes at $f_{z,R}$. Since this function is convex and $E_z(\lambda)$ is an unconstrained problem, we conclude that $f_{z,R}$ is the minimizer of $E_z(\lambda) = E_z(\Gamma_z(R))$. ∎

Proposition 8.24 Λ_z *is a decreasing global homeomorphism with inverse* Γ_z.

Proof. Since K is a Mercer kernel, the matrix $K[\mathbf{x}]$ is positive definite by the invertibility assumption. So there exist an orthogonal matrix P and a diagonal matrix D such that $K[\mathbf{x}] = PDP^{-1}$. Moreover, the main diagonal entries $d_1, \ldots, d_m$ of D are positive. Let $\mathbf{y}' = P^{-1}\mathbf{y}$. Then, by Proposition 8.7, $f_{\mathbf{z},\gamma} = \sum_{i=1}^{m} a_i K_{x_i}$ with a satisfying $(\gamma m \mathrm{Id} + D)P^{-1}a = P^{-1}\mathbf{y} = \mathbf{y}'$. It follows that

$$P^{-1}a = \left(\frac{y'_i}{\gamma m + d_i} \right)_{i=1}^{m}$$

and, using $P^{\mathrm{T}} = P^{-1}$,

$$\Lambda_{\mathbf{z}}(\gamma) = \|f_{\mathbf{z},\gamma}\|_K = a^{\mathrm{T}} K[\mathbf{x}]a = (P^{-1}a)^{\mathrm{T}} DP^{-1}a = \sqrt{\sum_{i=1}^{m} d_i \left(\frac{y'_i}{\gamma m + d_i} \right)^2},$$

which is positive for all $\gamma \in [0, +\infty)$ since $y \neq 0$ by assumption. Differentiating with respect to γ,

$$\Lambda'_{\mathbf{z}}(\gamma) = -\frac{1}{\|f_{\mathbf{z},\gamma}\|_K} \sum_{i=1}^{m} d_i \frac{y'^2_i}{(\gamma m + d_i)^3}.$$

This expression is negative for all $\gamma \in [0, +\infty)$. This shows that $\Lambda_{\mathbf{z}}$ is strictly decreasing in its domain. The first statement now follows since $\Lambda_{\mathbf{z}}$ is continuous, $\Lambda_{\mathbf{z}}(0) = R_0$, and $\Lambda_{\mathbf{z}}(\gamma) \to 0$ when $\gamma \to \infty$.

To prove the second statement, consider $\gamma > 0$. Then, by Proposition 8.23(i) and (ii),

$$f_{\mathbf{z},\gamma} = f_{\mathbf{z},\Lambda_{\mathbf{z}}(\gamma)} = f_{\mathbf{z},\Gamma_{\mathbf{z}}(\Lambda_{\mathbf{z}}(\gamma))}.$$

To prove that $\gamma = \Gamma_{\mathbf{z}}(\Lambda_{\mathbf{z}}(\gamma))$, it is thus enough to prove that for $\gamma, \gamma' \in (0, +\infty)$, if $f_{\mathbf{z},\gamma} = f_{\mathbf{z},\gamma'}$, then $\gamma = \gamma'$. To do so, let i be such that $y'_i \neq 0$ (such an i exists since $\mathbf{y} \neq 0$). Since the coefficient vectors for $f_{\mathbf{z},\gamma}$ and $f_{\mathbf{z},\gamma'}$ are the same a with $P^{-1}a = \left(y'_i/(\gamma m + d_i) \right)_{i=1}^{m}$, we have in particular

$$\frac{y'_i}{\gamma m + d_i} = \frac{y'_i}{\gamma' m + d_i},$$

whence it follows that $\gamma = \gamma'$.   ∎

Corollary 8.25 *For all $R < R_0$, the minimizer $f_{\mathbf{z},R}$ of $\overline{E_{\mathbf{z}}}(R)$ is unique.*

Proof. Let $\gamma = \Gamma_{\mathbf{z}}(R)$. Then $f_{\mathbf{z},R} = f_{\mathbf{z},\gamma}$ by Proposition 8.23(ii). Now use that $f_{\mathbf{z},\gamma}$ is unique.   ∎

Theorem 8.21 now follows from Propositions 8.23 and 8.24 and Corollary 8.25.

Remark 8.26 Let $E(\gamma)$ and $\overline{E}(R)$ be the problems

$$\min \quad \int (f(x) - y)^2 \, d\rho + \gamma \|f\|_K^2$$

$$\text{s.t.} \quad f \in \mathcal{H}_K$$

and

$$\min \quad \int (f(x) - y)^2 \, d\rho$$

$$\text{s.t.} \quad f \in B(\mathcal{H}_K, R),$$

respectively, where $R, \gamma > 0$. Denote by f_γ and f_R their minimizers, respectively. Also, let $R_1 = \|f_\rho\|_K$ if $f_\rho \in \mathcal{H}_K$ and $R_1 = \infty$ otherwise. A development similar to the one in this section shows the existence of a decreasing global homeomorphism

$$\Lambda : (0, +\infty) \to (0, R_1)$$

satisfying

(i) for all $\gamma > 0, f_\gamma$ is the minimizer of $\overline{E}(\Lambda(\gamma))$, and
(ii) for all $R \in (0, R_1), f_R$ is the minimizer of $E(\Lambda^{-1}(R))$.

Here f_R is the target function $f_{\mathcal{H}}$ when $\mathcal{H}$ is $I_K(B_R)$.

8.7 References and additional remarks

The problem of approximating a function from sparse data is often ill posed. A standard approach to dealing with ill-posedness is regularization theory [36, 54, 64, 102, 130]. Regularization schemes with RKHSs were introduced to learning theory in [137] using spline kernels and in [53, 134, 133] using general Mercer kernels. A key feature of RKHSs is ensuring that the minimizer of the regularization scheme can be found in the subspace spanned by $\{K_{x_i}\}_{i=1}^m$. Hence, the minimization over the possibly infinite-dimensional function space $\mathcal{H}_K$ is reduced to minimization over a finite-dimensional space [50]. This follows from the reproducing property in RKHSs. We have devoted Section 2.8 to this feature. It is extended to other contexts in the next two chapters.

The error analysis for the least squares regularization scheme was considered in [38] in terms of covering numbers. The distance between $f_{z,\gamma}$ and f_γ was studied in [24] using stability analysis. In [153], using leave-one-out techniques, it was proved that

$$\mathop{\mathbf{E}}_{\mathbf{z}\in Z^m}\big(\mathcal{E}(f_{\mathbf{z},\gamma})\big) \le \left(1 + \frac{2\mathbf{C}_K^2}{m\gamma}\right)^2 \inf_{f\in\mathcal{H}_K}\left\{\mathcal{E}(f) + \gamma\,\|f\|_K^2\right\}.$$

In [42, 43] a functional analysis approach was employed to show that for any $0 < \delta < 1$, with confidence $1 - \delta$,

$$\big|\mathcal{E}(f_{\mathbf{z},\gamma}) - \mathcal{E}(f_\gamma)\big| \le \frac{M\,\mathbf{C}_K^2}{\sqrt{m}}\left(1 + \frac{\mathbf{C}_K}{\sqrt{\gamma}}\right)\left(1 + \sqrt{2\log(2/\delta)}\right).$$

Parts (i) and (ii) of Proposition 8.5 were given in [39]. Part (iii) with $1 < \theta \le 2$ was proved in [115], and the extension to $2 < \theta \le 3$ was shown by Mihn in the appendix to [116]. In [115], a modified McDiarmid inequality was used to derive error bounds in the metric induced by $\|\ \|_K$. If f_ρ is in the range of L_K, then, for any $0 < \delta < 1$ with confidence $1 - \delta$,

$$\|f_{\mathbf{z},\gamma} - f_\rho\|_K^2 \le \widetilde{C}\left(\frac{\big(\log(4/\delta)\big)^2}{m}\right)^{1/3} \quad \text{by taking} \quad \gamma = \left(\frac{\big(\log(4/\delta)\big)^2}{m}\right)^{1/3}$$

holds, where $\widetilde{C}$ is a constant independent of m and δ. In [116] a Bennett inequality for vector-valued random variables with values in Hilbert spaces is applied, which yields better error bounds. If f_ρ is in the range of L_K, then we have

$$\|f_{\mathbf{z},\gamma} - f_\rho\|_{\mathscr{L}^2_{\rho_X}}^2 \le \widetilde{C}\big(\log(4/\delta)\big)^2 (1/m)^{2/3} \quad \text{by taking} \quad \gamma = \log(4/\delta)\,(1/m)^{1/3}.$$

These results are capacity-independent error bounds. The error analysis presented in this chapter is capacity dependent and was mainly done in [143]. When $f_\rho \in \mathcal{H}_K$ and $s^* < 2$, the learning rate given by Theorem 8.1 is better than capacity-independent ones.

A proof of Proposition 8.20 can be found, for instance, in [11].

For some applications, such as signal processing, inverse problems, and numerical analysis, the data $(x_i)_{i=1}^m$ may be deterministic, not randomly drawn according to ρ_X. Then the regularization scheme $\inf_{f\in\mathcal{H}_K}\mathcal{E}_{\mathbf{z},\gamma}(f)$ involves only the random data $(y_i)_{i=1}^m$. For active learning [33, 81], the data $(x_i)_{i=1}^m$ are drawn according to a user-defined distribution that is different from ρ_X. Such schemes and their connections to richness of data have been studied in [114].

9
Support vector machines for classification

In the previous chapters we have dealt with the problem of learning a function $f : X \to Y$ when $Y = \mathbb{R}$. We have described algorithms producing an approximation $f_{\mathbf{z}}$ of f from a given sample $\mathbf{z} \in Z^m$ and we have measured the quality of this approximation with the generalization error $\mathcal{E}$ as a ruler.

Although this setting applies to a good number of situations arising in practice, there are quite a few that can be better approached. One paramount example is that described in Case 1.5. Recall that in this case we dealt with a space Y consisting of two elements (in Case 1.5 they were 0 and 1). Problems consisting of learning a binary (or finitely) valued function are called *classification problems*. They occur frequently in practice (e.g., the determination, from a given sample of clinical data, of whether a patient suffers a certain disease), and they will be the subject of this (and the next) chapter.

A *binary classifier* on a compact metric space X is a function $f : X \to \{1, -1\}$. To provide some continuity in our notation, we denote $Y = \{-1, 1\}$ and keep $Z = X \times Y$. Classification problems thus consist of learning binary classifiers. To measure the quality of our approximations, an appropriate notion of error is essential.

Definition 9.1 Let ρ be a probability distribution on $Z := X \times Y$. The *misclassification error* $\mathcal{R}(f)$ for a classifier $f : X \to Y$ is defined to be the probability of a wrong prediction, that is, the measure of the event $\{f(x) \neq y\}$,

$$\mathcal{R}(f) := \Prob_{z \in Z} \{f(x) \neq y\} = \int_X \Prob_{y \in Y}(y \neq f(x) \mid x) \, d\rho_X. \tag{9.1}$$

Our target concept (in the sense of Case 1.5) is the set $T := \{x \in X \mid \Prob\{y = 1 \mid x\} \geq \frac{1}{2}\}$, since the conditional distribution at x is a binary distribution.

One goal of this chapter is to describe an approach to producing classifiers from samples (and an RKHS $\mathcal{H}_K$) known as support vector machines.

157

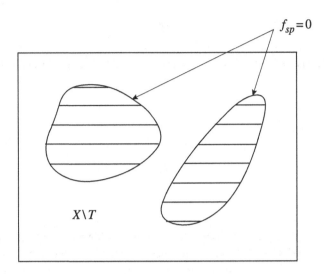

Figure 9.1

Needless to say, we are interested in bounding the misclassification error of the classifiers obtained in this way. We do so for a certain class of noise-free measures that we call *weakly separable*. Roughly speaking (a formal definition follows in Section 9.7), these are measures for which there exists a function $f_{\mathsf{sp}} \in \mathcal{H}_K$ such that $x \in T \iff f_{\mathsf{sp}}(x) \geq 0$ and satisfy a decay condition near the boundary of T. The situation is as in Figure 9.1, where the rectangle is the set X, the dashed regions represent the set T, and their boundaries are the zero set of f_{sp}.

Support vector machines produce classifiers from a sample $\mathbf{z}$, a real number $\gamma > 0$ (a regularization parameter as in Chapter 8), and an RKHS $\mathcal{H}_K$. Let us denote by $F_{\mathbf{z},\gamma}$ such a classifier. One major result in this chapter is the following (a more detailed statement is given in Theorem 9.26 below).

Theorem 9.2 *Assume ρ is weakly separable by $\mathcal{H}_K$. Let B_1 denote the unit ball in $\mathcal{H}_K$.*

(i) *If $\log \mathcal{N}(B_1, \eta) \leq C_0(1/\eta)^p$ for some $p, C_0 > 0$ and all $\eta > 0$, then, taking $\gamma = m^{-\beta}$ (for some $\beta > 0$), we have, with confidence $1 - \delta$,*

$$\mathcal{R}(F_{\mathbf{z},\gamma}) \leq \widetilde{C}\left(\left(\frac{1}{m}\right)^r \left(\log\frac{2}{\delta}\right)^2\right),$$

where r and $\widetilde{C}$ are positive constants independent of m and δ.

(ii) *If* $\log \mathcal{N}(B_1, \eta) \leq C_0 (\log(1/\eta))^p$ *for some* $p, C_0 > 0$ *and all* $0 < \eta < 1$, *then, for sufficiently large m and some* $\beta > 0$, *taking* $\gamma = m^{-\beta}$, *we have, with confidence* $1 - \delta$,

$$\mathcal{R}(F_{\mathbf{z},\gamma}) \leq \tilde{C} \left(\left(\frac{1}{m} \right)^r (\log m)^p \left(\log \frac{2}{\delta} \right)^2 \right),$$

where r and $\tilde{C}$ *are positive constants independent of m and* δ.

We note that the exponent β in both parts of Theorem 9.2, unfortunately, depends on p. Details of this dependence are made explicit in Section 9.7, where Theorem 9.26 is proved.

9.1 Binary classifiers

Just as in Chapter 1, where we saw that the real-valued function minimizing the error is the regression function f_ρ, we may wonder which binary classifier minimizes $\mathcal{R}$. The answer is simple. For a function $f : X \to \mathbb{R}$ define

$$\text{sgn}(f)(x) = \begin{cases} 1 & \text{if } f(x) \geq 0 \\ -1 & \text{if } f(x) < 0. \end{cases}$$

Also, let $K_\rho := \{x \in X : f_\rho(x) = 0\}$ and $\kappa_\rho = \rho_X(K_\rho)$.

Proposition 9.3
 (i) *For any classifier* f,

$$\mathcal{R}(f) = \tfrac{1}{2} \kappa_\rho + \int_{X \setminus K_\rho} \underset{Y}{\text{Prob}}(y \neq f(x)|x) \, d\rho_X.$$

(ii) $\mathcal{R}$ *is minimized by any classifier coinciding on* $X \setminus K_\rho$ *with*

$$f_c := \text{sgn}(f_\rho).$$

Proof. Since $Y = \{1, -1\}$, we have $f_\rho(x) = \text{Prob}_Y(y = 1 \mid x) - \text{Prob}_Y(y = -1 \mid x)$. This means that

$$f_c(x) = \text{sgn}(f_\rho)(x) = \begin{cases} 1 & \text{if } \text{Prob}_Y(y = 1 \mid x) \geq \text{Prob}_Y(y = -1 \mid x) \\ -1 & \text{if } \text{Prob}_Y(y = 1 \mid x) < \text{Prob}_Y(y = -1 \mid x). \end{cases}$$

$$(9.2)$$

For any classifier f and any $x \in K_\rho$, we have $\text{Prob}_Y(y \neq f(x)|x) = \tfrac{1}{2}$. Hence statement (i) holds.

For the second statement, we observe that for $x \in X \setminus K_\rho$, $\text{Prob}_Y(y = f_c(x) \mid x) > \text{Prob}_Y(y \neq f_c(x) \mid x)$. Then, for any classifier f, we have either $f(x) = f_c(x)$ or $\text{Prob}(y \neq f(x) \mid x) = \text{Prob}(y = f_c(x) \mid x) > \text{Prob}(y \neq f_c(x) \mid x)$. Hence $\mathcal{R}(f) \geq \mathcal{R}(f_c)$, and equality holds if and only if f and f_c are equal almost everywhere on X/K_ρ. ∎

The classifier f_c is called *Bayes rule*.

Remark 9.4 The role played by the quantity κ_ρ is reminiscent of that played by σ_ρ^2 in the regression setting. Note that κ_ρ depends only on ρ. Therefore, its occurrence in Proposition 9.3(i) – just as that of σ_ρ^2 in Proposition 1.8 – is independent of f. In this sense, it yields a lower bound for the misclassification error and is, again, a measure of how well conditioned ρ is.

As ρ is unknown, the best classifier f_c cannot be found directly. The goal of *classification algorithms* is to find classifiers that approximate Bayes rule f_c from samples $\mathbf{z} \in Z^m$.

A possible strategy to achieve this goal could consist of fixing an RKHS $\mathcal{H}_K$ and a $\gamma > 0$, finding the minimizer $f_{\mathbf{z},\gamma}$ of the regularized empirical error (as described in Chapter 8), that is,

$$f_{\mathbf{z},\gamma} = \underset{f \in \mathcal{H}_K}{\text{argmin}} \frac{1}{m} \sum_{i=1}^{m} (f(x_i) - y_i)^2 + \gamma \|f\|_K^2, \tag{9.3}$$

and then taking the function $\text{sgn}(f_{\mathbf{z},\gamma})$ as an approximation of f_c. Note that this strategy minimizes a functional on a set of real-valued continuous functions and then applies the sgn function to the computed minimizer to obtain a classifier.

A different strategy consists of first taking signs to obtain the set $\{\text{sgn}(f) \mid f \in \mathcal{H}_K\}$ of classifiers and then minimizing an empirical error over this set. To see which empirical error we want to minimize, note that for a classifier $f : X \to Y$,

$$\mathcal{R}(f) = \int_Z \chi_{\{f(x) \neq y\}} \, d\rho = \int_Z \chi_{\{yf(x) = -1\}} \, d\rho.$$

Then, for $f \in \mathcal{H}_K$ satisfying $f(x) \neq 0$ almost everywhere,

$$\mathcal{R}(\text{sgn}(f)) = \int_Z \chi_{\{\text{sgn}(f(x))y = -1\}} \, d\rho = \int_Z \chi_{\{\text{sgn}(yf(x)) = -1\}} \, d\rho = \int_Z \chi_{\{yf(x) < 0\}} \, d\rho.$$

By discretizing the integral into a sum, given the sample $\mathbf{z} = \{(x_i, y_i)\}_{i=1}^m \in Z^m$, one might consider a binary classifier $\text{sgn}(f)$, where f is a solution of

$$\underset{\substack{f \in \mathcal{H}_K \\ f(x) \neq 0 \text{ a.e.}}}{\text{argmin}} \frac{1}{m} \sum_{i=1}^m \chi_{\{y_i f(x_i) < 0\}},$$

or, dropping the restriction that $f(x) \neq 0$ a.e. for simplicity,

$$\underset{f \in \mathcal{H}_K}{\text{argmin}} \frac{1}{m} \sum_{i=1}^m \chi_{\{y_i f(x_i) < 0\}}. \tag{9.4}$$

Note that in practical terms, we are again minimizing over $\mathcal{H}_K$. But we are now minimizing a different functional.

It is clear, however, that if f is any minimizer of (9.4), so is αf for all $\alpha > 0$. This shows that the regularized version of (9.4) (regularized by adding the term $\gamma \|f\|_K^2$ to the functional to be minimized) has no solution. It also shows that we can take as minimizer a function with norm 1. We conclude that we can approximate the Bayes rule by $\text{sgn}(f_\mathbf{z}^0)$, where $f_\mathbf{z}^0$ is given by

$$f_\mathbf{z}^0 := \underset{\substack{f \in \mathcal{H}_K \\ \|f\|_K = 1}}{\text{argmin}} \frac{1}{m} \sum_{i=1}^m \chi_{\{y_i f(x_i) < 0\}}. \tag{9.5}$$

We show in the next section that although we can reduce the computation of $f_\mathbf{z}^0$ to a nonlinear programming problem, the problem is not a convex one. Hence, we do not possess efficient algorithms to find $f_\mathbf{z}^0$ (cf. Section 2.7). We also introduce a third approach that lies somewhere in between those leading to problems (9.3) and (9.5). This new approach then occupies us for the remainder of this (and the next) chapter. We focus on its geometric background, error analysis, and algorithmic features.

9.2 Regularized classifiers

A *loss (function)* is a function $\phi : \mathbb{R} \to \mathbb{R}_+$. For $(x, y) \in Z$ and $f : X \to \mathbb{R}$, the quantity $\phi(yf(x))$ measures the local error (w.r.t. ϕ). Recall from Chapter 1 that this is the error resulting from the use of f as a model for the process producing y from x. Global errors are obtained by averaging over Z and empirical errors by averaging over a sample $\mathbf{z} \in Z^m$.

Definition 9.5 The *generalization error* associated with the loss ϕ is defined as

$$\mathcal{E}^{\phi}(f) := \int_{Z} \phi(yf(x)) \, d\rho.$$

The *empirical error* associated with the loss ϕ and a sample $\mathbf{z} \in Z^m$ is defined as

$$\mathcal{E}_{\mathbf{z}}^{\phi}(f) := \frac{1}{m} \sum_{i=1}^{m} \phi(y_i f(x_i)).$$

If $f \in \mathcal{H}_K$ for some Mercer kernel K, then we can define regularized versions of these errors. For $\gamma > 0$, we define the *regularized error*

$$\mathcal{E}_{\gamma}^{\phi}(f) := \int_{Z} \phi(yf(x)) \, d\rho + \gamma \|f\|_{K}^{2}$$

and the *regularized empirical error*

$$\mathcal{E}_{\mathbf{z},\gamma}^{\phi}(f) := \frac{1}{m} \sum_{i=1}^{m} \phi(y_i f(x_i)) + \gamma \|f\|_{K}^{2}.$$

Examples of loss functions are the *misclassification loss*

$$\phi_0(t) = \begin{cases} 0 & \text{if } t \geq 0 \\ 1 & \text{if } t < 0 \end{cases}$$

and the *least-squares loss* $\phi_{\text{ls}} = (1 - t)^2$. Note that for functions $f : X \to \mathbb{R}$ and points $x \in X$ such that $f(x) \neq 0$, $\phi_0(yf(x)) = \chi_{\{y \neq \text{sgn}(f(x))\}}$; that is, the local error is 1 if y and $f(x)$ have different signs and 0 when the signs are the same.

Proposition 9.6 *Restricted to binary classifiers, the generalization error w.r.t. ϕ_0 is the misclassification error $\mathcal{R}$; that is, for all classifiers f,*

$$\mathcal{R}(f) = \mathcal{E}^{\phi_0}(f).$$

In addition, the generalization error w.r.t. ϕ_{ls} is the generalization error $\mathcal{E}$. Similar statements hold for the empirical errors.

Proof. The first statement follows from the equalities

$$\mathcal{R}(f) = \int_{Z} \chi_{\{yf(x)=-1\}} \, d\rho = \int_{Z} \phi_0(yf(x)) \, d\rho = \mathcal{E}^{\phi_0}(f).$$

For the second statement, note that the generalization error $\mathcal{E}(f)$ of f satisfies

$$\mathcal{E}(f) = \int_Z (y - f(x))^2 \, d\rho = \int_Z (1 - yf(x))^2 \, d\rho = \mathcal{E}^{\phi_{ls}}(f),$$

since elements $y \in Y = \{-1, 1\}$ satisfy $y^2 = 1$ and therefore

$$(y - f(x))^2 = (y - y^2 f(x))^2 = y^2 (1 - yf(x))^2 = (1 - yf(x))^2. \qquad \blacksquare$$

Recall that $\mathcal{H}_{K,\mathbf{z}}$ is the finite-dimensional subspace of $\mathcal{H}_K$ spanned by $\{K_{x_1}, \ldots, K_{x_m}\}$ and $P : \mathcal{H}_K \to \mathcal{H}_{K,\mathbf{z}}$ is the orthogonal projection. Corollary 2.26 showed that when $\mathcal{H} = I_K(B_R)$, the empirical target function $f_{\mathbf{z}}$ for the regression problem can be chosen in $\mathcal{H}_{K,\mathbf{z}}$. Proposition 8.7 gave a similar statement for the regularized empirical target function $f_{\mathbf{z},\gamma}$ (and exhibited explicit expressions for the coefficients of $f_{\mathbf{z},\gamma}$ as a linear combination of $\{K_{x_1}, \ldots, K_{x_m}\}$). The proofs of Proposition 2.25 and Corollary 2.26 readily extend to show the following result.

Proposition 9.7 *Let K be a Mercer kernel on X, and ϕ a loss function. Let also $B \subseteq \mathcal{H}_K$, $\gamma > 0$, and $\mathbf{z} \in Z^m$. If $f \in \mathcal{H}_K$ is a minimizer of $\mathcal{E}_{\mathbf{z}}^\phi$ in B, then $P(f)$ is a minimizer of $\mathcal{E}_{\mathbf{z}}^\phi$ in $P(B)$. If, in addition, $P(B) \subseteq B$ and $\mathcal{E}_{\mathbf{z}}^\phi$ can be minimized in B, then such a minimizer can be chosen in $P(B)$. Similar statements hold for $\mathcal{E}_{\mathbf{z},\gamma}^\phi$.* $\qquad \blacksquare$

We can use Proposition 9.7 to state the problem of computing $f_{\mathbf{z}}^0$ as a nonlinear programming problem.

Corollary 9.8 *We can take*

$$f_{\mathbf{z}}^0 := \underset{\substack{f \in \mathcal{H}_{K,\mathbf{z}} \\ \|f\|_K = 1}}{\operatorname{argmin}} \frac{1}{m} \sum_{i=1}^m \chi_{\{y_i f(x_i) < 0\}}.$$

Proof. Let f_* be a minimizer of $\mathcal{E}_{\mathbf{z}}^{\phi_0}(f) = \frac{1}{m} \sum_{i=1}^m \chi_{\{y_i f(x_i) < 0\}}$ in $\mathcal{H}_K \cap \{f \mid \|f\|_K = 1\}$. By Proposition 9.7, $P(f_*) \in \mathcal{H}_{K,\mathbf{z}}$ satisfies $\mathcal{E}_{\mathbf{z}}^{\phi_0}(f_*) = \mathcal{E}_{\mathbf{z}}^{\phi_0}(P(f_*))$.

If $P(f_*) \neq 0$, we thus have $\mathcal{E}_{\mathbf{z}}^{\phi_0}(f_*) = \mathcal{E}_{\mathbf{z}}^{\phi_0}(P(f_*)/\|P(f_*)\|_K)$, showing that a minimizer exists in $\mathcal{H}_{K,\mathbf{z}} \cap \{f \mid \|f\|_K = 1\}$.

If $P(f_*) = 0$, then $\mathcal{E}_{\mathbf{z}}^{\phi_0}(f_*) = \mathcal{E}_{\mathbf{z}}^{\phi_0}(0) = 1$, the maximal possible error. This means that for all $f \in \mathcal{H}_K$, $\mathcal{E}_{\mathbf{z}}^{\phi_0}(f) = 1$, so we may take any function in $\mathcal{H}_{K,\mathbf{z}} \cap \{f \mid \|f\|_K = 1\}$ as a minimizer of $\mathcal{E}_{\mathbf{z}}^{\phi_0}$. $\qquad \blacksquare$

Proposition 9.7 (and Corollary 9.8 when $\phi = \phi_0$) places the problem of finding minimizers of $\mathcal{E}_z^\phi$ or $\mathcal{E}_{z,\gamma}^\phi$ in the setting of the general nonlinear programming problem. But we would actually like to deal with a programming problem for which efficient algorithms exist – for instance, a convex programming problem. This is not the case, unfortunately, for the loss function ϕ_0.

Remark 9.9 Take $\phi = \phi_0$ and consider the problem of minimizing $\mathcal{E}_z^\phi$ on $\{f \in \mathcal{H}_K \mid \|f\|_K = 1\}$. By Corollary 9.8, we can minimize on $\{f \in \mathcal{H}_{K,z} \mid \|f\|_K = 1\}$ and take

$$f_z^0 = \sum_{j=1}^m c_{z,j} K_{x_j},$$

where

$$c_z = (c_{z,1}, \ldots, c_{z,m}) = \underset{\substack{c \in \mathbb{R}^m \\ c^{\mathrm{T}} K[\mathbf{x}]c=1}}{\operatorname{argmin}} \frac{1}{m} \sum_{i=1}^m \chi_{\left\{\sum_{j=1}^m c_j y_i K(x_i,x_j) < 0\right\}}.$$

Since $S_K^{m-1} = \{c \in \mathbb{R}^m \mid c^{\mathrm{T}} K[\mathbf{x}]c = 1\}$ is not a convex subset of $\mathbb{R}^m$ and $\chi_{\{\sum_{j=1}^m c_j y_i K(x_i,x_j) < 0\}}$ may not be a convex function of $c \in S_K^{m-1}$, the optimization problem of computing c_z is not, in general, a convex programming problem.

We would like thus to replace the loss ϕ_0 by a loss ϕ that, on one hand, approximates Bayes rule – for which we will require that ϕ is close to the misclassification loss ϕ_0 – and, on the other hand, leads to a convex programming problem. Although we could do so in the setting described in Chapter 1 (we actually did it with f_z^0 above), we instead consider the regularized setting of Chapter 8.

Definition 9.10 Let K be a Mercer kernel, ϕ a loss function, $z \in Z^m$, and $\gamma > 0$. The *regularized classifier* associated with K, ϕ, z, and γ is defined as $\operatorname{sgn}(f_{z,\gamma}^\phi)$, where

$$f_{z,\gamma}^\phi := \underset{f \in \mathcal{H}_K}{\operatorname{argmin}} \left\{ \frac{1}{m} \sum_{i=1}^m \phi\big(y_i f(x_i)\big) + \gamma \|f\|_K^2 \right\}. \tag{9.6}$$

Note that (9.6) is a regularization scheme like those described in Chapter 8. The constant $\gamma > 0$ is called the *regularization parameter*, and it is often selected as a function of m, $\gamma = \gamma(m)$.

Proposition 9.11 *If $\phi : \mathbb{R} \to \mathbb{R}_+$ is convex, then the optimization problem induced by (9.6) is a convex programming one.*

Proof. According to Proposition 9.7, $f_{\mathbf{z},\gamma}^{\phi} = \sum_{j=1}^{m} c_{\mathbf{z},j} K_{x_j}$, where

$$c_{\mathbf{z}} = (c_{\mathbf{z},1}, \ldots, c_{\mathbf{z},m}) = \operatorname*{argmin}_{c \in \mathbb{R}^m} \frac{1}{m} \sum_{i=1}^{m} \phi \left(\sum_{j=1}^{m} y_i K(x_i, x_j) c_j \right)$$

$$+ \gamma \sum_{i,j=1}^{m} c_i K(x_i, x_j) c_j.$$

For each $i = 1, \ldots, m$, $\phi \left(\sum_{j=1}^{m} y_i K(x_i, x_j) c_j \right) = \phi(\mathbf{y}^{\mathrm{T}} K[\mathbf{x}] c)$ is a convex function of $c \in \mathbb{R}^m$. In addition, since K is a Mercer kernel, the Gramian matrix $K[\mathbf{x}]$ is positive semidefinite. Therefore, the function $c \mapsto c^{\mathrm{T}} K[\mathbf{x}] c$ is convex. Thus, $c_{\mathbf{z}}$ is the minimizer of a convex function. ∎

Regularized classifiers associated with general loss functions are discussed in the next chapter. In particular, we show there that the least squares loss ϕ_{ls} yields a satisfactory algorithm from the point of view of convergence rates in its error analysis. Here we restrict our exposition to a special loss, called *hinge loss*,

$$\phi_{\mathsf{h}}(t) = (1 - t)_+ = \max\{1 - t, 0\}. \tag{9.7}$$

The regularized classifier associated with the hinge loss, the *support vector machine*, has been used extensively and appears to have a small misclassification error in practice. One nice property of the hinge loss ϕ_{h}, not possessed by the least squares loss ϕ_{ls}, is the elimination of the local error when $yf(x) > 1$. This property often makes the solution $f_{\mathbf{z},\gamma}^{\phi_{\mathsf{h}}}$ of (9.6) sparse in the representation $f_{\mathbf{z},\gamma}^{\phi_{\mathsf{h}}} = \sum_{i=1}^{m} c_{\mathbf{z},i} K_{x_i}$. That is, most coefficients $c_{\mathbf{z},i}$ in this representation vanish. Hence the computation of $f_{\mathbf{z},\gamma}^{\phi_{\mathsf{h}}}$ can, in practice, be very fast. We return to this issue at the end of Section 9.4.

Although the definition of the hinge loss may not suggest at a first glance any particular reason for inducing good classifiers, it turns out that there is some geometry to explain why it may do so. We next disgress on this geometry.

9.3 Optimal hyperplanes: the separable case

Suppose $X \subseteq \mathbb{R}^n$ and $\mathbf{z} = (z_1, \ldots, z_m)$ is a sample set with $z_i = (x_i, y_i)$, $i = 1, \ldots, m$. Then $\mathbf{z}$ consists of two classes with the following sets of indices: $\mathbf{I} = \{i \mid y_i = 1\}$ and $\mathbf{II} = \{i \mid y_i = -1\}$. Let H be a hyperplane given by $w \cdot x = b$ with $w \in \mathbb{R}^n$, $\|w\| = 1$, and $b \in \mathbb{R}$. We say that $\mathbf{I}$ and $\mathbf{II}$ are *separable* by H when, for $i = 1, \ldots, m$,

$$\begin{cases} w \cdot x_i > b & \text{if } i \in \mathbf{I} \\ w \cdot x_i < b & \text{if } i \in \mathbf{II}. \end{cases}$$

That is, points x_i corresponding to $\mathbf{I}$ and $\mathbf{II}$ lie on different sides of H. We say that $\mathbf{I}$ and $\mathbf{II}$ are *separable* (or that $\mathbf{z}$ is so) when there exists a hyperplane H separating them. As shown in Figure 9.2, if w is a unit vector in $\mathbb{R}^n$, then the distance from a point $x^* \in \mathbb{R}^n$ to the plane $w \cdot x = 0$ is $\|x^*\| |\cos \theta| = \|w\| \|x^*\| |\cos \theta| = |w \cdot x^*|$. For any $b \in \mathbb{R}$, the hyperplane H given by $w \cdot x = b$ is parallel to $w \cdot x = 0$ and the distance from the point x^* to H is $|w \cdot x^* - b|$. When $w \cdot x^* - b < 0$, the point x^* lies on the side of H opposite to the direction w.

If $\mathbf{I}$ and $\mathbf{II}$ are separable by H, points x_i with $i \in \mathbf{I}$ satisfy $w \cdot x_i - b > 0$ and the point(s) in this set closest to H is (are) at a distance $b_{\mathbf{I}}(w) := \min_{i \in \mathbf{I}} \{w \cdot x_i - b\} = \min_{i \in \mathbf{I}} w \cdot x_i - b$. Similarly, points x_i with $i \in \mathbf{II}$ satisfy $w \cdot x_i - b < 0$ and the points in this set closest to H is (are) at a distance $b_{\mathbf{II}}(w) := -\max_{i \in \mathbf{II}} \{w \cdot x_i - b\} = b - \max_{i \in \mathbf{II}} w \cdot x_i$.

If we shift the separating hyperplane to $w \cdot x = c(w)$ with

$$c(w) = \tfrac{1}{2} \left\{ \min_{i \in \mathbf{I}} w \cdot x_i + \max_{i \in \mathbf{II}} w \cdot x_i \right\},$$

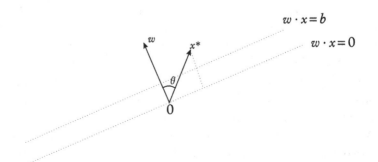

Figure 9.2

these distances become the same and equal to

$$\Delta(w) = \frac{1}{2} \left\{ \min_{i \in \mathbf{I}} w \cdot x_i - \max_{i \in \mathbf{II}} w \cdot x_i \right\}$$
$$= \frac{1}{2} \{ b_{\mathbf{I}}(w) + b + (b_{\mathbf{II}}(w) - b) \}$$
$$= \frac{1}{2} \{ b_{\mathbf{I}}(w) + b_{\mathbf{II}}(w) \} > 0.$$

Therefore, the two classes of points are separated by the hyperplane $w \cdot x = c(w)$ and satisfy

$$\begin{cases} w \cdot x_i - c(w) \geq \min_{i \in \mathbf{I}} w \cdot x_i - c(w) \\ \quad = \frac{1}{2} \{ \min_{i \in \mathbf{I}} w \cdot x_i - \max_{i \in \mathbf{II}} w \cdot x_i \} = \Delta(w) & \text{if } i \in \mathbf{I} \\ w \cdot x_i - c(w) \leq \max_{i \in \mathbf{II}} w \cdot x_i - c(w) \\ \quad = \frac{1}{2} \{ \max_{i \in \mathbf{II}} w \cdot x_i - \min_{i \in \mathbf{I}} w \cdot x_i \} = -\Delta(w) & \text{if } i \in \mathbf{II}. \end{cases}$$

Moreover, there exist points from $\mathbf{z}$ on both hyperplanes $w \cdot x = c(w) \pm \Delta(w)$ (see Figure 9.3).

The quantity $\Delta(w)$ is called the *margin* associated with the direction w, and the set $\{x \mid w \cdot x = c(w)\}$ is the associated *separating hyperplane*.

Different directions w induce different separating hyperplanes. In Figure 9.3, one can rotate w such that a hyperplane with smaller angle still separates the data, and such a separating hyperplane will have a larger margin (see Figure 9.4).

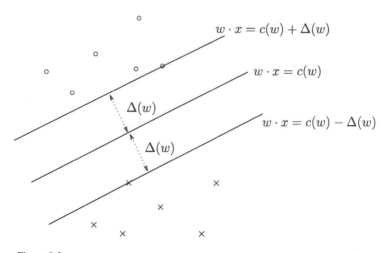

Figure 9.3

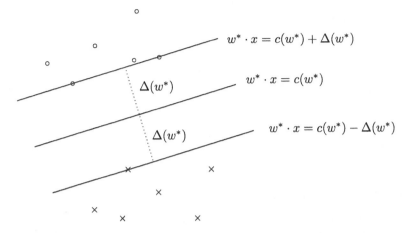

Figure 9.4

Any hyperplane in $\mathbb{R}^n$ induces a classifier. If its equation is $w \cdot x - b = 0$, then the function $x \mapsto \operatorname{sgn}(w \cdot x - b)$ is such a classifier. This reasoning suggests that the best classifier among those induced in this way may be that for which the direction w yields a separating hyperplane with the largest possible margin $\Delta(w)$. Given $\mathbf{z}$, such a direction is obtained by solving the optimization problem

$$\max_{\|w\|=1} \Delta(w)$$

or, in other words,

$$\max_{\|w\|=1} \frac{1}{2}\left\{\min_{y_i=1} w \cdot x_i - \max_{y_i=-1} w \cdot x_i\right\}. \tag{9.8}$$

If w^* is a maximizer of (9.8) with $\Delta(w^*) > 0$, then $w^* \cdot x = c(w^*)$ is called the *optimal hyperplane* and $\Delta(w^*)$ is called the (maximal) *margin of the sample*.

Theorem 9.12 *If $\mathbf{z}$ is separable with* **I** *and* **II** *both nonempty, then the optimization problem (9.8) has a unique solution w^*, $\Delta(w^*) > 0$, and the optimal separating hyperplane is given by $w^* \cdot x = c(w^*)$.*

Proof. The function $\Delta : \mathbb{R}^n \to \mathbb{R}$ defined by

$$\Delta(w) = \frac{1}{2}\left\{\min_{y_i=1} w \cdot x_i - \max_{y_i=-1} w \cdot x_i\right\}$$
$$= \min_{i \in \mathbf{I}} w \cdot \widetilde{x}_i + \min_{i \in \mathbf{II}} w \cdot \widetilde{x}_i,$$

where

$$\widetilde{x}_i = \begin{cases} \frac{1}{2}x_i & \text{if } y_i = 1 \\ -\frac{1}{2}x_i & \text{if } y_i = -1, \end{cases}$$

is continuous. Therefore, Δ achieves a maximum value over the compact set $\{w \in \mathbb{R}^n \mid \|w\| \leq 1\}$. The maximum cannot be achieved in the interior of this set; for w^* with $\|w^*\| < 1$, we have

$$\Delta \left(\frac{w^*}{\|w^*\|} \right) = \min_{i \in \mathbf{I}} \frac{w^*}{\|w^*\|} \cdot \widetilde{x}_i + \min_{i \in \mathbf{II}} \frac{w^*}{\|w^*\|} \cdot \widetilde{x}_i = \frac{1}{\|w^*\|} \Delta(w^*) > \Delta(w^*).$$

Furthermore, the maximum cannot be attained at two different points. Otherwise, for two maximizers $w_1^* \neq w_2^*$, we would have, for any $i \in \mathbf{I}$ and $j \in \mathbf{II}$,

$$w_1^* \cdot \widetilde{x}_i + w_1^* \cdot \widetilde{x}_j \geq \Delta(w_1^*), \quad w_2^* \cdot \widetilde{x}_i + w_2^* \cdot \widetilde{x}_j \geq \Delta(w_2^*) = \Delta(w_1^*),$$

which implies

$$\left(\tfrac{1}{2}w_1^* + \tfrac{1}{2}w_2^* \right) \cdot \widetilde{x}_i + \left(\tfrac{1}{2}w_1^* + \tfrac{1}{2}w_2^* \right) \cdot \widetilde{x}_j \geq \Delta(w_1^*).$$

That is, $\left(\tfrac{1}{2}w_1^* + \tfrac{1}{2}w_2^* \right)$ would be another maximizer, lying in the interior, which is not possible. ∎

For the optimal hyperplane $w^* \cdot x = c(w^*)$, all the vectors x_i satisfy

$$y_i(w^* \cdot x_i - c(w^*)) \geq \Delta(w^*)$$

no matter whether $y_i = 1$ or $y_i = -1$. The vectors x_i for which equality holds are called *support vectors*. From Figure 9.4, we see that these are points lying on the two separating hyperplanes $w^* \cdot x = c(w^*) \pm \Delta(w^*)$. The classifier $\mathbb{R}^n \to Y$ associated with w^* is given by

$$x \mapsto \text{sgn} \left(w^* \cdot x - c(w^*) \right).$$

9.4 Support vector machines

When $\mathbf{z}$ is separable, we can obtain a classifier by solving (9.8) and then taking, if w^* is the computed solution, the classifier $x \mapsto \text{sgn}(w^* \cdot x - c(w^*))$. We can also solve an equivalent form of (9.8).

Theorem 9.13 *Assume (9.8) has a solution w^* with $\Delta(w^*) > 0$. Then $w^* = \widetilde{w}/\|\widetilde{w}\|$, where $\widetilde{w}$ is a solution of*

$$\min_{w \in \mathbb{R}^n, b \in \mathbb{R}} \|w\|^2 \tag{9.9}$$
$$\text{s.t.} \qquad y_i(w \cdot x_i - b) \geq 1, \quad i = 1, \ldots, m.$$

Moreover, $\Delta(w^*) = 1/\|\widetilde{w}\|$ is the margin.

Proof. A minimizer $(\widetilde{w}, \widetilde{b})$ of the quadratic function $\|w\|^2$ subject to the linear constraints exists. Recall that

$$\Delta(w) = \tfrac{1}{2} \left\{ \min_{y_i=1} w \cdot x_i - \max_{y_i=-1} w \cdot x_i \right\}.$$

Then

$$\Delta\left(\frac{\widetilde{w}}{\|\widetilde{w}\|}\right) = \frac{1}{2} \left\{ \min_{y_i=1} \left(\frac{\widetilde{w}}{\|\widetilde{w}\|} \cdot x_i - \frac{\widetilde{b}}{\|\widetilde{w}\|}\right) \right.$$
$$\left. - \max_{y_j=-1} \left(\frac{\widetilde{w}}{\|\widetilde{w}\|} \cdot x_j - \frac{\widetilde{b}}{\|\widetilde{w}\|}\right) \right\} \geq \frac{1}{\|\widetilde{w}\|},$$

since $\widetilde{w} \cdot x_i - \widetilde{b} \geq 1$ when $y_i = 1$, and $\widetilde{w} \cdot x_j - \widetilde{b} \leq -1$ when $y_j = -1$.

We claim that $\Delta(w_0) \leq 1/\|\widetilde{w}\|$ for each unit vector w_0. If this is so, we can conclude from Theorem 9.12 that $\Delta(\widetilde{w}/\|\widetilde{w}\|) = 1/\|\widetilde{w}\| = \Delta(w^*)$ and $w^* = \widetilde{w}/\|\widetilde{w}\|$.

Suppose, to the contrary, that for some unit vector $w_0 \in \mathbb{R}^n$, $\Delta(w_0) > 1/\|\widetilde{w}\|$ holds. Consider the vector $\bar{w} = w_0/\Delta(w_0)$ together with

$$b = \frac{\tfrac{1}{2}\left(\min_{y_i=1} w_0 \cdot x_i + \max_{y_j=-1} w_0 \cdot x_j\right)}{\Delta(w_0)}.$$

They satisfy

$$\bar{w} \cdot x_i - b = \frac{w_0 \cdot x_i - \tfrac{1}{2}\left(\min_{y_i=1} w_0 \cdot x_i + \max_{y_j=-1} w_0 \cdot x_j\right)}{\Delta(w_0)}$$
$$\geq 1 \quad \text{if } y_i = 1$$

and

$$\bar{w} \cdot x_j - b = \frac{w_0 \cdot x_j - \tfrac{1}{2}\left(\min_{y_i=1} w_0 \cdot x_i + \max_{y_j=-1} w_0 \cdot x_j\right)}{\Delta(w_0)}$$
$$\leq -1 \quad \text{if } y_j = -1.$$

But $\|\bar{w}\|^2 = \|w_0\|^2/\Delta(w_0)^2 = 1/\Delta(w_0)^2 < \|\tilde{w}\|^2$, which is in contradiction with $\tilde{w}$ being a minimizer of (9.9). ∎

Thus, in the separable case, we can proceed by solving either the optimization problem (9.8) or that given by (9.9). The resulting classifier is called the *hard margin classifier*, and its margin is given by $\Delta(w^*)$ with w^* the solution of (9.8) or by $1/\|\tilde{w}\|$ with $\tilde{w}$ the solution of (9.9).

It follows from Theorem 9.12 that there are at least n support vectors. In most applications of the support vector machine, the number of support vectors is much smaller than the sample size m. This makes the algorithm solving (9.9) run faster.

Support vector machines (SVMs) consist of a family of efficient classification algorithms: the SVM hard margin classifier (9.9), which works for separable data, the SVM soft margin classifier (9.10) for nonseparable data (see next section), and the general SVM algorithm (9.6) associated with the hinge loss ϕ_h and a general Mercer kernel K. The first two classifiers can be expressed in terms of the linear kernel $K(x,y) = x \cdot y + 1$, whereas the general SVM involves general Mercer kernels: the polynomial kernel $(x \cdot y + 1)^d$ with $d \in \mathbb{N}$ or Gaussians $\exp\{-\|x - y\|^2/\sigma^2\}$ with $\sigma > 0$. These SVM algorithms share a special feature caused by the hinge loss ϕ_h: the solution $f_{\mathbf{z},\gamma}^{\phi_h} = \sum_{i=1}^m c_{\mathbf{z},i} K_{x_i}$ often has a sparse vector of coefficients $c_{\mathbf{z}} = (c_{\mathbf{z},1}, \ldots, c_{\mathbf{z},m})$, which makes the algorithm computing $c_{\mathbf{z}}$ run faster.

9.5 Optimal hyperplanes: the nonseparable case

In the nonseparable situation, there are no $w \in \mathbb{R}^n$ and $b \in \mathbb{R}$ such that the points in $\mathbf{z}$ can be separated in to two classes with $y_i = 1$ and $y_i = -1$ by the hyperplane $w \cdot x = b$. In this case, we look for the *soft margin classifier*. This is defined by introducing slack variables $\xi = (\xi_1, \ldots, \xi_m)$ and considering the problem

$$
\begin{aligned}
\min_{w \in \mathbb{R}^n, b \in \mathbb{R}, \xi \in \mathbb{R}^m} \quad & \|w\|^2 + \frac{1}{\gamma m} \sum_{i=1}^m \xi_i \\
\text{s.t.} \quad & y_i(w \cdot x_i - b) \geq 1 - \xi_i \\
& \xi_i \geq 0, \quad i = 1, \ldots, m.
\end{aligned}
\tag{9.10}
$$

Here $\gamma > 0$ is a regularization parameter. If $(\tilde{w}, \tilde{b}, \tilde{\xi})$ is a solution of (9.10), then its associated soft margin classifier is defined by $x \mapsto \text{sgn}(\tilde{w} \cdot x - \tilde{b})$.

The hard margin problem (9.9) in the separable case can be seen as a special case of the soft margin one (9.10) corresponding to $\frac{1}{\gamma} = \infty$, in which case all solutions have $\tilde{\xi} = 0$.

We claimed at the end of Section 9.2 that the regularized classifier associated with the hinge loss was related to our previous discussion of margins and separating hyperplanes. To see why this is so we next show that the soft margin classifier is a special example of (9.6). Recall that the *hinge loss* ϕ_h is defined by

$$\phi_h(t) = (1 - t)_+ = \max\{1 - t, 0\}.$$

If $(\widetilde{w}, \widetilde{b}, \widetilde{\xi})$ is a solution of (9.10), then we must have $\widetilde{\xi}_i = (1 - y_i(\widetilde{w} \cdot x_i - \widetilde{b}))_+$; that is, $\widetilde{\xi}_i = \phi_h(y_i(\widetilde{w} \cdot x_i - \widetilde{b}))$. Hence, (9.10) can be expressed by means of the loss ϕ_h as

$$\min_{w \in \mathbb{R}^n, b \in \mathbb{R}} \frac{1}{m} \sum_{i=1}^{m} \phi_h(y_i(w \cdot x_i - b)) + \gamma \|w\|^2.$$

If we consider the linear Mercer kernel K on $\mathbb{R}^n \times \mathbb{R}^n$ given by $K(x, y) = x \cdot y$, then $\mathcal{H}_K = \{w \cdot x \mid x \in \mathbb{R}^n\}$, $\|K_w\|_K^2 = \|w\|^2$, and (9.10) can be written as

$$\min_{f \in \mathcal{H}_K, b \in \mathbb{R}} \frac{1}{m} \sum_{i=1}^{m} \phi_h(y_i(f(x_i) - b)) + \gamma \|f\|_K^2. \tag{9.11}$$

The scheme (9.11) is the same as (9.6) with the linear kernel except for the constant term b, called *offset*. [1]

One motivation to consider scheme (9.6) with an arbitrary Mercer kernel is the expectation of separating data by surfaces instead of hyperplanes only. Let f be a function on $\mathbb{R}^n$, and $f(x) = 0$ the corresponding surface. The two classes **I** and **II** are separable by this surface if, for $i = 1, \ldots, m$,

$$\begin{cases} f(x_i) > 0 & \text{if } i \in \mathbf{I} \\ f(x_i) < 0 & \text{if } i \in \mathbf{II}; \end{cases}$$

that is, if $y_i f(x_i) > 0$ for $i = 1, \ldots, m$. This set of inequalities is an empirical version of the separation condition "$yf(x) > 0$ almost surely" for the probability distribution ρ on Z. Such a separation condition is more general than the separation by hyperplanes. In order to find such a separating surface using efficient algorithms (convex optimization), we require that the function f lies in an RKHS $\mathcal{H}_K$. Under such a separation condition, one may take $\gamma = 0$ and algorithm (9.6) corresponds to a hard margin classifier. This is the context of the next two sections, on error analysis.

---

[1] We could have considered the scheme (9.11) with offset. We did not do so for simplicity of exposition. References to work on the general case can be found in Section 9.8.

9.6 Error analysis for separable measures

In this section we present an error analysis for scheme (9.6) with the hinge loss $\phi_h(t) = (1 - t)_+$ for separable distributions.

Definition 9.14 Let $\mathcal{H}_K$ be an RKHS of functions on X, and ρ a probability measure on $Z = X \times Y$. We say that ρ is *strictly separable by* $\mathcal{H}_K$ *with margin* $\Delta > 0$ if there is some $f_{sp} \in \mathcal{H}_K$ such that $\|f_{sp}\|_K = 1$ and $yf_{sp}(x) \geq \Delta$ almost surely.

Remark 9.15
(i) Even under the weaker condition that $yf_{sp}(x) > 0$ almost surely (which we consider in the next section), we have $y = \text{sgn}(f_{sp}(x))$ almost surely. Hence, the variance σ_ρ^2 vanishes (i.e., ρ is noise free) and so does κ_ρ.
(ii) As a consequence of (i), $f_c = \text{sgn}(f_{sp})$.
(iii) Since f_{sp} is continuous and $|f_{sp}(x)| \geq \Delta$ almost surely, it follows that if ρ is strictly separable, then

$$\rho_X \left(\overline{T} \cap \overline{X \setminus T} \right) = 0,$$

where $T = \{x \in X \mid f_c(x) = 1\}$. This implies that if X is connected, $\rho_X(T) > 0$, and $\rho_X(X \setminus T) > 0$, then ρ is degenerate. The situation would be as in Figure 9.5, where the two dashed regions represent the support of the set T, those with dots represent the support of $X \setminus T$, and the remainder of the rectangle has measure zero (for the measure ρ_X).

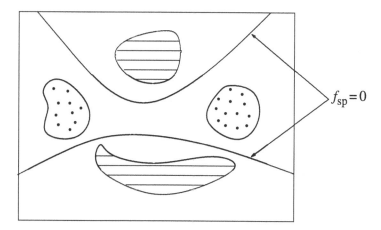

Figure 9.5

Theorem 9.16 *If ρ is strictly separable by $\mathcal{H}_K$ with margin Δ, then, for almost every $\mathbf{z} \in Z^m$,*

$$\mathcal{E}_{\mathbf{z}}^{\phi_{\mathsf{h}}} (f_{\mathbf{z},\gamma}^{\phi_{\mathsf{h}}}) \leq \frac{\gamma}{\Delta^2}$$

and $\|f_{\mathbf{z},\gamma}^{\phi_{\mathsf{h}}}\|_K \leq \frac{1}{\Delta}$.

Proof. Since $f_{\mathsf{sp}}/\Delta \in \mathcal{H}_K$, we see from the definition of $f_{\mathbf{z},\gamma}^{\phi_{\mathsf{h}}}$ that

$$\mathcal{E}_{\mathbf{z}}^{\phi_{\mathsf{h}}} (f_{\mathbf{z},\gamma}^{\phi_{\mathsf{h}}}) + \gamma \|f_{\mathbf{z},\gamma}^{\phi_{\mathsf{h}}}\|_K^2 \leq \mathcal{E}_{\mathbf{z}}^{\phi_{\mathsf{h}}} \left(\frac{f_{\mathsf{sp}}}{\Delta}\right) + \gamma \left\|\frac{f_{\mathsf{sp}}}{\Delta}\right\|_K^2.$$

But $y(f_{\mathsf{sp}}(x)/\Delta) \geq 1$ almost surely, that is, $1 - y(f_{\mathsf{sp}}(x)/\Delta) \leq 0$, so we have $\phi_{\mathsf{h}} \left(y(f_{\mathsf{sp}}(x)/\Delta)\right) = 0$ almost surely. It follows that $\mathcal{E}_{\mathbf{z}}^{\phi_{\mathsf{h}}} \left(f_{\mathsf{sp}}/\Delta\right) = 0$. Since $\left\|f_{\mathsf{sp}}/\Delta\right\|_K^2 = 1/\Delta^2$,

$$\mathcal{E}_{\mathbf{z}}^{\phi_{\mathsf{h}}} (f_{\mathbf{z},\gamma}^{\phi_{\mathsf{h}}}) + \gamma \|f_{\mathbf{z},\gamma}^{\phi_{\mathsf{h}}}\|_K^2 \leq \frac{\gamma}{\Delta^2}$$

holds and the statement follows.   ∎

The results in Chapter 8 lead us to expect the solution $f_{\mathbf{z},\gamma}^{\phi_{\mathsf{h}}}$ of (9.6) to satisfy $\mathcal{E}^{\phi_{\mathsf{h}}}(f_{\mathbf{z},\gamma}^{\phi_{\mathsf{h}}}) \rightarrow \mathcal{E}^{\phi_{\mathsf{h}}}(f_\rho^{\phi_{\mathsf{h}}})$, where $f_\rho^{\phi_{\mathsf{h}}}$ is a minimizer of $\mathcal{E}^{\phi_{\mathsf{h}}}$. We next show that this is indeed the case. To this end, we first characterize $f_\rho^{\phi_{\mathsf{h}}}$. For $x \in X$, let $\eta_x := \mathrm{Prob}_Y(y = 1 \mid x)$.

Theorem 9.17 *For any measurable function $f : X \rightarrow \mathbb{R}$*

$$\mathcal{E}^{\phi_{\mathsf{h}}}(f) \geq \mathcal{E}^{\phi_{\mathsf{h}}}(f_c)$$

holds.

That is, the Bayes rule f_c is a minimizer $f_\rho^{\phi_{\mathsf{h}}}$ of $\mathcal{E}^{\phi_{\mathsf{h}}}$.

Proof. Write $\mathcal{E}^{\phi_{\mathsf{h}}}(f) = \int_X \Phi_{\mathsf{h},x}(f(x)) \, d\rho_X$, where

$$\Phi_{\mathsf{h},x}(t) = \int_Y \phi_{\mathsf{h}}(yt) \, d\rho(y \mid x) = \phi_{\mathsf{h}}(t)\eta_x + \phi_{\mathsf{h}}(-t)(1 - \eta_x).$$

When $t = f_c(x) \in \{1, -1\}$, for $y = f_c(x)$ one finds that $yt = 1$ and $\phi_{\mathsf{h}}(yt) = 0$, whereas for $y = -f_c(x) \neq f_c(x)$, $yt = -1$ and $\phi_{\mathsf{h}}(yt) = 2$. So $\int_Y \phi_{\mathsf{h}}(yt) \, d\rho(y \mid x) = 2 \, \mathrm{Prob}(y \neq f_c(x) \mid x)$ and $\Phi_{\mathsf{h},x}(f_c(x)) = 2 \, \mathrm{Prob}(y \neq f_c(x) \mid x)$.

According to (9.2), $\mathrm{Prob}(y \neq f_c(x) \mid x) \leq \mathrm{Prob}(y = s \mid x)$ for $s = \pm 1$. Hence, $\Phi_{\mathsf{h},x}(f_c(x)) \leq 2 \, \mathrm{Prob}(y = s \mid x)$ for any $s \in \{1, -1\}$.

If $t \geq 1$, then $\phi_{\mathsf{h}}(t) = 0$ and $\Phi_{\mathsf{h},x}(t) = (1 + t)(1 - \eta_x) \geq 2(1 - \eta_x) \geq \Phi_{\mathsf{h},x}(f_c(x))$.

If $t \le -1$, then $\phi_h(-t) = 0$ and $\Phi_{h,x}(t) = (1-t)\eta_x \ge 2\eta_x \ge \Phi_{h,x}(f_c(x))$. If $-1 < t < 1$, then $\Phi_{h,x}(t) = (1-t)\eta_x + (1+t)(1-\eta_x) \ge (1-t)\frac{1}{2}\Phi_{h,x}(f_c(x)) + (1+t)\frac{1}{2}\Phi_{h,x}(f_c(x)) = \Phi_{h,x}(f_c(x))$.
Thus, we have $\Phi_{h,x}(t) \ge \Phi_{h,x}(f_c(x))$ for all $t \in \mathbb{R}$. In particular,

$$\mathcal{E}^{\phi_h}(f) = \int_X \Phi_{h,x}(f(x))\,d\rho_X \ge \int_X \Phi_{h,x}(f_c(x))\,d\rho_X = \mathcal{E}^{\phi_h}(f_c). \qquad \blacksquare$$

When ρ is strictly separable by $\mathcal{H}_K$, we see that $\mathrm{sgn}(yf_{sp}(x)) = 1$ and hence $y = \mathrm{sgn}(f_{sp}(x))$ almost surely. This means $f_c(x) = \mathrm{sgn}(f_{sp}(x))$ and $y = f_c(x)$ almost surely. In this case, we have $\mathcal{E}^{\phi_h}(f_c) = \int_Z (1 - yf_c(x))_+ = 0$. Therefore, we expect $\mathcal{E}^{\phi_h}(f_{\mathbf{z},\gamma}^{\phi_h}) \to 0$. To get error bounds showing that this is the case we write

$$\mathcal{E}^{\phi_h}(f_{\mathbf{z},\gamma}^{\phi_h}) = \mathcal{E}^{\phi_h}(f_{\mathbf{z},\gamma}^{\phi_h}) - \mathcal{E}_{\mathbf{z}}^{\phi_h}(f_{\mathbf{z},\gamma}^{\phi_h}) + \mathcal{E}_{\mathbf{z}}^{\phi_h}(f_{\mathbf{z},\gamma}^{\phi_h}) \le \mathcal{E}^{\phi_h}(f_{\mathbf{z},\gamma}^{\phi_h}) - \mathcal{E}_{\mathbf{z}}^{\phi_h}(f_{\mathbf{z},\gamma}^{\phi_h}) + \frac{\gamma}{\Delta^2}.$$
$$(9.12)$$

Here we have used the first inequality in Theorem 9.16. The second inequality of that theorem tells us that $f_{\mathbf{z},\gamma}^{\phi_h}$ lies in the set $\{f \in \mathcal{H}_K \mid \|f\|_K \le 1/\Delta\}$. So it is sufficient to estimate $\mathcal{E}^{\phi_h}(f) - \mathcal{E}_{\mathbf{z}}^{\phi_h}(f)$ for functions f in this set in some uniform way. We can use the same idea we used in Lemmas 3.18 and 3.19.

Lemma 9.18 *Suppose a random variable ξ satisfies $0 \le \xi \le M$. Denote $\mu = \mathbf{E}(\xi)$. For every $\varepsilon > 0$ and $0 < \alpha \le 1$,*

$$\mathrm{Prob}_{\mathbf{z} \in Z^m} \left\{ \frac{\mu - \frac{1}{m}\sum_{i=1}^m \xi(z_i)}{\sqrt{\mu + \varepsilon}} \ge \alpha\sqrt{\varepsilon} \right\} \le \exp\left\{ -\frac{3\alpha^2 m\varepsilon}{8M} \right\}$$

holds.

Proof. The proof follows from Lemma 3.18, since the assumption $0 \le \xi \le M$ implies $|\xi - \mu| \le M$ and $\mathbf{E}(\xi^2) \le M\mathbf{E}(\xi)$. $\blacksquare$

Lemma 9.19 *Let $\mathcal{F}$ be a subset of $\mathcal{C}(X)$ such that $\|f\|_{\mathcal{C}(X)} \le B$ for all $f \in \mathcal{F}$. Then, for every $\varepsilon > 0$ and $0 < \alpha \le 1$, we have*

$$\mathrm{Prob}_{\mathbf{z} \in Z^m} \left\{ \sup_{f \in \mathcal{F}} \frac{\mathcal{E}^{\phi_h}(f) - \mathcal{E}_{\mathbf{z}}^{\phi_h}(f)}{\sqrt{\mathcal{E}^{\phi_h}(f) + \varepsilon}} \ge 4\alpha\sqrt{\varepsilon} \right\} \le \mathcal{N}(\mathcal{F}, \alpha\varepsilon) \exp\left\{ -\frac{3\alpha^2 m\varepsilon}{8(1+B)} \right\}.$$

Proof. Let $\{f_1, \dots, f_N\}$ be an $\alpha\varepsilon$-net for $\mathcal{F}$ with $N = \mathcal{N}(\mathcal{F}, \alpha\varepsilon)$. Then, for each $f \in \mathcal{F}$, there is some $j \le N$ such that $\|f - f_j\|_{\mathcal{C}(X)} \le \alpha\varepsilon$. Since ϕ_h is Lipschitz, $|\phi_h(t) - \phi_h(t')| \le |t - t'|$ for all $t, t' \in \mathbb{R}$. Therefore,

$|\phi_{\mathsf{h}}(yf(x)) - \phi_{\mathsf{h}}(yf_j(x))| \leq \|f - f_j\|_{\mathscr{C}(X)} \leq \alpha\varepsilon$. It follows that $|\mathcal{E}^{\phi_{\mathsf{h}}}(f) - \mathcal{E}^{\phi_{\mathsf{h}}}(f_j)| \leq \alpha\varepsilon$ and $|\mathcal{E}_{\mathbf{z}}^{\phi_{\mathsf{h}}}(f) - \mathcal{E}_{\mathbf{z}}^{\phi_{\mathsf{h}}}(f_j)| \leq \alpha\varepsilon$. Hence,

$$\frac{|\mathcal{E}^{\phi_{\mathsf{h}}}(f) - \mathcal{E}^{\phi_{\mathsf{h}}}(f_j)|}{\sqrt{\mathcal{E}^{\phi_{\mathsf{h}}}(f) + \varepsilon}} \leq \alpha\sqrt{\varepsilon} \quad \text{and} \quad \frac{|\mathcal{E}_{\mathbf{z}}^{\phi_{\mathsf{h}}}(f) - \mathcal{E}_{\mathbf{z}}^{\phi_{\mathsf{h}}}(f_j)|}{\sqrt{\mathcal{E}^{\phi_{\mathsf{h}}}(f) + \varepsilon}} \leq \alpha\sqrt{\varepsilon}.$$

Also,

$$\sqrt{\mathcal{E}^{\phi_{\mathsf{h}}}(f_j) + \varepsilon} \leq \sqrt{\mathcal{E}^{\phi_{\mathsf{h}}}(f) + \varepsilon + \mathcal{E}^{\phi_{\mathsf{h}}}(f_j) - \mathcal{E}^{\phi_{\mathsf{h}}}(f)}$$

$$\leq \sqrt{\mathcal{E}^{\phi_{\mathsf{h}}}(f) + \varepsilon} + \sqrt{|\mathcal{E}^{\phi_{\mathsf{h}}}(f_j) - \mathcal{E}^{\phi_{\mathsf{h}}}(f)|}.$$

Since $\alpha \leq 1$, we have $|\mathcal{E}^{\phi_{\mathsf{h}}}(f) - \mathcal{E}^{\phi_{\mathsf{h}}}(f_j)| \leq \varepsilon \leq \varepsilon + \mathcal{E}^{\phi_{\mathsf{h}}}(f)$ and then

$$\sqrt{\mathcal{E}^{\phi_{\mathsf{h}}}(f_j) + \varepsilon} \leq 2\sqrt{\mathcal{E}^{\phi_{\mathsf{h}}}(f) + \varepsilon}. \tag{9.13}$$

Therefore,

$$\frac{\mathcal{E}^{\phi_{\mathsf{h}}}(f) - \mathcal{E}_{\mathbf{z}}^{\phi_{\mathsf{h}}}(f)}{\sqrt{\mathcal{E}^{\phi_{\mathsf{h}}}(f) + \varepsilon}} = \frac{\mathcal{E}^{\phi_{\mathsf{h}}}(f) - \mathcal{E}^{\phi_{\mathsf{h}}}(f_j)}{\sqrt{\mathcal{E}^{\phi_{\mathsf{h}}}(f) + \varepsilon}} + \frac{\mathcal{E}_{\mathbf{z}}^{\phi_{\mathsf{h}}}(f_j) - \mathcal{E}_{\mathbf{z}}^{\phi_{\mathsf{h}}}(f)}{\sqrt{\mathcal{E}^{\phi_{\mathsf{h}}}(f) + \varepsilon}}$$

$$+ \frac{\mathcal{E}^{\phi_{\mathsf{h}}}(f_j) - \mathcal{E}_{\mathbf{z}}^{\phi_{\mathsf{h}}}(f_j)}{\sqrt{\mathcal{E}^{\phi_{\mathsf{h}}}(f) + \varepsilon}} \leq 2\alpha\sqrt{\varepsilon} + \frac{\mathcal{E}^{\phi_{\mathsf{h}}}(f_j) - \mathcal{E}_{\mathbf{z}}^{\phi_{\mathsf{h}}}(f_j)}{\sqrt{\mathcal{E}^{\phi_{\mathsf{h}}}(f) + \varepsilon}}.$$

It follows that if $(\mathcal{E}^{\phi_{\mathsf{h}}}(f) - \mathcal{E}_{\mathbf{z}}^{\phi_{\mathsf{h}}}(f))/\sqrt{\mathcal{E}^{\phi_{\mathsf{h}}}(f) + \varepsilon} \geq 4\alpha\sqrt{\varepsilon}$ for some $f \in \mathcal{F}$, then

$$\frac{\mathcal{E}^{\phi_{\mathsf{h}}}(f_j) - \mathcal{E}_{\mathbf{z}}^{\phi_{\mathsf{h}}}(f_j)}{\sqrt{\mathcal{E}^{\phi_{\mathsf{h}}}(f) + \varepsilon}} \geq 2\alpha\sqrt{\varepsilon}.$$

This, together with (9.13), tells us that

$$\frac{\mathcal{E}^{\phi_{\mathsf{h}}}(f_j) - \mathcal{E}_{\mathbf{z}}^{\phi_{\mathsf{h}}}(f_j)}{\sqrt{\mathcal{E}^{\phi_{\mathsf{h}}}(f_j) + \varepsilon}} \geq \alpha\sqrt{\varepsilon}.$$

Thus,

$$\text{Prob}_{\mathbf{z} \in Z^m} \left\{ \sup_{f \in \mathcal{F}} \frac{\mathcal{E}^{\phi_{\mathsf{h}}}(f) - \mathcal{E}_{\mathbf{z}}^{\phi_{\mathsf{h}}}(f)}{\sqrt{\mathcal{E}^{\phi_{\mathsf{h}}}(f) + \varepsilon}} \geq 4\alpha\sqrt{\varepsilon} \right\}$$

$$\leq \sum_{j=1}^{N} \text{Prob}_{\mathbf{z} \in Z^m} \left\{ \frac{\mathcal{E}^{\phi_{\mathsf{h}}}(f_j) - \mathcal{E}_{\mathbf{z}}^{\phi_{\mathsf{h}}}(f_j)}{\sqrt{\mathcal{E}^{\phi_{\mathsf{h}}}(f_j) + \varepsilon}} \geq \alpha\sqrt{\varepsilon} \right\}.$$

The statement now follows from Lemma 9.18 applied to the random variable $\xi = \phi_h(yf_j(x))$, for $j = 1,\ldots,N$, which satisfies $0 \le \xi \le 1 + \|f_j\|_{\mathscr{C}(X)} \le 1 + B$. ∎

We can now derive error bounds for strictly separable measures. Recall that B_1 denotes the unit ball of $\mathcal{H}_K$ as a subset of $\mathscr{C}(X)$.

Theorem 9.20 *If ρ is strictly separable by $\mathcal{H}_K$ with margin Δ, then, for any $0 < \delta < 1$,*

$$\mathcal{E}^{\phi_h}(f_{z,\gamma}^{\phi_h}) \le 2\varepsilon^*(m,\delta) + \frac{2\gamma}{\Delta^2}$$

with confidence at least $1 - \delta$, where $\varepsilon^*(m,\delta)$ is the smallest positive solution of the inequality in ε

$$\log \mathcal{N}\left(B_1, \frac{\Delta\varepsilon}{4}\right) - \frac{3m\varepsilon}{128(1 + \mathbf{C}_K/\Delta)} \le \log \delta.$$

In addition,

(i) *If $\log \mathcal{N}(B_1, \eta) \le C_0(1/\eta)^p$ for some $p, C_0 > 0$, and all $\eta > 0$, then*

$$\varepsilon^*(m,\delta) \le 86(1 + \mathbf{C}_K/\Delta) \max \left\{ \frac{\log(1/\delta)}{m}, C_0^{1/(1+p)} \right.$$
$$\left. \left(\frac{4}{\Delta}\right)^{p/(1+p)} \left(\frac{1}{m}\right)^{1/(1+p)} \right\}.$$

(ii) *If $\log \mathcal{N}(B_1, \eta) \le C_0(\log(1/\eta))^p$ for some $p, C_0 > 0$ and all $0 < \eta < 1$, then, for $m \ge \max\{4/\Delta, 3\}$,*

$$\varepsilon^*(m,\delta) \le \frac{(\log m)^p}{m} \left\{ 1 + 43(1 + \mathbf{C}_K/\Delta)(2^p C_0 + \log(1/\delta)) \right\}.$$

Proof. We apply Lemma 9.19 to the set $\mathcal{F} = \{f \in \mathcal{H}_K \mid \|f\|_K \le \frac{1}{\Delta}\}$. Each function $f \in \mathcal{F}$ satisfies $\|f\|_{\mathscr{C}(X)} \le \mathbf{C}_K\|f\|_K \le \mathbf{C}_K/\Delta$. By Lemma 9.19, for any $0 < \alpha \le 1$, with confidence at least $1 - \mathcal{N}(\mathcal{F}, \alpha\varepsilon) \exp\{-3\alpha^2 m\varepsilon/(8(1 + \mathbf{C}_K/\Delta))\}$, we have

$$\sup_{f \in \mathcal{F}} \frac{\mathcal{E}^{\phi_h}(f) - \mathcal{E}_z^{\phi_h}(f)}{\sqrt{\mathcal{E}^{\phi_h}(f) + \varepsilon}} \le 4\alpha\sqrt{\varepsilon}.$$

In particular, the function $f_{\mathbf{z},\gamma}^{\phi_h}$, which belongs to $\mathcal{F}$ by Theorem 9.16, satisfies

$$\frac{\mathcal{E}^{\phi_h}(f_{\mathbf{z},\gamma}^{\phi_h}) - \mathcal{E}_{\mathbf{z}}^{\phi_h}(f_{\mathbf{z},\gamma}^{\phi_h})}{\sqrt{\mathcal{E}^{\phi_h}(f_{\mathbf{z},\gamma}^{\phi_h}) + \varepsilon}} \le 4\alpha\sqrt{\varepsilon}.$$

This, together with (9.12), yields

$$\mathcal{E}^{\phi_h}(f_{\mathbf{z},\gamma}^{\phi_h}) \le 4\alpha\sqrt{\varepsilon}\sqrt{\mathcal{E}^{\phi_h}(f_{\mathbf{z},\gamma}^{\phi_h}) + \varepsilon} + \frac{\gamma}{\Delta^2}.$$

If we denote $t = \sqrt{\mathcal{E}^{\phi_h}(f_{\mathbf{z},\gamma}^{\phi_h}) + \varepsilon}$, this inequality becomes

$$t^2 - 4\alpha\sqrt{\varepsilon}\,t - \left(\varepsilon + \frac{\gamma}{\Delta^2}\right) \le 0.$$

Solving the associated quadratic equation and taking into account that $t = \sqrt{\mathcal{E}^{\phi_h}(f_{\mathbf{z},\gamma}^{\phi_h}) + \varepsilon} \ge 0$, we deduce that

$$0 \le t \le 2\alpha\sqrt{\varepsilon} + \sqrt{(2\alpha\sqrt{\varepsilon})^2 + \varepsilon + \frac{\gamma}{\Delta^2}}.$$

Hence, using the elementary inequality $(a + b)^2 \le 2a^2 + 2b^2$, we obtain

$$\mathcal{E}^{\phi_h}(f_{\mathbf{z},\gamma}^{\phi_h}) = t^2 - \varepsilon \le 2(4\alpha^2\varepsilon) + 2\left((2\alpha\sqrt{\varepsilon})^2 + \varepsilon + \frac{\gamma}{\Delta^2}\right) - \varepsilon = 16\alpha^2\varepsilon + \varepsilon + \frac{2\gamma}{\Delta^2}.$$

Set $\alpha = \frac{1}{4}$ and $\varepsilon = \varepsilon^*(m,\delta)$ as in the statement. Then the confidence $1 - \mathcal{N}(\mathcal{F}, \alpha\varepsilon)\exp\{-3\alpha^2 m\varepsilon/(8(1 + \mathbf{C}_K/\Delta))\}$ is at least $1 - \delta$, and with this confidence we have

$$\mathcal{E}^{\phi_h}(f_{\mathbf{z},\gamma}^{\phi_h}) \le 2\varepsilon^*(m,\delta) + \frac{2\gamma}{\Delta^2}.$$

(i) If $\log\mathcal{N}(B_1, \eta) \le C_0(1/\eta)^p$, then $\varepsilon^*(m,\delta) \le \varepsilon^*$, where ε^* satisfies

$$C_0\left(\frac{4}{\Delta\varepsilon}\right)^p - \frac{3m\varepsilon}{128(1 + \mathbf{C}_K/\Delta)} = \log\delta.$$

This equation can be written as

$$\varepsilon^{1+p} - \frac{128(1 + \mathbf{C}_K/\Delta)}{3m}\log\frac{1}{\delta}\varepsilon^p - \frac{128(1 + \mathbf{C}_K/\Delta)C_0}{3m}\left(\frac{4}{\Delta}\right)^p = 0.$$

Then Lemma 7.2 with $d = 2$ yields

$$\varepsilon^*(m, \delta) \leq \varepsilon^* \leq \max \left\{ \frac{256(1 + \mathbf{C}_K/\Delta)}{3m} \log \frac{1}{\delta}, \right.$$
$$\left. \left(\frac{256(1 + \mathbf{C}_K/\Delta)C_0}{3} \right)^{1/(1+p)} \left(\frac{4}{\Delta} \right)^{p/(1+p)} m^{-1/(1+p)} \right\}.$$

(ii) If $\log \mathcal{N}(B_1, \eta) \leq C_0(\log(1/\eta))^p$, then $\varepsilon^*(m, \delta) \leq \varepsilon^*$, where ε^* satisfies

$$C_0 \left(\log \frac{4}{\Delta \varepsilon} \right)^p - \frac{3m\varepsilon}{128(1 + \mathbf{C}_K/\Delta)} = \log \delta.$$

The function $h : \mathbb{R}_+ \to \mathbb{R}$ defined by $h(\varepsilon) = C_0 \left(\log \frac{4}{\Delta \varepsilon} \right)^p - 3m\varepsilon / (128(1 + \mathbf{C}_K/\Delta))$ is decreasing. Take

$$A = \max \left\{ \frac{(2^p C_0 + \log(1/\delta))128(1 + \mathbf{C}_K/\Delta)}{3}, 1 \right\}$$

and $\varepsilon = A(\log m)^p / m$. Then, for $m \geq \max\{4/\Delta, 3\}$,

$$\frac{A(\log m)^p}{m} \geq \frac{1}{m} \quad \text{and} \quad \log \frac{4}{\Delta} + \log m \leq 2 \log m.$$

It follows that

$$h \left(\frac{A(\log m)^p}{m} \right) \leq C_0 \left(\log \frac{4}{\Delta} + \log m \right)^p - (\log m)^p \left(2^p C_0 + \log \frac{1}{\delta} \right)$$
$$\leq -(\log m)^p \log \frac{1}{\delta} \leq \log \delta.$$

Since h is decreasing, we have

$$\varepsilon^*(m, \delta) \leq \varepsilon^* \leq \frac{A(\log m)^p}{m}. \qquad \blacksquare$$

In order to obtain estimates for the misclassification error from the generalization error, we need to compare $\mathcal{R}$ with $\mathcal{E}^{\phi_h}$. This is simple.

Theorem 9.21 *For any measure ρ and any measurable function $f : X \to \mathbb{R}$,*

$$\mathcal{R}(\mathrm{sgn}(f)) - \mathcal{R}(f_c) \leq \mathcal{E}^{\phi_h}(f) - \mathcal{E}^{\phi_h}(f_c).$$

Proof. Denote $X_c = \{x \in X \mid \text{sgn}(f)(x) \neq f_c(x)\}$. By the definition of the misclassification error,

$$\mathcal{R}(\text{sgn}(f)) - \mathcal{R}(f_c) = \int_{X_c} \Pr_{Y}(y \neq \text{sgn}(f)(x) \mid x)$$
$$- \Pr_{Y}(y \neq f_c(x) \mid x) \, d\rho_X.$$

For a point $x \in X_c$, we know that $\text{Prob}_Y(y \neq \text{sgn}(f)(x) \mid x) = \text{Prob}_Y(y = f_c(x) \mid x)$. Hence, $\text{Prob}_Y(y \neq \text{sgn}(f)(x) \mid x) - \text{Prob}_Y(y \neq f_c(x) \mid x) = f_\rho(x)$ or $-f_\rho(x)$ according to whether $f_\rho(x) \geq 0$. It follows that $|f_\rho(x)| = \text{Prob}_Y(y = f_c(x) \mid x) - \text{Prob}_Y(y \neq f_c(x) \mid x)$ and, therefore,

$$\mathcal{R}(\text{sgn}(f)) - \mathcal{R}(f_c) = \int_{X_c} |f_\rho(x)| \, d\rho_X. \tag{9.14}$$

By the definition of ϕ_h,

$$\phi_h(yf_c(x)) = (1 - yf_c(x))_+ = \begin{cases} 0 & \text{if } y = f_c(x) \\ 2 & \text{if } y \neq f_c(x). \end{cases}$$

Hence $\mathcal{E}^{\phi_h}(f_c) = \int_X \int_Y \phi_h(yf_c(x)) \, d\rho(y|x) \, d\rho_X = \int_X 2\,\text{Prob}_Y(y \neq f_c(x)|x) \, d\rho_X$. Furthermore, $\mathcal{E}^{\phi_h}(f) = \int_X \int_Y \phi_h(yf(x)) \, d\rho(y|x) \, d\rho_X$. Thus, it is sufficient for us to prove that

$$\int_Y \phi_h(yf(x)) \, d\rho(y|x) - 2\Pr_{Y}(y \neq f_c(x)|x) \geq |f_\rho(x)|, \quad \forall x \in X_c. \tag{9.15}$$

We prove (9.15) in two cases.

If $|f(x)| > 1$, then $x \in X_c$ implies that $\text{sgn}(f(x)) \neq f_c(x)$ and $\phi_h(-f_c(x) f(x)) = 0$. Hence,

$$\int_Y \phi_h(yf(x)) \, d\rho(y|x) = \phi_h(f_c(x)f(x)) \Pr_{Y}(y = f_c(x)|x)$$
$$= (1 - f_c(x)f(x)) \Pr_{Y}(y = f_c(x)|x).$$

Since $-f_c(x)f(x) = |f(x)| > 1$, it follows that

$$\int_Y \phi_h(yf(x)) \, d\rho(y|x) - 2\Pr_{Y}(y \neq f_c(x)|x)$$
$$\geq (1 + |f(x)|) \left(\Pr_{Y}(y = f_c(x)|x) - \Pr_{Y}(y \neq f_c(x)|x) \right)$$
$$= (1 + |f(x)|)|f_\rho(x)| \geq |f_\rho(x)|.$$

If $|f(x)| \leq 1$, then

$$\int_Y \phi_\mathsf{h}(yf(x)) \, d\rho(y|x) - 2 \operatorname*{Prob}_Y(y \neq f_c(x)|x)$$

$$= (1 - f_c(x)f(x)) \operatorname*{Prob}_Y(y = f_c(x)|x) + (1 + f_c(x)f(x)) \operatorname*{Prob}_Y(y \neq f_c(x)|x)$$

$$- 2 \operatorname*{Prob}_Y(y \neq f_c(x)|x)$$

$$= \operatorname*{Prob}_Y(y = f_c(x)|x) - \operatorname*{Prob}_Y(y \neq f_c(x)|x)$$

$$+ f_c(x)f(x) \left(\operatorname*{Prob}_Y(y \neq f_c(x)|x) - \operatorname*{Prob}_Y(y = f_c(x)|x) \right)$$

$$= (1 - f_c(x)f(x))|f_\rho(x)|.$$

But $x \in X_c$ implies that $f_c(x)f(x) \leq 0$ and $1 - f_c(x)f(x) = 1 + |f(x)|$. So in this case we have

$$\int_Y \phi_\mathsf{h}(yf(x)) \, d\rho(y|x) - 2 \operatorname*{Prob}_Y(y \neq f_c(x)|x) = (1 + |f(x)|)|f_\rho(x)| \geq |f_\rho(x)|.$$

∎

Combining Theorems 9.20 and 9.21, we can derive bounds for the misclassification error for the support vector machine soft margin classifier for strictly separable measures satisfying $\mathcal{R}(f_c) = \mathcal{E}^{\phi_\mathsf{h}}(f_c) = 0$.

Corollary 9.22 *Assume ρ is strictly separable by $\mathcal{H}_K$ with margin Δ.*

(i) *If $\log \mathcal{N}(B_1, \eta) \leq C_0(1/\eta)^p$ for some $p, C_0 > 0$ and all $\eta > 0$, then, with confidence $1 - \delta$,*

$$\mathcal{R}(\mathrm{sgn}(f_{\mathbf{z},\gamma}^{\phi_\mathsf{h}})) \leq \frac{2\gamma}{\Delta^2} + 172(1 + \mathbf{C}_K/\Delta)$$

$$\max \left\{ \frac{\log(1/\delta)}{m}, C_0^{1/(1+p)} \left(\frac{4}{\Delta} \right)^{p/(1+p)} \left(\frac{1}{m} \right)^{1/(1+p)} \right\}.$$

(ii) *If $\log \mathcal{N}(B_1, \eta) \leq C_0(\log(1/\eta))^p$ for some $p, C_0 > 0$ and all $0 < \eta < 1$, then, for $m \geq \max\{4/\Delta, 3\}$, with confidence $1 - \delta$,*

$$\mathcal{R}(\mathrm{sgn}(f_{\mathbf{z},\gamma}^{\phi_\mathsf{h}})) \leq \frac{2\gamma}{\Delta^2} + \frac{(\log m)^p}{m} \left\{ 2 + 86(1 + \mathbf{C}_K/\Delta)(2^p C_0 + \log(1/\delta)) \right\}. \quad ∎$$

It follows from Corollary 9.22 that for strictly separable measures, we may take $\gamma = 0$. In this case, the penalized term in (9.6) vanishes and the soft margin classifier becomes a hard margin one.

9.7 Weakly separable measures

We continue our discussion on separable measures. By abandoning the positive margin $\Delta > 0$ assumption, we consider weakly separable measures.

Definition 9.23 We say that ρ is *weakly separable by* $\mathcal{H}_K$ if there is some function $f_{\text{sp}} \in \mathcal{H}_K$ satisfying $\|f_{\text{sp}}\|_K = 1$ and $y f_{\text{sp}}(x) > 0$ almost surely. It has *separation triple* $(\theta, \Delta, C_0) \in (0, \infty] \times (0, \infty)^2$ if, for all $t > 0$,

$$\rho_X\{x \in X : |f_{\text{sp}}(x)| < \Delta t\} \leq C_0 t^\theta. \tag{9.16}$$

The largest θ for which there are positive constants Δ, C_0 such that (θ, Δ, C_0) is a separation triple is called the *separation exponent* of ρ (w.r.t. $\mathcal{H}_K$ and f_{sp}).

Remark 9.24 Note that when $\theta = \infty$, condition (9.16) is the same as $\rho_X\{x \in X : |f_{\text{sp}}(x)| < \Delta t\} = 0$ for all $0 < t < 1$. That is, $|f_{\text{sp}}(x)| \geq \Delta$ almost surely. But $y f_{\text{sp}}(x) > 0$ almost surely. So a weakly separable measure with $\theta = \infty$ is exactly a strictly separable measure with margin Δ.

Lemma 9.25 *Assume ρ is weakly separable by $\mathcal{H}_K$ with separation triple (θ, Δ, C_0). Take*

$$f_\gamma = \left(\frac{C_0}{\gamma}\right)^{1/(2+\theta)} \Delta^{-\theta/(2+\theta)} f_{\text{sp}}. \tag{9.17}$$

Then

$$\mathcal{E}^{\phi_h}(f_\gamma) + \gamma \|f_\gamma\|_K^2 \leq 2C_0^{2/(2+\theta)} \left(\frac{\gamma}{\Delta^2}\right)^{\theta/(2+\theta)}.$$

Proof. Write $f_\gamma = f_{\text{sp}}/\Delta t$, with $t > 0$ to be determined. Since $y f_{\text{sp}}(x) > 0$ almost surely, the same holds for $y f_\gamma(x) > 0$. Hence, $\phi_h(y f_\gamma(x)) < 1$ and $\phi_h(y f_\gamma(x)) > 0$ only if $y f_\gamma(x) < 1$, that is, if $|f_\gamma(x)| = |f_{\text{sp}}(x)/\Delta t| < 1$. Therefore,

$$\mathcal{E}^{\phi_h}(f_\gamma) = \int_Z \phi_h(y f_\gamma(x)) \, d\rho = \int_X \phi_h(|f_\gamma(x)|) \, d\rho_X$$

$$= \int_{|f_\gamma(x)| < 1} (1 - |f_\gamma(x)|) \, d\rho_X \leq \rho_X\{x \in X : |f_\gamma(x)|$$

$$< 1\} \leq \rho_X\{x \in X : |f_{\text{sp}}(x)| < \Delta t\} \leq C_0 t^\theta.$$

But $\gamma \|f_\gamma\|_K^2 = \gamma 1/(\Delta t)^2$. Setting $t = \left(\gamma/C_0\Delta^2\right)^{1/(2+\theta)}$ proves our statement. ∎

Theorem 9.26 *If ρ is weakly separable by $\mathcal{H}_K$ with a separation triple (θ, Δ, C_0) then, for any $0 < \delta < 1$, with confidence $1 - \delta$, we have*

$$\mathcal{R}(\mathrm{sgn}(f_{\mathbf{z},\gamma}^{\phi_h})) \leq 2\varepsilon^*(m, \delta, \gamma) + 8C_0^{2/(2+\theta)} \left(\frac{\gamma}{\Delta^2}\right)^{\theta/(2+\theta)} + \frac{3\log(2/\delta)}{m},$$

where $\varepsilon^*(m, \delta, \gamma)$ is the smallest positive solution of the inequality

$$\log \mathcal{N}\left(B_1, \frac{\varepsilon}{4R}\right) - \frac{3m\varepsilon}{128(1 + C_K R)} \leq \log \frac{\delta}{2}$$

with $R = 2C_0^{1/(2+\theta)} \Delta^{-\theta/(2+\theta)} \gamma^{-1/(2+\theta)} + \sqrt{\frac{2\log(2/\delta)}{m\gamma}}$.
In addition,

(i) *If $\log \mathcal{N}(B_1, \eta) \leq C_0(1/\eta)^p$ for some $p, C_0 > 0$ and all $\eta > 0$, then, taking $\gamma = m^{-\beta}$ with $0 < \beta < \frac{2+\theta}{\max\{\theta, 1+2p\}}$, we have, with confidence $1 - \delta$,*

$$\mathcal{R}(\mathrm{sgn}(f_{\mathbf{z},\gamma}^{\phi_h})) \leq \widetilde{C}\left(\left(\frac{1}{m}\right)^r \left(\log \frac{2}{\delta}\right)^2\right)$$

with $r = \min\left\{\frac{2+\theta-\beta}{2+\theta}, \frac{2+\theta-\beta-2p\beta}{(2+\theta)(1+p)}, \frac{\theta\beta}{2+\theta}\right\}$ and $\widetilde{C}$ a constant independent of m and δ.

(ii) *If $\log \mathcal{N}(B_1, \eta) \leq C_0(\log(1/\eta))^p$ for some $p, C_0 > 0$ and all $0 < \eta < 1$, then, for $m \geq \max\{4/\Delta, 3\}$, taking $\gamma = m^{-(2+\theta)/(1+\theta)}$, we have, with confidence $1 - \delta$,*

$$\mathcal{R}(\mathrm{sgn}(f_{\mathbf{z},\gamma}^{\phi_h})) \leq \widetilde{C}\left(\left(\frac{1}{m}\right)^{\theta/(1+\theta)} (\log m)^p \left(\log \frac{2}{\delta}\right)^2\right).$$

Proof. By Theorem 9.21, it is sufficient to bound $\mathcal{E}^{\phi_h}(f_{\mathbf{z},\gamma}^{\phi_h})$ as stated, since $y = \mathrm{sgn}(f_{\mathrm{sp}}(x)) = f_c(x)$ almost surely and therefore $\mathcal{E}^{\phi_h}(f_c) = 0$ and $\mathcal{R}(f_c) = 0$.

Choose f_γ by (9.17). Decompose $\mathcal{E}^{\phi_h}(f_{\mathbf{z},\gamma}^{\phi_h}) + \gamma \|f_{\mathbf{z},\gamma}^{\phi_h}\|_K^2$ as

$$\mathcal{E}^{\phi_h}(f_{\mathbf{z},\gamma}^{\phi_h}) - \mathcal{E}_{\mathbf{z}}^{\phi_h}(f_{\mathbf{z},\gamma}^{\phi_h}) + \left\{\mathcal{E}_{\mathbf{z}}^{\phi_h}(f_{\mathbf{z},\gamma}^{\phi_h}) + \gamma \|f_{\mathbf{z},\gamma}^{\phi_h}\|_K^2 - \left(\mathcal{E}_{\mathbf{z}}^{\phi_h}(f_\gamma) + \gamma \|f_\gamma\|_K^2\right)\right\}$$

$$+ \mathcal{E}_{\mathbf{z}}^{\phi_h}(f_\gamma) + \gamma \|f_\gamma\|_K^2.$$

Since the middle term is at most zero by (9.6), we have

$$\mathcal{E}^{\phi_h}(f_{\mathbf{z},\gamma}^{\phi_h}) + \gamma \|f_{\mathbf{z},\gamma}^{\phi_h}\|_K^2 \leq \left\{\mathcal{E}^{\phi_h}(f_{\mathbf{z},\gamma}^{\phi_h}) - \mathcal{E}_{\mathbf{z}}^{\phi_h}(f_{\mathbf{z},\gamma}^{\phi_h})\right\} + \left\{\mathcal{E}_{\mathbf{z}}^{\phi_h}(f_\gamma) + \gamma \|f_\gamma\|_K^2\right\}.$$

To bound the last term, consider the random variable $\xi = \phi_h(yf_\gamma(x))$. Since $yf_\gamma(x) > 0$ almost surely, we have $0 \leq \xi \leq 1$. Also, $\sigma^2(\xi) \leq \mathbf{E}(\xi) = \mathcal{E}^{\phi_h}(f_\gamma)$. Apply the one-side Bernstein inequality to ξ and deduce, for each $\varepsilon > 0$,

$$\operatorname*{Prob}_{\mathbf{z} \in Z^m} \left\{ \mathcal{E}_{\mathbf{z}}^{\phi_h}(f_\gamma) - \mathcal{E}^{\phi_h}(f_\gamma) \leq \varepsilon \right\} \geq 1 - \exp\left\{ -\frac{m\varepsilon^2}{2(\mathcal{E}^{\phi_h}(f_\gamma) + \frac{1}{3}\varepsilon)} \right\}.$$

By solving the quadratic equation

$$-\frac{m\varepsilon^2}{2(\mathcal{E}^{\phi_h}(f_\gamma) + \frac{1}{3}\varepsilon)} = \log\frac{\delta}{2},$$

we conclude that, with confidence $1 - \frac{\delta}{2}$,

$$\mathcal{E}_{\mathbf{z}}^{\phi_h}(f_\gamma) - \mathcal{E}^{\phi_h}(f_\gamma) \leq \frac{\frac{1}{3}\log\frac{2}{\delta} + \sqrt{\left(\frac{1}{3}\log\frac{2}{\delta}\right)^2 + 2m\mathcal{E}^{\phi_h}(f_\gamma)\log\frac{2}{\delta}}}{m}$$

$$\leq \frac{7\log\frac{2}{\delta}}{6m} + \mathcal{E}^{\phi_h}(f_\gamma).$$

Thus, there exists a subset U_1 of Z^m with $\rho(U_1) \geq 1 - \frac{\delta}{2}$ such that

$$\mathcal{E}_{\mathbf{z}}^{\phi_h}(f_\gamma) \leq 2\mathcal{E}^{\phi_h}(f_\gamma) + \frac{7\log\frac{2}{\delta}}{6m}, \quad \forall \mathbf{z} \in U_1.$$

Then, by Lemma 9.25, for all $\mathbf{z} \in U_1$,

$$\mathcal{E}_{\mathbf{z}}^{\phi_h}(f_{\mathbf{z},\gamma}^{\phi_h}) + \gamma\|f_{\mathbf{z},\gamma}^{\phi_h}\|_K^2 \leq \mathcal{E}_{\mathbf{z}}^{\phi_h}(f_\gamma) + \gamma\|f_\gamma\|_K^2$$

$$\leq 4C_0^{2/(2+\theta)}\left(\frac{\gamma}{\Delta^2}\right)^{\theta/(2+\theta)} + \frac{7\log\frac{2}{\delta}}{6m}. \tag{9.18}$$

In particular, taking $R = 2C_0^{1/(2+\theta)}\Delta^{-\theta/(2+\theta)}\gamma^{-1/(2+\theta)} + \sqrt{\frac{2\log(2/\delta)}{m\gamma}}$, we have, for all $\mathbf{z} \in U_1$,

$$f_{\mathbf{z},\gamma}^{\phi_h} \in \mathcal{F} = \{f \in \mathcal{H}_K : \|f\|_K \leq R\}, \quad \forall \mathbf{z} \in U_1.$$

Now apply Lemma 9.19 to the set $\mathcal{F}$ with $\alpha = \frac{1}{4}$. We have $\mathcal{N}(\mathcal{F}, \frac{\varepsilon}{4}) = \mathcal{N}(B_1, \frac{\varepsilon}{4R})$ and $\|f\|_{\mathscr{C}(X)} \leq C_K R$ for all $f \in \mathcal{F}$. By Lemma 9.19, we can find a subset U_2 of Z^m with $\rho(U_2) \geq 1 - \mathcal{N}(B_1, \frac{\varepsilon}{4R}) \exp\{-3m\varepsilon/(128(1 + \mathbf{C}_K R))\}$ such that, for all $f \in \mathcal{F}$,

$$\frac{\mathcal{E}^{\phi_h}(f) - \mathcal{E}_{\mathbf{z}}^{\phi_h}(f)}{\sqrt{\mathcal{E}^{\phi_h}(f) + \varepsilon}} \leq \sqrt{\varepsilon}.$$

In particular, when $\mathbf{z} \in U_1 \cap U_2$, we have $f_{\mathbf{z},\gamma}^{\phi_h} \in \mathcal{F}$ and, hence,

$$\mathcal{E}^{\phi_h}(f_{\mathbf{z},\gamma}^{\phi_h}) - \mathcal{E}_{\mathbf{z}}^{\phi_h}(f_{\mathbf{z},\gamma}^{\phi_h}) \leq \sqrt{\varepsilon}\sqrt{\mathcal{E}^{\phi_h}(f_{\mathbf{z},\gamma}^{\phi_h}) + \varepsilon} \leq \tfrac{1}{2}\mathcal{E}^{\phi_h}(f_{\mathbf{z},\gamma}^{\phi_h}) + \varepsilon.$$

Take $\varepsilon = \varepsilon^*(m, \delta, \gamma)$ to be the smallest positive solution of

$$\log \mathcal{N}\left(B_1, \frac{\varepsilon}{4R}\right) - \frac{3m\varepsilon}{128(1 + \mathbf{C}_K R)} \leq \log \frac{\delta}{2}.$$

Then $\rho(U_2) \geq 1 - \frac{\delta}{2}$ and, for $\mathbf{z} \in U_1 \cap U_2$, (9.18) implies

$$\mathcal{E}^{\phi_h}(f_{\mathbf{z},\gamma}^{\phi_h}) \leq \frac{1}{2}\mathcal{E}^{\phi_h}(f_{\mathbf{z},\gamma}^{\phi_h}) + \varepsilon^*(m, \delta, \gamma) + 4C_0^{2/(2+\theta)}\left(\frac{\gamma}{\Delta^2}\right)^{\theta/(2+\theta)} + \frac{7 \log \frac{2}{\delta}}{6m}.$$

It follows that

$$\mathcal{E}^{\phi_h}(f_{\mathbf{z},\gamma}^{\phi_h}) \leq 2\varepsilon^*(m, \delta, \gamma) + 8C_0^{2/(2+\theta)}\left(\frac{\gamma}{\Delta^2}\right)^{\theta/(2+\theta)} + \frac{7 \log \frac{2}{\delta}}{3m}.$$

Since $\rho(U_1 \cap U_2) \geq 1 - \delta$, our first statement holds. The rest of the result, statements (i) and (ii), follows from Theorem 9.20 after replacing Δ by $\frac{1}{R}$ and δ by $\frac{\delta}{2}$. ∎

9.8 References and additional remarks

The support vector machine was introduced by Vapnik and his collaborators. It appeared in [20] with polynomial kernels $K(x, y) = (1 + x \cdot y)^d$ and in [35] with general Mercer kernels. More details about the algorithm for solving optimization problem (9.11) can be found in [37, 107, 134, 152].

Proposition 9.3 and some other properties of the Bayes rule can be found in [44]. Proposition 9.7 (a representer theorem) can be found in [137]. The material in Sections 9.3–9.5 is taken from [134].

Theorem 9.17 was proved in [138] and Theorem 9.21 in [154]. The idea of comparing excess errors also appeared in [80].

The error analysis for support vector machines and strictly separable distributions was already well understood in the early works on support vector machines (see [134, 37]). The concept of weakly separable distribution was introduced, and the error analysis for such a distribution was performed, in [31].

When the support vector machine soft margin classifier contains an offset term b as in (9.11), the algorithm is more flexible and more general data can be separated. But the error analysis is more complex than for scheme (9.6), which has no offset. The bound for $\|f_{\mathbf{z},\gamma}^{\phi_h}\|_K$ becomes larger than those shown in Theorem 9.16 and (9.18). But the approach we have used for scheme (9.6) can be applied as well and a similar error analysis can be performed. For details, see [31].

10

General regularized classifiers

In Chapter 9 we saw that solving classification problems amounts to approximating the Bayes rule f_c (w.r.t. the misclassification error) and we described a learning algorithm, the support vector machine, producing such aproximations from a sample $\mathbf{z}$, a Mercer kernel K, and a regularization parameter $\gamma > 0$. The main result in Chapter 9 estimated the quality of the approximations obtained under a separability hypothesis on ρ. The classifier produced by the support vector machine is the regularized classifier associated with $\mathbf{z}$, K, γ, and a particular loss function, the hinge loss ϕ_h. Recall that for a loss function ϕ, this regularized classifier is given by $\mathrm{sgn}(f_{\mathbf{z},\gamma}^\phi)$, with

$$f_{\mathbf{z},\gamma}^\phi := \underset{f \in \mathcal{H}_K}{\mathrm{argmin}} \left\{ \frac{1}{m} \sum_{i=1}^{m} \phi(y_i f(x_i)) + \gamma \|f\|_K^2 \right\}. \tag{10.1}$$

In this chapter we extend this development in two ways. First, we remove the separability assumption. Second, we replace the hinge loss ϕ_h by arbitrary loss functions within a certain class. Note that it would not be of interest to consider completely arbitrary loss functions, since many such functions would lead to optimization problems (10.1) for which no efficient algorithm is known. The following definition yields an intermediate class of loss functions.

Definition 10.1 We say that $\phi : \mathbb{R} \to \mathbb{R}_+$ is a *classifying loss (function)* if it is convex and differentiable at 0 with $\phi'(0) < 0$, and if the smallest zero of ϕ is 1.

Examples of classifying loss functions are the least squares loss $\phi_{ls}(t)$, the hinge loss ϕ_h, and, for $1 \leq q < \infty$, the *q-norm (support vector machine) loss* defined by $\phi_q(t) := (\phi_h(t))^q$.

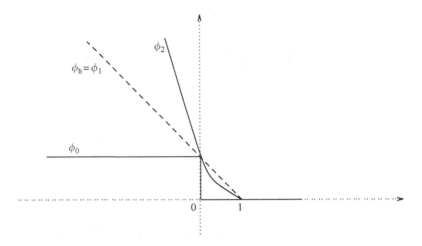

Figure 10.1

Note that Proposition 9.11 implies that optimization problem (10.1) for a classifying loss function is a convex programming problem. One special feature shared by ϕ_{ls}, ϕ_h, and $\phi_2 = (\phi_h)^2$ is that their associated convex programming problems are quadratic programming problems. This allows for many efficient algorithms to be applied when computing a solution of (10.1). Note that ϕ_{ls} differs from ϕ_2 by the addition of a symmetric part on the right of 1.

Figure 10.1 shows the shape of some of these classifying loss functions (together with that of ϕ_0).

Our goal, as in previous chapters, is to understand how close $\mathrm{sgn}(f^{\phi}_{\mathbf{z},\gamma})$ is to f_c (w.r.t. the misclassification error). In other words, we want to estimate the *excess misclassification error* $\mathcal{R}(\mathrm{sgn}(f^{\phi}_{\mathbf{z},\gamma})) - \mathcal{R}(f_c)$. Note that in Chapter 9 we had $\mathcal{R}(f_c) = 0$ because of the separability assumption. This is no longer the case. The main result in this chapter, Theorem 10.24, this goal achieves for various kernels K and classifying loss functions.

The following two theorems, easily derived from Theorem 10.24, become specific for $\mathscr{C}^{\infty}$ kernels and the hinge loss ϕ_h and the least squares loss ϕ_{ls}, respectively.

Theorem 10.2 *Assume that $X \subseteq \mathbb{R}^n$, K is $\mathscr{C}^{\infty}$ in $X \times X$, and, for some $\beta > 0$,*

$$\inf_{f \in \mathcal{H}_K} \{ \| f - f_c \|_{\mathscr{L}^1_{\rho_X}} + \gamma \| f \|_K^2 \} = \mathcal{O}(\gamma^{\beta}). \tag{10.2}$$

Choose $\gamma = m^{-2/(1+\beta)}$. *Then, for any* $0 < \varepsilon < \frac{1}{2}$ *and* $0 < \delta < 1$, *with confidence* $1 - \delta$,

$$\mathcal{R}(\text{sgn}(f_{z,\gamma}^{\phi_h})) - \mathcal{R}(f_c) \leq \tilde{C} \log \frac{2}{\delta} \left(\frac{1}{m}\right)^{\theta}$$

holds, where $\theta = \min\left\{\dfrac{2\beta}{1+\beta}, \dfrac{1}{2} - \varepsilon\right\}$ *and* $\tilde{C}$ *is a constant depending on* ε *but not on* m *or* δ.

Condition (10.2) measures how quickly f_c is approximated by functions from $\mathcal{H}_K$ in the metric $\mathscr{L}^1_{\rho_X}$. When $\mathcal{H}_K$ is dense in $\mathscr{L}^1_{\rho_X}(X)$, the quantity on the left-hand side of (10.2) tends to zero as $\gamma \to 0$. What (10.2) requires is a certain decay for this convergence. This can be stated as some interpolation space condition for the function f_c.

Theorem 10.3 *Assume that* K *is* $\mathscr{C}^\infty$ *in* $X \times X$ *and that for some* $\beta > 0$,

$$\inf_{f \in \mathcal{H}_K} \{\|f - f_\rho\|^2_{\mathscr{L}^2_{\rho_X}} + \gamma \|f\|^2_K\} = \mathcal{O}(\gamma^\beta).$$

Choose $\gamma = \frac{1}{m}$. *Then, for any* $0 < \varepsilon < \frac{1}{2}$ *and* $0 < \delta < 1$, *with confidence* $1 - \delta$,

$$\mathcal{R}(\text{sgn}(f_{z,\gamma}^{\phi_{ls}})) - \mathcal{R}(f_c) \leq \tilde{C} \left(\log \frac{2}{\delta}\right)^{1/2} \left(\frac{1}{m}\right)^{(1/2)\min\{\beta, 1-\varepsilon\}}$$

holds, where $\tilde{C}$ *is a constant depending on* ε *but not on* m *or* δ.

Again, the exponents β in Theorems 10.2 and 10.3 depend on the measure ρ. We note, however, that in the latter this exponent occurs only in the bounds; that is, the regularization parameter γ can be chosen without knowing β and, actually, without any knowledge about ρ.

10.1 Bounding the misclassification error in terms of the generalization error

The classification algorithm induced by (10.1) is a regularization scheme. Thus, we expect that our knowledge from Chapter 8 can be used in its analysis. Note, however, that the minimized errors – the generalization error in Chapter 8 and the error with respect to the loss ϕ here – are different and, naturally enough, so are their minimizers.

Definition 10.4 Denote by $f_\rho^\phi : X \to \mathbb{R}$ any measurable function minimizing the generalization error with respect to ϕ for example, for almost all $x \in X$,

$$f_\rho^\phi(x) := \underset{t \in \mathbb{R}}{\operatorname{argmin}} \int_Y \phi(yt) \, d\rho(y \mid x) = \underset{t \in \mathbb{R}}{\operatorname{argmin}} \, \phi(t)\eta_x + \phi(-t)(1 - \eta_x).$$

Our goal in this chapter is to show that under some mild conditions, for any classifying loss ϕ satisfying $\phi''(0) > 0$, we have $\mathcal{E}^\phi(f_{\mathbf{z},\gamma}^\phi) - \mathcal{E}^\phi(f_\rho^\phi) \to 0$ with high confidence as $m \to \infty$ and $\gamma = \gamma(m) \to 0$. We saw in Chapter 9 that this is the case for $\phi = \phi_\mathsf{h}$ and weakly separable measures. We begin in this section by extending Theorem 9.21.

Theorem 10.5 *Let ϕ be a classifying loss such that $\phi''(0)$ exists and is positive. Then there is a constant $c_\phi > 0$ such that for all measurable functions $f : X \to \mathbb{R}$,*

$$\mathcal{R}(\operatorname{sgn}(f)) - \mathcal{R}(f_c) \le c_\phi \sqrt{\mathcal{E}^\phi(f) - \mathcal{E}^\phi(f_\rho^\phi)}.$$

If $\mathcal{R}(f_c) = 0$, then the bound can be improved to

$$\mathcal{R}(\operatorname{sgn}(f)) - \mathcal{R}(f_c) \le c_\phi \left\{ \mathcal{E}^\phi(f) - \mathcal{E}^\phi(f_\rho^\phi) \right\}.$$

To prove Theorem 10.5, we want to understand the behavior of f_ρ^ϕ. To this end, we introduce an auxiliary function. In what follows, fix a classifying loss ϕ.

Definition 10.6 Define the *localizing function* $\Phi = \Phi_x : (\mathbb{R} \cup \{\pm\infty\}) \to \mathbb{R}_+$ to be the function associated with ϕ, ρ, and x given by

$$\Phi(t) = \eta_x \phi(t) + (1 - \eta_x)\phi(-t). \tag{10.3}$$

The following property of classifying loss functions follows immediately from their convexity. Denote by ϕ'_- (respectively, ϕ'_+) the left derivative (respectively, right derivative) of ϕ.

Figure 10.2 shows the localizing functions corresponding to ϕ_0, ϕ_1, and ϕ_2 for $\eta(x) = 0.75$.

Lemma 10.7 *A classifying loss ϕ is strictly decreasing on $(-\infty, 1]$ and nondecreasing on $(1, +\infty)$. It satisfies $\phi'_+(t) < 0$ for $t \in (-\infty, 1)$ and $\phi'_-(t) \ge 0$ for $t \in (1, +\infty)$.* ∎

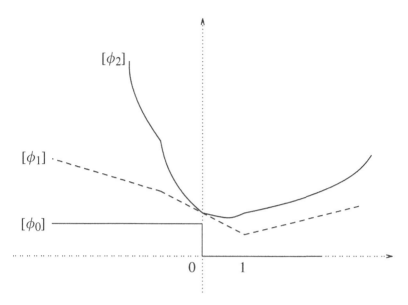

Figure 10.2

Note that if ϕ is a classifying loss, then, for all $x \in X$, Φ_x is convex and $\mathcal{E}^\phi(f) = \int_X \Phi_x(f(x)) \, d\rho_X$. Denote

$$f_\rho^-(x) := \sup\{t \in \mathbb{R} \mid \Phi_-'(t) = \eta_x \phi_-'(t) - (1 - \eta_x)\phi_+'(-t) < 0\}$$

and

$$f_\rho^+(x) := \inf\{t \in \mathbb{R} \mid \Phi_+'(t) = \eta_x \phi_+'(t) - (1 - \eta_x)\phi_-'(-t) > 0\}.$$

The convexity of Φ implies that $f_\rho^-(x) \le f_\rho^+(x)$.

Theorem 10.8 *Let ϕ be a classifying loss function and $x \in X$.*

(i) *The convex function Φ_x is strictly decreasing on $(-\infty, f_\rho^-(x)]$, strictly increasing on $[f_\rho^+(x), +\infty)$, and constant on $[f_\rho^-(x), f_\rho^+(x)]$.*

(ii) *$f_\rho^\phi(x)$ is a minimizer of Φ_x and can be taken to be any value in $[f_\rho^-(x), f_\rho^+(x)]$.*

(iii) *The following holds:* $\begin{cases} 0 \le f_\rho^-(x) \le f_\rho^\phi(x) & \text{if } f_\rho(x) > 0 \\ f_\rho^\phi(x) \le f_\rho^+(x) \le 0 & \text{if } f_\rho(x) < 0 \\ f_\rho^-(x) \le 0 \le f_\rho^+(x) & \text{if } f_\rho(x) = 0. \end{cases}$

(iv) *$f_\rho^-(x) \le 1$ and $f_\rho^+(x) \ge -1$.*

Proof.

(i) Since $\Phi = \Phi_x$ is convex, its one-side derivatives are both well defined and nondecreasing, and $\Phi'_-(t) \leq \Phi'_+(t)$ for every $t \in \mathbb{R}$. Then Φ is strictly decreasing on the interval $\left(-\infty, f_\rho^-(x)\right]$, since $\Phi'_-(t) < 0$ on this interval. In the same way, $\Phi'_+(t) > 0$ for $t > f_\rho^+(x)$, so Φ is strictly increasing on $\left[f_\rho^+(x), \infty\right)$.

For $t \in [f_\rho^-(x), f_\rho^+(x)]$, we have $0 \leq \Phi'_-(t) \leq \Phi'_+(t) \leq 0$. Hence Φ is constant on $[f_\rho^-(x), f_\rho^+(x)]$ and its value on this interval is its minimum.

(ii) Let $x \in X$. If we denote

$$\mathcal{E}^\phi(f \mid x) := \int_Y \phi(yf(x))\,d\rho(y \mid x) = \eta_x\phi(f(x)) + (1 - \eta_x)\phi(-f(x)),$$

$$(10.4)$$

then $\mathcal{E}^\phi(f \mid x) = \Phi(f(x))$. It follows that $f_\rho^\phi(x)$, which minimizes $\mathcal{E}^\phi(\cdot \mid x)$, is also a minimizer of Φ.

(iii) Observe that

$$f_\rho(x) = \eta_x - (1 - \eta_x) = 2\left(\eta_x - \tfrac{1}{2}\right).$$

$$(10.5)$$

Since ϕ is differentiable at 0, so is Φ, and $\Phi'(0) = (2\eta_x - 1)$ $\phi'(0) = f_\rho(x)\phi'(0)$. We now reason by cases and use that $\phi'(0) < 0$. When $f_\rho(x) > 0$, we have $\Phi'_-(0) = \Phi'(0) < 0$ and $f_\rho^-(x) \geq 0$. When $f_\rho(x) < 0$, $\Phi'_+(0) = \Phi'(0) > 0$ and $f_\rho^+(x) \leq 0$ hold. Finally, when $f_\rho(x) = 0$, we have $\Phi'_-(0) = \Phi'_+(0) = 0$, which implies $f_\rho^-(x) \leq 0$ and $f_\rho^+(x) \geq 0$.

(iv) When $t > 1$, Lemma 10.7 tells us that $\phi'_-(t) \geq 0$ and $\phi'_+(-t) < 0$. Hence $\Phi'_-(t) \geq 0$ and $f_\rho^-(x) \leq 1$. In the same way, $f_\rho^+(x) \geq -1$ follows from $\phi'_+(t) < 0$ and $\phi'_-(-t) \geq 0$ for $t < -1$. ∎

Assumption 10.9 It follows from Theorem 10.8 that f_ρ^ϕ can be chosen to satisfy

$$|f_\rho^\phi(x)| \leq 1 \quad \text{and} \quad f_\rho^\phi(x) = 0 \text{ if } f_\rho(x) = 0.$$

$$(10.6)$$

In the remainder of this chapter we assume, without loss of generality, that f_ρ^ϕ satisfies (10.6).

Lemma 10.10 *Let ϕ be a classifying loss such that $\phi''(0)$ exists and is positive. Then there is a constant $C = C_\phi > 0$ such that for all $x \in X$,*

$$\Phi(0) - \Phi(f_\rho^\phi(x)) \geq C\left(\eta_x - \tfrac{1}{2}\right)^2.$$

Proof. By the definition of $\phi''(0)$, there exists some $\frac{1}{2} \geq c_0 > 0$ such that for all $t \in [-c_0, c_0]$,

$$\left| \frac{\phi'(t) - \phi'(0)}{t} - \phi''(0) \right| \leq \frac{\phi''(0)}{2}.$$

This implies that

$$\phi'(0) + \phi''(0)t - \frac{\phi''(0)}{2}|t| \leq \phi'(t) \leq \phi'(0) + \phi''(0)t$$

$$+ \frac{\phi''(0)}{2}|t|, \quad \forall t \in [-c_0, c_0]. \quad (10.7)$$

Let $x \in X$. Consider the case $\eta_x > \frac{1}{2}$ first.
Denote $\Delta = \min\{\frac{-\phi'(0)}{\phi''(0)}(\eta_x - \frac{1}{2}), c_0\}$. For $0 \leq t \leq c_0$,

$$\Phi'(t) = \eta_x \phi'(t) - (1 - \eta_x)\phi'(-t) \leq (2\eta_x - 1)\phi'(0) + \phi''(0)t + \frac{\phi''(0)}{2}t.$$

Thus, for $0 \leq t \leq \Delta \leq \frac{-\phi'(0)}{\phi''(0)}(\eta_x - \frac{1}{2})$, we have

$$\Phi'(t) \leq (2\eta_x - 1)\phi'(0) + \frac{3}{2}\phi''(0)\left\{ \frac{-\phi'(0)}{\phi''(0)}\left(\eta_x - \frac{1}{2}\right) \right\}$$

$$\leq \frac{\phi'(0)}{2}\left(\eta_x - \frac{1}{2}\right) < 0.$$

Therefore Φ is strictly decreasing on the interval $[0, \Delta]$. But $f_\rho^\phi(x)$ is its minimal point, so

$$\Phi(0) - \Phi(f_\rho^\phi(x)) \geq \Phi(0) - \Phi(\Delta) \geq -\frac{\phi'(0)}{2}\left(\eta_x - \frac{1}{2}\right)\Delta.$$

When $\frac{-\phi'(0)}{\phi''(0)}\left(\eta_x - \frac{1}{2}\right) \leq c_0$, we have $\Delta = \frac{-\phi'(0)}{\phi''(0)}\left(\eta_x - \frac{1}{2}\right)$. When $\frac{-\phi'(0)}{\phi''(0)}\left(\eta_x - \frac{1}{2}\right) > c_0$, we have $\Delta = c_0 \geq 2c_0\,(\eta_x - \frac{1}{2})$. In both cases, we have

$$\Phi(0) - \Phi(f_\rho^\phi(x)) \geq \frac{-\phi'(0)}{2}\left(\eta_x - \frac{1}{2}\right)^2 \min\left\{ \frac{-\phi'(0)}{\phi''(0)}, 2c_0 \right\}.$$

That is, the desired inequality holds with

$$C = \min\left\{ -\phi'(0)c_0, \frac{(-\phi'(0))^2}{2\phi''(0)} \right\}.$$

The proof for $\eta_x < \frac{1}{2}$ is similar: one estimates the upper bound of $\Phi(t)$ for $t < 0$.   ∎

Proof of Theorem 10.5   Denote $X_c = \{x \in X \mid \mathrm{sgn}(f)(x) \neq f_c(x)\}$. Recall (9.14). Applying the Cauchy–Schwarz inequality and the fact that ρ_X is a probability measure on X, we get

$$\mathcal{R}(\mathrm{sgn}(f)) - \mathcal{R}(f_c) \leq \left\{ \int_{X_c} |f_\rho(x)|^2 \, d\rho_X \right\}^{1/2} \left\{ \int_{X_c} 1 \, d\rho_X \right\}^{1/2}$$

$$\leq \left\{ \int_{X_c} |f_\rho(x)|^2 \, d\rho_X \right\}^{1/2}.$$

We then use Lemma 10.10 and (10.5) to find that

$$\mathcal{R}(\mathrm{sgn}(f)) - \mathcal{R}(f_c) \leq 2 \left\{ \frac{1}{C} \int_{X_c} \left\{ \Phi(0) - \Phi(f_\rho^\phi(x)) \right\} d\rho_X \right\}^{1/2}.$$

Let $x \in X_c$. If $f_\rho(x) > 0$, then $f_c(x) = 1$ and $f(x) < 0$. By Theorem 10.8, $f_\rho^-(x) \geq 0$ and Φ is strictly decreasing on $(-\infty, 0]$. So $\Phi(f(x)) > \Phi(0)$ in this case. In the same way, if $f_\rho(x) < 0$, then $f(x) \geq 0$. By Theorem 10.8, $f_\rho^+(x) \leq 0$ and Φ is strictly increasing on $[0, +\infty)$. So $\Phi(f(x)) \geq \Phi(0)$. Finally, if $f_\rho(x) = 0$, by (10.6), $f_\rho^\phi(x) = 0$ and then $\Phi(0) - \Phi(f_\rho^\phi(x)) = 0$.

In all three cases we have $\Phi(0) - \Phi(f_\rho^\phi(x)) \leq \Phi(f(x)) - \Phi(f_\rho^\phi(x))$. Hence,

$$\int_{X_c} \left\{ \Phi(0) - \Phi(f_\rho^\phi(x)) \right\} d\rho_X \leq \int_{X_c} \left\{ \Phi(f(x)) - \Phi(f_\rho^\phi(x)) \right\} d\rho_X$$

$$\leq \int_X \left\{ \Phi(f(x)) - \Phi(f_\rho^\phi(x)) \right\} d\rho_X = \mathcal{E}^\phi(f) - \mathcal{E}^\phi(f_\rho^\phi).$$

This proves the first desired bound with $c_\phi = 2/\sqrt{C}$.

If $\mathcal{R}(f_c) = 0$, then $y = f_c(x)$ almost surely and $\eta_x = 1$ or 0 almost everywhere. This means that $|f_\rho(x)| = 1$ and $|f_\rho(x)| = |f_\rho(x)|^2$ almost everywhere. Using this with respect to relation (9.14), we see that $\mathcal{R}(\mathrm{sgn}(f)) - \mathcal{R}(f_c) = \int_{X_c} |f_\rho(x)|^2 \, d\rho_X$. Then the above procedure yields the second bound with $c_\phi = \frac{4}{C}$.   ∎

10.2 Projection and error decomposition

Since regularized classifiers are obtained by composing the sgn function with a real-valued function $f : X \to \mathbb{R}$, we may improve the error estimates by replacing image values of f by their projection onto $[-1, 1]$. This section develops this idea.

Definition 10.11 The *projection operator* π on the space of measurable functions $f : X \to \mathbb{R}$ is defined by

$$\pi(f)(x) = \begin{cases} 1 & \text{if } f(x) > 1 \\ -1 & \text{if } f(x) < -1 \\ f(x) & \text{if } -1 \leq f(x) \leq 1. \end{cases}$$

Trivially, $\text{sgn}(\pi(f)) = \text{sgn}(f)$. Lemma 10.7 tells us that $\phi(y(\pi(f))(x)) \leq \phi(yf(x))$. Then

$$\mathcal{E}^\phi(\pi(f)) \leq \mathcal{E}^\phi(f) \quad \text{and} \quad \mathcal{E}^\phi_{\mathbf{z}}(\pi(f)) \leq \mathcal{E}^\phi_{\mathbf{z}}(f). \tag{10.8}$$

Together with Theorem 10.5, this implies that if $\phi''(0) > 0$,

$$\mathcal{R}(\text{sgn}(f^\phi_{\mathbf{z},\gamma})) - \mathcal{R}(f_c) \leq c_\phi \sqrt{\mathcal{E}^\phi(\pi(f^\phi_{\mathbf{z},\gamma})) - \mathcal{E}^\phi(f^\phi_\rho)}.$$

Thus the analysis for the excess misclassification error of $f^\phi_{\mathbf{z},\gamma}$ is reduced into that for the excess generalization error $\mathcal{E}^\phi(\pi(f^\phi_{\mathbf{z},\gamma})) - \mathcal{E}^\phi(f^\phi_\rho)$. We carry out the latter analysis in the next two sections.

The following result is similar to Theorem 8.3.

Theorem 10.12 *Let ϕ be a classifying loss, $f^\phi_{\mathbf{z},\gamma}$ be defined by (10.1), and $f_\gamma \in \mathcal{H}_K$. Then $\mathcal{E}^\phi(\pi(f^\phi_{\mathbf{z},\gamma})) - \mathcal{E}^\phi(f^\phi_\rho)$ is bounded by*

$$\mathcal{E}^\phi(\pi(f^\phi_{\mathbf{z},\gamma})) - \mathcal{E}^\phi(f^\phi_\rho) + \gamma \|f^\phi_{\mathbf{z},\gamma}\|^2_K \leq \left\{ \mathcal{E}^\phi(f_\gamma) - \mathcal{E}^\phi(f^\phi_\rho) + \gamma \|f_\gamma\|^2_K \right\}$$

$$+ \left\{ \left[\left[\mathcal{E}^\phi_{\mathbf{z}}(f_\gamma) - \mathcal{E}^\phi_{\mathbf{z}}(f^\phi_\rho) \right] - \left[\mathcal{E}^\phi(f_\gamma) - \mathcal{E}^\phi(f^\phi_\rho) \right] \right\} \tag{10.9}$$

$$+ \left\{ \left[\mathcal{E}^\phi(\pi(f^\phi_{\mathbf{z},\gamma})) - \mathcal{E}^\phi(f^\phi_\rho) \right] - \left[\mathcal{E}^\phi_{\mathbf{z}}(\pi(f^\phi_{\mathbf{z},\gamma})) - \mathcal{E}^\phi_{\mathbf{z}}(f^\phi_\rho) \right] \right\} \right\}.$$

Proof. The proof follows from (10.8) using the procedure in the proof of Theorem 8.3. ∎

The function $f_\gamma \in \mathcal{H}_K$ in Theorem 10.12 is called a *regularizing function*. It is arbitrarily chosen and depends on γ. One standard choice is the function

$$\widetilde{f}^\phi_\gamma := \underset{f \in \mathcal{H}_K}{\text{argmin}} \left\{ \mathcal{E}^\phi(f) + \gamma \|f\|^2_K \right\}. \tag{10.10}$$

The first term on the right-hand side of (10.9) is estimated in the next section. It is the regularized error (w.r.t. ϕ) of f_γ,

$$\mathcal{D}(\gamma, \phi) := \mathcal{E}^\phi(f_\gamma) - \mathcal{E}^\phi(f_\rho^\phi) + \gamma \|f_\gamma\|_K^2. \tag{10.11}$$

The second and third terms on the right-hand side of (10.9) decompose the sample error $\mathcal{E}^\phi(\pi(f_{\mathbf{z},\gamma}^\phi)) - \mathcal{E}^\phi(f_\gamma)$. The second term is about a single random variable involving only one function f_γ and is easy to handle; we bound it in Section 10.4. The third term is more complex. In the form presented, the function $\pi(f_{\mathbf{z},\gamma}^\phi)$ is projected from $f_{\mathbf{z},\gamma}^\phi$. This projection maintains the misclassification error: $\mathcal{R}(\mathrm{sgn}(\pi(f_{\mathbf{z},\gamma}^\phi))) = \mathcal{R}(\mathrm{sgn}(f_{\mathbf{z},\gamma}^\phi))$. However, it causes the random variable $\phi(y\pi(f_{\mathbf{z},\gamma}^\phi)(x))$ to be bounded by $\phi(-1)$, a bound that is often smaller than that for $\phi(yf_{\mathbf{z},\gamma}^\phi(x))$. This allows for improved bounds for the sample error for classification algorithms. We bound this third term in Section 10.5.

10.3 Bounds for the regularized error $\mathcal{D}(\gamma, \phi)$ of f_γ

In this section we estimate the regularized error $\mathcal{D}(\gamma, \phi)$ of f_γ. This estimate follows from estimates for $\mathcal{E}^\phi(f_\gamma) - \mathcal{E}^\phi(f_\rho^\phi)$. Define the function $\Psi : \mathbb{R}_+ \to \mathbb{R}_+$ by

$$\Psi(t) = \max\{|\phi_-'(t)|, |\phi_+'(t)|, |\phi_-'(-t)|, |\phi_+'(-t)|\}.$$

Theorem 10.13 *Let ϕ be a classifying loss. For any measurable function f,*

$$\mathcal{E}^\phi(f) - \mathcal{E}^\phi(f_\rho^\phi) \le \int_X \Psi(|f(x)|) \left| f(x) - f_\rho^\phi(x) \right| d\rho_X.$$

If, in addition, $\phi \in \mathcal{C}^2(\mathbb{R})$, then we have

$$\mathcal{E}^\phi(f) - \mathcal{E}^\phi(f_\rho^\phi) \le \frac{1}{2} \int_X \big\{ \|\phi''\|_{\mathscr{L}^\infty_{[-1,1]}} \\ + \|\phi''\|_{\mathscr{L}^\infty_{[-|f(x)|,|f(x)|]}} \big\} \left| f(x) - f_\rho^\phi(x) \right|^2 d\rho_X.$$

Proof. It follows from (10.3) and (10.4) that

$$\mathcal{E}^\phi(f) - \mathcal{E}^\phi(f_\rho^\phi) = \int_X \Phi(f(x)) - \Phi(f_\rho^\phi(x)) \, d\rho_X.$$

By Theorem 10.8, Φ is constant on $[f_\rho^-(x), f_\rho^+(x)]$. So we need only bound for those points x for which the value $f(x)$ is outside this interval.

If $f(x) > f_\rho^+(x)$, then, by Theorem 10.8 and since $f_\rho^+(x) \geq -1$, Φ is strictly increasing on $[f_\rho^+(x), f(x)]$. Moreover, the convexity of Φ implies that

$$\Phi(f(x)) - \Phi(f_\rho^\phi(x)) \leq \Phi_-'(f(x))\left(f(x) - f_\rho^\phi(x)\right)$$
$$\leq \max\left\{\phi_-'(f(x)), |\phi_+'(-f(x))|\right\} \left|f(x) - f_\rho^\phi(x)\right|$$
$$\leq \Psi(|f(x)|) \left|f(x) - f_\rho^\phi(x)\right|.$$

Similarly, if $f(x) < f_\rho^-(x)$, then, by Theorem 10.8 again and since $f_\rho^-(x) \leq 1$, Φ is strictly decreasing on $[f(x), f_\rho^-(x)]$, and

$$\Phi(f(x)) - \Phi(f_\rho^\phi(x)) \leq \Phi_+'(f(x))\left(f(x) - f_\rho^\phi(x)\right) \leq \Psi(|f(x)|) \left|f(x) - f_\rho^\phi(x)\right|.$$

Thus, we have

$$\Phi(f(x)) - \Phi(f_\rho^\phi(x)) \leq \Psi(|f(x)|) \left|f(x) - f_\rho^\phi(x)\right|.$$

This gives the first bound.

If $\phi \in \mathcal{C}^2(\mathbb{R})$, so is Φ. Then $\Phi'(f_\rho^\phi(x)) = 0$ since $f_\rho^\phi(x)$ is a minimum of Φ. When $f(x) > f_\rho^+(x)$, using Taylor's expansion,

$$\Phi(f(x)) - \Phi(f_\rho^\phi(x)) = \Phi'(f_\rho^\phi(x))\left(f(x) - f_\rho^\phi(x)\right) + \int_{f_\rho^\phi(x)}^{f(x)} (f(x) - t)\,\Phi''(t)\,dt$$
$$\leq \|\Phi''\|_{\mathscr{L}^\infty[f_\rho^\phi(x), f(x)]} \frac{1}{2} \left|f(x) - f_\rho^\phi(x)\right|^2.$$

Now, since $f_\rho^\phi(x) \in [-1, 1]$, use that

$$\|\Phi''\|_{\mathscr{L}^\infty[f_\rho^\phi(x), f(x)]} \leq \max\{\|\phi''\|_{\mathscr{L}^\infty[-1,1]}, \|\phi''\|_{\mathscr{L}^\infty[-|f(x)|,|f(x)|]}\}$$

to get the desired result.

The case $f(x) < f_\rho^-(x)$ is dealt with in the same way.   ∎

Corollary 10.14 *Let ϕ be a classifying loss with $\|\phi'\|_\infty < \infty$. For any measurable function f,*

$$\mathcal{E}^\phi(f) - \mathcal{E}^\phi(f_\rho^\phi) \leq \|\phi'\|_\infty \|f - f_\rho^\phi\|_{\mathscr{L}^1_{\rho_X}}.$$

If $\phi \in \mathscr{C}^2(\mathbb{R})$ and $\|\phi''\|_\infty < \infty$, then we have

$$\mathcal{E}^\phi(f) - \mathcal{E}^\phi(f_\rho^\phi) \le \|\phi''\|_\infty \|f - f_\rho^\phi\|_{\mathscr{L}^2_{\rho_X}}^2. \qquad \blacksquare$$

10.4 Bounds for the sample error term involving f_γ

In this section, we bound the sample error term in (10.9) involving f_γ, that is, $\left[\mathcal{E}_\mathbf{z}^\phi(f_\gamma) - \mathcal{E}_\mathbf{z}^\phi(f_\rho^\phi)\right] - \left[\mathcal{E}^\phi(f_\gamma) - \mathcal{E}^\phi(f_\rho^\phi)\right]$. This can be written as $\frac{1}{m}\sum_{i=1}^m \xi(z_i) - \mathbf{E}(\xi)$ with ξ the random variable on (Z, ρ) given by $\xi(z) = \phi(yf_\gamma(x)) - \phi(yf_\rho^\phi(x))$. To bound this quantity using the Bernstein inequalities, we need to control the variance. We do so by means of the following constant determined by ϕ and ρ.

Definition 10.15 The *variancing power* $\tau = \tau_{\phi,\rho}$ of the pair (ϕ, ρ) is defined to be the maximal number τ in $[0, 1]$ such that for some constant $C_1 > 0$ and any measurable function $f : X \to [-1, 1]$,

$$\mathbf{E}\left\{\left(\phi(yf(x)) - \phi(yf_\rho^\phi(x))\right)^2\right\} \le C_1 \left(\mathcal{E}^\phi(f) - \mathcal{E}^\phi(f_\rho^\phi)\right)^\tau. \qquad (10.12)$$

Since (10.12) always holds with $\tau = 0$ and $C_1 = (\phi(-1))^2$, the variancing power $\tau_{\phi,\rho}$ is well defined.

Example 10.16 For $\phi_{\mathsf{ls}}(t) = (1 - t)^2$ we have $\tau_{\phi,\rho} = 1$ for any probability measure ρ.

Proof. For $\phi_{\mathsf{ls}}(t) = (1 - t)^2$ we know that $\phi_{\mathsf{ls}}(yf(x)) = (y - f(x))^2$ and $f_\rho^\phi = f_\rho$. Hence (10.12) is valid with $\tau = 1$ and $C_1 = \sup_{(x,y)\in Z} \left(y - f(x) + y - f_\rho(x)\right)^2 \le 16$. $\qquad \blacksquare$

In general, τ depends on the convexity of ϕ and on how much noise ρ contains.

In particular, for the q-norm loss ϕ_q, $\tau_{\phi_q,\rho} = 1$ when $1 < q \le 2$ and $0 < \tau_{\phi_q,\rho} < 1$ when $q > 2$.

Lemma 10.17 *Let $q, q^* > 1$ be such that $\frac{1}{q} + \frac{1}{q^*} = 1$. Then*

$$a \cdot b \le \frac{1}{q}a^q + \frac{1}{q^*}b^{q^*}, \quad \forall a, b > 0.$$

Proof. Let $b > 0$. Define a function $f : \mathbb{R}_+ \to \mathbb{R}$ by $f(a) = a \cdot b - \frac{1}{q}a^q - \frac{1}{q^*}b^{q^*}$. This satisfies

$$f'(a) = b - a^{q-1}, \qquad f''(a) = -(q-1)a^{q-2} < 0, \quad \forall a > 0.$$

Hence f is a concave function on $\mathbb{R}_+$ and takes its maximum value at the unique point $a^* = b^{1/(q-1)}$ where $f'(a^*) = 0$. But $q^* = \frac{q}{q-1}$ and

$$f(a^*) = a^* \cdot b - \frac{1}{q}(a^*)^q - \frac{1}{q^*}b^{q^*} = b^{q/(q-1)} - \frac{1}{q}b^{q/(q-1)} - \frac{1}{q^*}b^{q/(q-1)} = 0.$$

Therefore, $f(a) \le f(a^*) = 0$ for all $a \in \mathbb{R}_+$. This is true for any $b > 0$. So the inequality holds. ∎

Proposition 10.18 *Let $\tau = \tau_{\phi,\rho}$ and $B_\gamma := \max\{\phi(\|f_\gamma\|_\infty), \phi(-\|f_\gamma\|_\infty)\}$. For any $0 < \delta < 1$, with confidence $1 - \delta$, the quantity $\left[\mathcal{E}_\mathbf{z}^\phi(f_\gamma) - \mathcal{E}_\mathbf{z}^\phi(f_\rho^\phi)\right] - \left[\mathcal{E}^\phi(f_\gamma) - \mathcal{E}^\phi(f_\rho^\phi)\right]$ is bounded by*

$$\frac{5B_\gamma + 2\phi(-1)}{3m}\log\left(\frac{2}{\delta}\right) + \left(\frac{2C_1 \log(2/\delta)}{m}\right)^{1/(2-\tau)} + \mathcal{E}^\phi(f_\gamma) - \mathcal{E}^\phi(f_\rho^\phi).$$

Proof. Write the random variable $\xi(z) = \phi(yf_\gamma(x)) - \phi(yf_\rho^\phi(x))$ on (Z, ρ) as $\xi = \xi_1 + \xi_2$, where

$$\xi_1 := \phi(yf_\gamma(x)) - \phi(y\pi(f_\gamma)(x)), \quad \xi_2 := \phi(y\pi(f_\gamma)(x)) - \phi(yf_\rho^\phi(x)). \tag{10.13}$$

The first part ξ_1 is a random variable satisfying $0 \le \xi_1 \le B_\gamma$. Applying the one-side Bernstein inequality to ξ_1, we obtain, for any $\varepsilon > 0$,

$$\operatorname*{Prob}_{\mathbf{z} \in Z^m}\left\{\frac{1}{m}\sum_{i=1}^m \xi_1(z_i) - \mathbf{E}(\xi_1) > \varepsilon\right\} \le \exp\left\{-\frac{m\varepsilon^2}{2\left(\sigma^2(\xi_1) + \frac{1}{3}B_\gamma\varepsilon\right)}\right\}.$$

Solving the quadratic equation for ε given by

$$\frac{m\varepsilon^2}{2\left(\sigma^2(\xi_1) + \frac{1}{3}B_\gamma\varepsilon\right)} = \log\frac{2}{\delta},$$

we see that for any $0 < \delta < 1$, there exists a subset U_1 of Z^m with measure at least $1 - \frac{\delta}{2}$ such that for every $\mathbf{z} \in U_1$,

$$\frac{1}{m}\sum_{i=1}^{m}\xi_1(z_i) - \mathbf{E}(\xi_1)$$

$$\leq \frac{\frac{1}{3}B_\gamma\log(2/\delta) + \sqrt{\left(\frac{1}{3}B_\gamma\log(2/\delta)\right)^2 + 2m\sigma^2(\xi_1)\log(2/\delta)}}{m}.$$

But $\sigma^2(\xi_1) \leq \mathbf{E}(\xi_1^2) \leq B_\gamma\mathbf{E}(\xi_1)$. Therefore,

$$\frac{1}{m}\sum_{i=1}^{m}\xi_1(z_i) - \mathbf{E}(\xi_1) \leq \frac{5B_\gamma\log(2/\delta)}{3m} + \mathbf{E}(\xi_1), \quad \forall \mathbf{z} \in U_1.$$

Next we consider ξ_2. This is a random variable bounded by $\phi(-1)$. Applying the one-side Bernstein inequality as above, we obtain another subset U_2 of Z^m with measure at least $1 - \frac{\delta}{2}$ such that for every $\mathbf{z} \in U_2$,

$$\frac{1}{m}\sum_{i=1}^{m}\xi_2(z_i) - \mathbf{E}(\xi_2) \leq \frac{2\phi(-1)\log(2/\delta)}{3m} + \sqrt{\frac{2\log(2/\delta)\sigma^2(\xi_2)}{m}}.$$

The definition of the variancing power τ gives $\sigma^2(\xi_2) \leq \mathbf{E}(\xi_2^2) \leq C_1\{\mathbf{E}(\xi_2)\}^\tau$. Applying Lemma 10.17 to $q = \frac{2}{2-\tau}, q^* = \frac{2}{\tau}, a = \sqrt{2\log(2/\delta)C_1/m}$, and $b = \sqrt{\{\mathbf{E}(\xi_2)\}^\tau}$, we obtain

$$\sqrt{\frac{2\log(2/\delta)\sigma^2(\xi_2)}{m}} \leq \left(1 - \frac{\tau}{2}\right)\left(\frac{2\log(2/\delta)C_1}{m}\right)^{1/(2-\tau)} + \frac{\tau}{2}\mathbf{E}(\xi_2).$$

Hence, for all $\mathbf{z} \in U_2$,

$$\frac{1}{m}\sum_{i=1}^{m}\xi_2(z_i) - \mathbf{E}(\xi_2) \leq \frac{2\phi(-1)\log(2/\delta)}{3m} + \left(\frac{2\log(2/\delta)C_1}{m}\right)^{1/(2-\tau)} + \mathbf{E}(\xi_2).$$

Combining these inequalities for ξ_1 and ξ_2 with the fact that $\mathbf{E}(\xi_1) + \mathbf{E}(\xi_2) = \mathbf{E}(\xi) = \mathcal{E}^\phi(f_\gamma) - \mathcal{E}^\phi(f_\rho^\phi)$, we conclude that for all $\mathbf{z} \in U_1 \cap U_2$,

$$
\left[\mathcal{E}_{\mathbf{z}}^\phi(f_\gamma) - \mathcal{E}_{\mathbf{z}}^\phi(f_\rho^\phi) \right] - \left[\mathcal{E}^\phi(f_\gamma) - \mathcal{E}^\phi(f_\rho^\phi) \right]
$$
$$
\leq \frac{5B_\gamma \log(2/\delta) + 2\phi(-1) \log(2/\delta)}{3m} + \left(\frac{2 \log(2/\delta) C_1}{m} \right)^{(1/2 - \tau)}
$$
$$
+ \mathcal{E}^\phi(f_\gamma) - \mathcal{E}^\phi(f_\rho^\phi).
$$

Since the measure of $U_1 \cap U_2$ is at least $1 - \delta$, this bound holds with the confidence claimed. ∎

10.5 Bounds for the sample error term involving $f_{z,\gamma}^\phi$

The other term of the sample error in (10.9), $\left[\mathcal{E}^\phi(\pi(f_{z,\gamma}^\phi)) - \mathcal{E}^\phi(f_\rho^\phi) \right] - \left[\mathcal{E}_{\mathbf{z}}^\phi(\pi(f_{z,\gamma}^\phi)) - \mathcal{E}_{\mathbf{z}}^\phi(f_\rho^\phi) \right]$, involves the function $f_{z,\gamma}^\phi$ and thus runs over a set of functions. To bound it, we use – as we have already done in similar cases – a probability inequality for a function set in terms of the covering numbers of the set.

The following probability inequality can be proved using the one-side Bernstein inequality as in Lemma 3.18.

Lemma 10.19 *Let ξ be a random variable on Z with mean μ and variance σ^2. Assume that $\mu \geq 0$, $|\xi - \mu| \leq B$ almost everywhere, and $\sigma^2 \leq c\mu^\tau$ for some $0 \leq \tau \leq 2$ and $c, B \geq 0$. Then, for every $\varepsilon > 0$,*

$$
\operatorname*{Prob}_{\mathbf{z} \in Z^m} \left\{ \frac{\mu - \frac{1}{m} \sum_{i=1}^m \xi(z_i)}{\sqrt{\mu^\tau + \varepsilon^\tau}} > \varepsilon^{1 - \frac{\tau}{2}} \right\} \leq \exp\left\{ -\frac{m\varepsilon^{2-\tau}}{2(c + \frac{1}{3}B\varepsilon^{1-\tau})} \right\}
$$

holds. ∎

Also, the following inequality for a function set can be proved in the same way as Lemma 3.19.

Lemma 10.20 *Let $0 \leq \tau \leq 1$, $c, B \geq 0$, and $\mathcal{G}$ be a set of functions on Z such that for every $g \in \mathcal{G}$, $\mathbf{E}(g) \geq 0$, $\|g - \mathbf{E}(g)\|_{\mathscr{L}_\rho^\infty} \leq B$, and $\mathbf{E}(g^2) \leq c(\mathbf{E}(g))^\tau$.*

Then, for all $\varepsilon > 0$,

$$\Prob_{z \in Z^m} \left\{ \sup_{g \in \mathcal{G}} \frac{\mathbf{E}(g) - \frac{1}{m} \sum_{i=1}^m g(z_i)}{\sqrt{(\mathbf{E}(g))^\tau + \varepsilon^\tau}} > 4\varepsilon^{1-\tau/2} \right\}$$

$$\leq \mathcal{N}(\mathcal{G}, \varepsilon) \exp \left\{ -\frac{m\varepsilon^{2-\tau}}{2(c + \frac{1}{3}B\varepsilon^{1-\tau})} \right\}. \qquad \blacksquare$$

We can now derive the sample error bounds along the same lines we followed in the previous chapter for the regression problem.

Lemma 10.21 *Let $\tau = \tau_{\phi,\rho}$. For any $R > 0$ and any $\varepsilon > 0$.*

$$\Prob_{z \in Z^m} \left\{ \sup_{f \in B_R} \frac{\left\{ \mathcal{E}^\phi(\pi(f)) - \mathcal{E}^\phi(f_\rho^\phi) \right\} - \left\{ \mathcal{E}_z^\phi(\pi(f)) - \mathcal{E}_z^\phi(f_\rho^\phi) \right\}}{\sqrt{\left(\mathcal{E}^\phi(\pi(f)) - \mathcal{E}^\phi(f_\rho^\phi) \right)^\tau + \varepsilon^\tau}} \leq 4\varepsilon^{1-\tau/2} \right\}$$

$$\geq 1 - \mathcal{N}\left(B_1, \frac{\varepsilon}{R|\phi'(-1)|} \right) \exp \left\{ -\frac{m\varepsilon^{2-\tau}}{2C_1 + \frac{4}{3}\phi(-1)\varepsilon^{1-\tau}} \right\}$$

holds.

Proof. Apply Lemma 10.20 to the function set

$$\mathcal{F}_R^\phi = \left\{ \phi(y(\pi f)(x)) - \phi(yf_\rho^\phi(x)) : f \in B_R \right\}. \qquad (10.14)$$

Each function $g \in \mathcal{F}_R^\phi$ satisfies $\mathbf{E}(g^2) \leq c\,(\mathbf{E}(g))^\tau$ for $c = C_1$ and $\|g - \mathbf{E}(g)\|_{\mathscr{L}_\rho^\infty} \leq B := 2\phi(-1)$. Therefore, to draw our conclusion from Lemma 10.20, we need only bound the covering number $\mathcal{N}(\mathcal{F}_R^\phi, \varepsilon)$. To do so, we note that for $f_1, f_2 \in B_R$ and $(x, y) \in Z$, we have

$$\left| \left\{ \phi(y(\pi f_1)(x)) - \phi(yf_\rho^\phi(x)) \right\} - \left\{ \phi(y(\pi f_2)(x)) - \phi(yf_\rho^\phi(x)) \right\} \right|$$

$$= |\phi(y(\pi f_1)(x)) - \phi(y(\pi f_2)(x))| \leq |\phi'(-1)| \|f_1 - f_2\|_\infty.$$

Therefore

$$\mathcal{N}\left(\mathcal{F}_R^\phi, \varepsilon \right) \leq \mathcal{N}\left(B_R, \frac{\varepsilon}{|\phi'(-1)|} \right),$$

proving the statement. $\qquad \blacksquare$

Define $\varepsilon^*(m, R, \delta)$ to be the smallest positive number ε satisfying

$$\log \mathcal{N}\left(B_1, \frac{\varepsilon}{R|\phi'(-1)|}\right) - \frac{m\varepsilon^{2-\tau}}{2C_1 + \frac{4}{3}\phi(-1)\varepsilon^{1-\tau}} \leq \log \delta. \tag{10.15}$$

Then the confidence for the error $\varepsilon = \varepsilon^*(m, R, \delta)$ in Lemma 10.21 is at least $1 - \delta$.

For $R > 0$, denote

$$\mathcal{W}(R) = \left\{\mathbf{z} \in Z^m : \|f_{\mathbf{z},\gamma}^{\phi}\|_K \leq R\right\}.$$

Proposition 10.22 *For all $0 < \delta < 1$ and $R > 0$, there is a subset V_R of Z^m with measure at most δ such that for all $\mathbf{z} \in \mathcal{W}(R) \setminus V_R$, the quantity $\mathcal{E}^{\phi}(\pi(f_{\mathbf{z},\gamma}^{\phi})) - \mathcal{E}^{\phi}(f_{\rho}^{\phi}) + \gamma \|f_{\mathbf{z},\gamma}^{\phi}\|_K^2$ is bounded by*

$$4\mathcal{D}(\gamma, \phi) + 24\varepsilon^*(m, R, \delta/2) + \frac{10B_\gamma + 4\phi(-1)}{3m} \log (4/\delta)$$

$$+ 2\left(\frac{2C_1 \log (4/\delta)}{m}\right)^{1/(2-\tau)}.$$

Proof. Lemma 10.17 implies that for $0 < \tau < 1$,

$$\sqrt{\left(\mathcal{E}^{\phi}(\pi(f)) - \mathcal{E}^{\phi}(f_{\rho}^{\phi})\right)^{\tau} + \varepsilon^{\tau}} \cdot 4\varepsilon^{1-\tau/2} \leq \frac{\tau}{2}\left(\mathcal{E}^{\phi}(\pi(f)) - \mathcal{E}^{\phi}(f_{\rho}^{\phi})\right)$$

$$+ (1 - \tau/2)4^{1/(1-\tau/2)}\varepsilon + 4\varepsilon$$

$$\leq \frac{1}{2}\left(\mathcal{E}^{\phi}(\pi(f)) - \mathcal{E}^{\phi}(f_{\rho}^{\phi})\right) + 12\varepsilon.$$

Putting this into Lemma 10.21 with $\varepsilon = \varepsilon^*(m, R, \delta/2)$, we deduce that for $\mathbf{z} \in \mathcal{W}(R)$, with confidence $1 - \frac{\delta}{2}$,

$$\left[\mathcal{E}^{\phi}(\pi(f_{\mathbf{z},\gamma}^{\phi})) - \mathcal{E}^{\phi}(f_{\rho}^{\phi})\right] - \left[\mathcal{E}_{\mathbf{z}}^{\phi}(\pi(f_{\mathbf{z},\gamma}^{\phi})) - \mathcal{E}_{\mathbf{z}}^{\phi}(f_{\rho}^{\phi})\right]$$

$$\leq \frac{1}{2}\left(\mathcal{E}^{\phi}(\pi(f_{\mathbf{z},\gamma}^{\phi})) - \mathcal{E}^{\phi}(f_{\rho}^{\phi})\right) + 12\varepsilon^*(m, R, \delta/2).$$

Proposition 10.18 with δ replaced by $\frac{\delta}{2}$ guarantees that with confidence $1 - \frac{\delta}{2}$, the quantity $\left[\mathcal{E}_{\mathbf{z}}^{\phi}(f_\gamma) - \mathcal{E}_{\mathbf{z}}^{\phi}(f_{\rho}^{\phi})\right] - \left[\mathcal{E}^{\phi}(f_\gamma) - \mathcal{E}^{\phi}(f_{\rho}^{\phi})\right]$ is bounded by

$$\frac{5B_\gamma + 2\phi(-1)}{3m} \log (4/\delta) + \left(\frac{2C_1 \log (4/\delta)}{m}\right)^{1/(2-\tau)} + \mathcal{E}^{\phi}(f_\gamma) - \mathcal{E}^{\phi}(f_{\rho}^{\phi}).$$

Combining these two bounds with (10.9), we see that for $\mathbf{z} \in \mathcal{W}(R)$, with confidence $1 - \delta$,

$$
\mathcal{E}^{\phi}(\pi(f_{\mathbf{z},\gamma}^{\phi})) - \mathcal{E}^{\phi}(f_{\rho}^{\phi}) + \gamma \|f_{\mathbf{z},\gamma}^{\phi}\|_K^2 \leq \mathcal{D}(\gamma, \phi) + \frac{1}{2}\left(\mathcal{E}^{\phi}(\pi(f_{\mathbf{z},\gamma}^{\phi})) - \mathcal{E}^{\phi}(f_{\rho}^{\phi})\right)
$$
$$
+ 12\varepsilon^*(m, R, \delta/2) + \frac{5B_\gamma + 2\phi(-1)}{3m}\log{(4/\delta)} + \left(\frac{2C_1 \log{(4/\delta)}}{m}\right)^{1/(2-\tau)}
$$
$$
+ \mathcal{E}^{\phi}(f_\gamma) - \mathcal{E}^{\phi}(f_{\rho}^{\phi}).
$$

This gives the desired bound. ∎

Lemma 10.23 *For all $\gamma > 0$ and $\mathbf{z} \in Z^m$,*

$$
\|f_{\mathbf{z},\gamma}^{\phi}\|_K \leq \sqrt{\phi(0)/\gamma}.
$$

Proof. Since $f_{\mathbf{z},\gamma}^{\phi}$ minimizes $\mathcal{E}_{\mathbf{z}}^{\phi}(f) + \gamma \|f\|_K^2$ in $\mathcal{H}_K$, choosing $f = 0$ implies that

$$
\gamma \|f_{\mathbf{z},\gamma}^{\phi}\|_K^2 \leq \mathcal{E}_{\mathbf{z}}^{\phi}(f_{\mathbf{z},\gamma}^{\phi}) + \gamma \|f_{\mathbf{z},\gamma}^{\phi}\|_K^2 \leq \mathcal{E}_{\mathbf{z}}^{\phi}(0) + 0 = \frac{1}{m}\sum_{i=1}^{m}\phi(0) = \phi(0).
$$

Therefore, $\|f_{\mathbf{z},\gamma}^{\phi}\|_K \leq \sqrt{\phi(0)/\gamma}$ for all $\mathbf{z} \in Z^m$. ∎

By Lemma 10.23, $\mathcal{W}(\sqrt{\phi(0)/\gamma}) = Z^m$. Taking $R := \sqrt{\phi(0)/\gamma}$, we can derive a weak error bound, as we did in Section 8.3. But we can do better. A bound for the norm $\|f_{\mathbf{z},\gamma}^{\phi}\|_K$ improving that of Lemma 10.23 can be shown to hold with high probability. To show this is the target of the next section. Note that we could now wrap the results in this and the two preceding sections into a single statement bounding the excess misclassification error $\mathcal{R}(\text{sgn}(f_{\mathbf{z},\gamma}^{\phi})) - \mathcal{R}(f_c)$. We actually do that, in Corollary 10.25, once we have obtained a better bound for the norm $\|f_{\mathbf{z},\gamma}^{\phi}\|_K$.

10.6 Stronger error bounds

In this section we derive bounds for $\mathcal{E}^{\phi}(\pi(f_{\mathbf{z},\gamma}^{\phi})) - \mathcal{E}^{\phi}(f_{\rho}^{\phi})$, improving those that would follow from the preceding sections, at the cost of a few mild assumptions.

Theorem 10.24 *Assume the following with positive constants p, C_0, C_{ϕ}', A, $q \geq 1$, and $\beta \leq 1$.*

(i) *K satisfies $\log \mathcal{N}(B_1, \eta) \leq C_0 (1/\eta)^p$.*

(ii) $\phi(t) \leq C'_\phi |t|^q$ for all $t \notin (-1, 1)$.

(iii) $\mathcal{D}(\gamma, \phi) \leq A\gamma^\beta$ for each $\gamma > 0$.

Choose $\gamma = m^{-\zeta}$ *with* $\zeta = \frac{1}{\beta + q(1-\beta)/2}$. *Then, for all* $0 < \eta \leq \frac{1}{2}$ *and all* $0 < \delta < 1$, *with confidence* $1 - \delta$,

$$\mathcal{E}^\phi(\pi(f_{\mathbf{z},\gamma}^\phi)) - \mathcal{E}^\phi(f_\rho^\phi) \leq \widetilde{C}_\eta \log \frac{2}{\delta} m^{-\theta},$$

where

$$\theta := \min \left\{ \frac{\beta}{\beta + q(1-\beta)/2}, \frac{1 - p\widetilde{r}}{2 - \tau + p} \right\}, \quad s := \frac{p}{2(1+p)},$$

$$\widetilde{r} := \max \left\{ \frac{\zeta - 1/(2 - \tau + p)}{2(1-s)} + \eta, \frac{1 - \beta}{2}\zeta, \zeta \left(\frac{1}{2} + \frac{q}{4}(1-\beta) \right) - \frac{1}{2} \right\}$$

and $\widetilde{C}_\eta$ *is a constant depending on* η *and the constants in conditions (i)–(iii), but not on* m *or* δ.

The following corollary follows from Theorems 10.5 and 10.24.

Corollary 10.25 *Under the hypothesis and with the notations of Theorem 10.24, if* $\phi''(0) > 0$, *then, with confidence at least* $1 - \delta$, *we have*

$$\mathcal{R}(\mathrm{sgn}(f_{\mathbf{z},\gamma}^\phi)) - \mathcal{R}(f_c) \leq c_\phi \sqrt{\widetilde{C}_\eta \log \frac{2}{\delta} m^{-\theta}}. \qquad \blacksquare$$

When the kernel is $\mathscr{C}^\infty$ on $X \subset \mathbb{R}^n$, we know (cf. Theorem 5.1(i)) that p in Theorem 10.24(i) can be arbitrarily small. We thus get the following result.

Corollary 10.26 *Assume that K is* $\mathscr{C}^\infty$ *on* $X \times X$ *and* $\phi(t) \leq C'_\phi |t|^q$ *for all* $t \notin (-1, 1)$ *and some* $q \geq 1$. *If* $\mathcal{D}(\gamma, \phi) \leq A\gamma^\beta$ *for all* $\gamma > 0$ *and some* $0 < \beta \leq 1$, *choose* $\gamma = m^{-\zeta}$ *with* $\zeta = \frac{1}{\beta + q(1-\beta)/2}$. *Then for any* $0 < \eta \leq \frac{1}{2}$ *and* $0 < \delta < 1$, *with confidence* $1 - \delta$,

$$\mathcal{E}^\phi(\pi(f_{\mathbf{z},\gamma}^\phi)) - \mathcal{E}^\phi(f_\rho^\phi) \leq \widetilde{C}_\eta \log \frac{2}{\delta} m^{-\theta},$$

where

$$\theta := \min \left\{ \frac{\beta}{\beta + q(1-\beta)/2}, \frac{1}{2 - \tau} - \eta \right\}$$

and $\widetilde{C}_\eta$ *is a constant depending on* η, *but not on* m *or* δ. $\qquad \blacksquare$

Theorem 10.2 follows from Corollary 10.26, Corollary 10.14, and Theorem 9.21 by taking $f_\gamma = \widetilde{f}_\gamma^\phi$ defined in (10.10).

Theorem 10.3 is a consequence of Corollary 10.26 and Theorem 10.5. In this case, in addition, we can take $q = 2$, which implies $\zeta = 1$.

The proof of Theorem 10.24 will follow from several lemmas. The idea is to find a radius R such that $\mathcal{W}(R)$ is close to Z^m with high probability.

First we establish a bound for the number $\varepsilon^*(m, R, \delta)$.

Lemma 10.27 *Assume K satisfies $\log \mathcal{N}(B_1, \eta) \le C_0(1/\eta)^p$ for some $p > 0$. Then for $R \ge 1$ and $0 < \delta \le \frac{1}{2}$, the quantity $\varepsilon^*(m, R, \delta)$ defined by (10.15) can be bounded by*

$$
\varepsilon^*(m, R, \delta) \le C_2 \left\{ \left(\frac{\log(1/\delta)}{m} \right)^{1/(2-\tau)} \right.
$$
$$
\left. + \max \left\{ \left(\frac{R^p}{m} \right)^{1/(1+p)}, \left(\frac{R^p}{m} \right)^{1/(2-\tau+p)} \right\} \right\},
$$

where $C_2 := (6\phi(-1) + 8C_1 + 1)(C_0 + 1)(|\phi'(-1)| + 1)$.

Proof. Using the covering number assumption, we see from (10.15) that $\varepsilon^*(m, R, \delta) \le \Delta$, where Δ is the unique positive number ϵ satisfying

$$
C_0 \left(\frac{R|\phi'(-1)|}{\epsilon} \right)^p - \frac{m\epsilon^{2-\tau}}{2C_1 + \frac{4}{3}\phi(-1)\epsilon^{1-\tau}} = \log \delta.
$$

We can rewrite this equation as

$$
\epsilon^{2-\tau+p} - \frac{4\phi(-1)\log(1/\delta)}{3m}\epsilon^{1-\tau+p} - \frac{2C_1 \log(1/\delta)}{m}\epsilon^p
$$
$$
- \frac{4\phi(-1)C_0}{3m}\left(R|\phi'(-1)|\right)^p \epsilon^{1-\tau} - \frac{2C_1 C_0}{m}\left(R|\phi'(-1)|\right)^p = 0.
$$

Applying Lemma 7.2 with $d = 4$ to this equation, we find that the solution Δ satisfies

$$
\Delta \le \max \left\{ \frac{16\phi(-1)\log(1/\delta)}{3m}, \left(\frac{8C_1 \log(1/\delta)}{m} \right)^{1/(2-\tau)}, \right.
$$
$$
\left. \left(\frac{16\phi(-1)C_0}{3m}\left(R|\phi'(-1)|\right)^p \right)^{1/(1+p)}, \left(\frac{8C_1 C_0}{m}\left(R|\phi'(-1)|\right)^p \right)^{1/(2-\tau+p)} \right\}.
$$

Therefore the desired bound for $\varepsilon^*(m, R, \delta)$ follows. ∎

The following lemma is a consequence of Lemma 10.27 and Proposition 10.22.

Lemma 10.28 *Under the assumptions of Theorem 10.24, choose $\gamma = m^{-\zeta}$ for some $\zeta > 0$. Then, for any $0 < \delta < 1$ and $R \geq 1$, there is a set $V_R \subseteq Z^m$ with measure at most δ such that for $m \geq (C_K^2 A)^{-1/(\zeta(1-\beta))}$,*

$$\mathcal{W}(R) \subseteq \mathcal{W}(\tilde{a}_m R^s + \tilde{b}_m) \cup V_R,$$

where $s := \frac{p}{2(1+p)}, \tilde{a}_m := 5\sqrt{C_2} m^{(\zeta - 1/(2-\tau+p))/2}$, and $\tilde{b}_m := C_3 \left(\log \frac{4}{\delta} \right)^{1/2} m^r$. Here the constants are

$$r := \max \left\{ \frac{\zeta - 1/(2-\tau)}{2}, \frac{1-\beta}{2}\zeta, \zeta \left(\frac{1}{2} + \frac{q}{4}(1-\beta) \right) - \frac{1}{2} \right\}$$

and

$$C_3 = 5\sqrt{C_2} + 2C_1^{1/(4-2\tau)} + \frac{2}{\sqrt{3}}\sqrt{\phi(-1)} + 2\sqrt{A} + 2\sqrt{C_\phi'} C_K^{q/2} A^{q/4}.$$

Proof. By Proposition 10.22, there is a set V_R with measure at most δ such that for all $\mathbf{z} \in \mathcal{W}(R) \setminus V_R$,

$$\gamma \| f_{\mathbf{z},\gamma}^\phi \|_K^2 \leq 4A\gamma^\beta + 24\varepsilon^*(m, R, \delta/2) + \frac{10B_\gamma + 4\phi(-1)}{3m} \log(4/\delta)$$
$$+ 2\left(\frac{2C_1 \log(4/\delta)}{m} \right)^{1/(2-\tau)}.$$

Since $\phi(t) \leq C_\phi' |t|^q$ for each $t \notin (-1,1)$, we see that $B_\gamma = \max\{\phi(\|f_\gamma\|_\infty), \phi(-\|f_\gamma\|_\infty)\}$ is bounded by $C_\phi' \left(\max\{\|f_\gamma\|_\infty, 1\} \right)^q$. But the assumption $\mathcal{D}(\gamma, \phi) \leq A\gamma^\beta$ implies that

$$\|f_\gamma\|_\infty \leq C_K \|f_\gamma\|_K \leq C_K \sqrt{\mathcal{D}(\gamma, \phi)/\gamma} \leq C_K \sqrt{A} \gamma^{(\beta-1)/2}.$$

Hence,

$$B_\gamma \leq C_\phi' C_K^q A^{q/2} \gamma^{q(\beta-1)/2}, \quad \text{when } C_K \sqrt{A} \gamma^{(\beta-1)/2} \geq 1. \tag{10.16}$$

Under this restriction it follows from Lemma 10.27 that $\mathbf{z} \in \mathcal{W}(\widetilde{R})$ for any $\widetilde{R}$ satisfying

$$
\widetilde{R} \geq \frac{1}{\sqrt{\gamma}} \left\{ 4A\gamma^{\beta} + \left(24C_2 + 4C_1^{1/(2-\tau)} + \frac{4}{3}\phi(-1) \right) \left(\frac{\log(4/\delta)}{m} \right)^{1/(2-\tau)} \right.
$$

$$
\left. + 24C_2 m^{-1/(2-\tau+p)} R^{p/(1+p)} + \frac{10}{3} C_\phi' \mathbf{C}_K^q A^{q/2} \gamma^{q(\beta-1)/2} \left(\frac{\log(4/\delta)}{m} \right)^{1/2} \right\}^{1/2}.
$$

Taking $\gamma = m^{-\zeta}$, we see that we can choose

$$
\widetilde{R} = 5\sqrt{C_2} m^{(\zeta - 1/(2-\tau+p))/2} r^{p/(2(1+p))} + C_3 \left(\log \frac{4}{\delta} \right)^{1/2} m^r.
$$

This proves the lemma.   ∎

Lemma 10.29 *Under the assumptions of Theorem 10.24, take* $\gamma = m^{-\zeta}$ *for some* $\zeta > 0$ *and let* $m \geq (\mathbf{C}_K^2 A)^{-1/(\zeta(1-\beta))}$. *Then, for any* $\eta > 0$ *and* $0 < \delta < 1$, *the set* $\mathcal{W}(R^*)$ *has measure at least* $1 - J_\eta \delta$, *where* $R^* = C_4 m^{r^*}$,

$$
J_\eta := \log_2 \max \left\{ \frac{\zeta}{2}, \frac{1}{(2-\tau+p)} \right\} + \log_2 \frac{1}{\eta} + 1,
$$

and

$$
r^* := \max \left\{ r, \frac{\zeta - 1/(2-\tau+p)}{2(1-s)} + \eta \right\}.
$$

The constant C_4 *is given by*

$$
C_4 = \left(5\sqrt{C_2} \right)^2 (\phi(0) + 1) + J_\eta \left(5\sqrt{C_2} \right)^2 C_3 \left(\log \frac{4}{\delta} \right)^{1/2}.
$$

Proof. Let J be a positive integer that will be determined later. Define a sequence $\{R^{(j)}\}_{j=0}^{J}$ by $R^{(0)} = \sqrt{\phi(0)/\gamma}$ and $R^{(j)} = \widetilde{a}_m \left(R^{(j-1)} \right)^s + \widetilde{b}_m$ for $1 \leq j \leq J$. Then we have

$$
R^{(J)} = (\widetilde{a}_m)^{1+s+s^2+\cdots+s^{J-1}} \left(R^{(0)} \right)^{s^J} + \sum_{j=0}^{J-1} (\widetilde{a}_m)^{1+s+s^2+\cdots+s^{j-1}} \left(\widetilde{b}_m \right)^{s^j}.
$$

(10.17)

The first term on the right-hand side of (10.17) equals

$$\left(5\sqrt{C_2}\right)^{\frac{1-s^J}{1-s}} m^{\frac{\zeta-1/(2-\tau+p)}{2}\cdot\frac{1-s^J}{1-s}} \left(\phi(0)\right)^{s^J/2} m^{\frac{\zeta}{2}s^J},$$

which, since $0 < s < \frac{1}{2}$, is bounded by

$$\left(5\sqrt{C_2}\right)^2 (\phi(0)+1)\, m^{\frac{\zeta-1/(2-\tau+p)}{2(1-s)}} \cdot m^{\left(\frac{\zeta}{2}-\frac{\zeta-1/(2-\tau+p)}{2(1-s)}\right)\cdot s^J}.$$

When $2^J \geq \max\{\frac{\zeta}{2\eta}, \frac{1}{(2-\tau+p)\eta}\}$, this upper bound is controlled by

$$\left(5\sqrt{C_2}\right)^2 (\phi(0)+1)\, m^{\frac{\zeta-1/(2-\tau+p)}{2(1-s)}+\eta}.$$

The second term on the right-hand side of (10.17) equals

$$\sum_{j=0}^{J-1} \left(5\sqrt{C_2}m^{\frac{\zeta-1/(2-\tau+p)}{2}}\right)^{\frac{1-s^j}{1-s}} \left(C_3\,(\log(4/\delta))^{1/2}\, m^r\right)^{s^j},$$

which is bounded by

$$m^{\frac{\zeta-1/(2-\tau+p)}{2(1-s)}} \sum_{j=0}^{J-1} \left(5\sqrt{C_2}\right)^2 C_3\,(\log(4/\delta))^{1/2}\, m^{\left(r-\frac{\zeta-1/(2-\tau+p)}{2(1-s)}\right)\cdot s^j}.$$

If $r \geq \frac{\zeta-1/(2-\tau+p)}{2(1-s)}$, this last expression is bounded by

$$m^{\frac{\zeta-1/(2-\tau+p)}{2(1-s)}} J \left(5\sqrt{C_2}\right)^2 C_3\,(\log(4/\delta))^{1/2}\, m^{r-\frac{\zeta-1/(2-\tau+p)}{2(1-s)}}$$

$$= J \left(5\sqrt{C_2}\right)^2 C_3\,(\log(4/\delta))^{1/2}\, m^r.$$

If $r < \frac{\zeta-1/(2-\tau+p)}{2(1-s)}$, an upper bound is easier:

$$m^{\frac{\zeta-1/(2-\tau+p)}{2(1-s)}} J \left(5\sqrt{C_2}\right)^2 C_3\,(\log(4/\delta))^{1/2}.$$

Thus, in either case, the second term has the upper bound $J \left(5\sqrt{C_2}\right)^2 C_3 (\log(4/\delta))^{1/2}\, m^{r^*}$.

Combining the bounds for the two terms, we have

$$R^{(J)} \leq \left\{\left(5\sqrt{C_2}\right)^2 (\phi(0)+1) + J \left(5\sqrt{C_2}\right)^2 C_3\,(\log(4/\delta))^{1/2}\right\} m^{r^*}.$$

Taking J to be J_η, we have $2^J > \max\{\frac{\zeta}{2\eta}, \frac{1}{(2-\tau+p)\eta}\}$ and we finish the proof. ∎

The proof of Theorem 10.24 follows from Lemmas 10.27 and 10.29 and Proposition 10.22. The constant $\tilde{C}_\eta$ can be explicitly obtained.

10.7 Improving learning rates by imposing noise conditions

There is a difference in the learning rates given by Theorem 10.2 (where the best rate is $\frac{1}{2} - \varepsilon$) and Theorem 9.26 (where the rate can be arbitrarily close to 1). This motivates the idea of improving the learning rates stated in this chapter by imposing some conditions on the measures. In this section we introduce one possible such condition.

Definition 10.30 Let $0 \le q \le \infty$. We say that ρ has *Tsybakov noise exponent* q if there exists a constant $c_q > 0$ such that for all $t > 0$,

$$\rho_X \left(\{x \in X : |f_\rho(x)| \le c_q t\} \right) \le t^q. \tag{10.18}$$

All distributions have at least noise exponent 0 since $t^0 = 1$. Deterministic distributions (which satisfy $|f_\rho(x)| \equiv 1$) have noise exponent $q = \infty$ with $c_\infty = 1$.

The Tsybakov noise condition improves the variancing power $\tau_{\phi,\rho}$. Let us show this for the hinge loss.

Lemma 10.31 *Let $0 \le q \le \infty$. If ρ has Tsybakov noise exponent q with (10.18) valid, then, for every function $f : X \to [-1, 1]$,*

$$\mathbf{E}\left\{(\phi_h(yf(x)) - \phi_h(yf_c(x)))^2\right\} \le 8 \left(\frac{1}{2c_q}\right)^{q/(q+1)} \left(\mathcal{E}^{\phi_h}(f) - \mathcal{E}^{\phi_h}(f_c)\right)^{q/(q+1)}$$

holds.

Proof. Since $f(x) \in [-1, 1]$, we have $\phi_h(yf(x)) - \phi_h(yf_c(x)) = y(f_c(x) - f(x))$. It follows that

$$\mathcal{E}^{\phi_h}(f) - \mathcal{E}^{\phi_h}(f_c) = \int_X (f_c(x) - f(x))f_\rho(x) \, d\rho_X = \int_X |f_c(x) - f(x)| \, |f_\rho(x)| \, d\rho_X$$

and

$$\mathbf{E}\left\{(\phi_h(yf(x)) - \phi_h(yf_c(x)))^2\right\} = \int_X |f_c(x) - f(x)|^2 \, d\rho_X.$$

Let $t > 0$ and separate the domain X into two sets: $X_t^+ := \{x \in X : |f_\rho(x)| > c_q t\}$ and $X_t^- := \{x \in X : |f_\rho(x)| \leq c_q t\}$. On X_t^+ we have $|f_c(x) - f(x)|^2 \leq 2|f_c(x) - f(x)| \frac{|f_\rho(x)|}{c_q t}$. On X_t^- we have $|f_c(x) - f(x)|^2 \leq 4$. It follows from (10.18) that

$$\int_X |f_c(x) - f(x)|^2 \, d\rho_X \leq \frac{2\left(\mathcal{E}^{\phi_h}(f) - \mathcal{E}^{\phi_h}(f_c)\right)}{c_q t} + 4\rho_X(X_t^-)$$

$$\leq \frac{2\left(\mathcal{E}^{\phi_h}(f) - \mathcal{E}^{\phi_h}(f_c)\right)}{c_q t} + 4t^q.$$

Choosing $t = \left\{(\mathcal{E}^{\phi_h}(f) - \mathcal{E}^{\phi_h}(f_c))/(2c_q)\right\}^{1/(q+1)}$ yields the desired bound. ∎

Lemma 10.31 tells us that the variancing power $\tau_{\phi_h,\rho}$ of the hinge loss equals $\frac{q}{q+1}$ when the measure ρ has Tsybakov noise exponent q. Combining this with Corollary 10.26 gives the following result on improved learning rates for measures satisfying the Tsybakov noise condition.

Theorem 10.32 *Under the assumption of Theorem 10.2, if ρ has Tsybakov noise exponent q with $0 \leq q \leq \infty$, then, for any $0 < \varepsilon < \frac{1}{2}$ and $0 < \delta < 1$, with confidence $1 - \delta$, we have*

$$\mathcal{R}(\mathrm{sgn}(f_{\mathbf{z},\gamma}^{\phi_h})) - \mathcal{R}(f_c) \leq \widetilde{C} \log \frac{2}{\delta} \left(\frac{1}{m}\right)^\theta$$

where $\theta = \min\left\{\frac{2\beta}{1+\beta}, \frac{q+1}{q+2} - \varepsilon\right\}$ and $\widetilde{C}$ is a constant independent of m and δ. ∎

In Theorem 10.32, the learning rate can be arbitrarily close to 1 when q is sufficiently large.

10.8 References and additional remarks

General expositions of convex loss functions for classification can be found in [14, 31]. Theorem 10.5, the use of the projection operator, and some estimates for the regularized error were provided in [31]. The error decomposition for regularization schemes was introduced in [145].

The convergence of the support vector machine (SVM) 1-norm soft margin classifier for general probability distributions (without separability conditions) was established in [121] when $\mathcal{H}_K$ is dense in $\mathscr{C}(X)$ (such a kernel K is called *universal*). Convergence rates in this situation were derived in [154]. For further results and references on convergence rates, see the thesis [140].

The error analysis in this chapter is taken from [142], where more technical and better error bounds are provided by means of the local Rademacher process, empirical covering numbers, and the entropy integral [84, 132]. The Tsybakov noise condition of Section 10.7 was introduced in [131].

The iteration technique used in the proof of Lemma 10.29 was given in [122] (see also [144]).

SVMs have many modifications for various purposes in different fields [134]. These include q-norm soft margin classifiers [31, 77], multiclass SVMs [4, 32, 75, 139], ν-SVMs [108], linear programming SVMs [26, 96, 98, 146], maximum entropy discrimination [65], and one-class SVMs [107, 128].

We conclude with some brief comments on current trends.

Learning theory is a rapidly growing field. Many people are working on both its foundations and its applications, from different points of view. This work develops the theory but also leaves many open questions. Here we mention some involving regularization schemes [48].

(i) Feature selection. One purpose is to understand structures of high-dimensional data. Topics include manifold learning or semisupervised learning [15, 23, 27, 34, 45, 97] and dimensionality reduction (see the introduction [55] of a special issue and references therein). Another purpose is to determine important features (variables) of functions defined on huge-dimensional spaces. Two approaches are the filter method and the wrapper method [69]. Regularization schemes for this purpose include those in [56, 58] and a least squares–type algorithm in [93] that learns gradients as vector-valued functions [89].

(ii) Multikernel regularization schemes. Let $K_\Sigma = \{K_\sigma : \sigma \in \Sigma\}$ be a set of Mercer kernels on X such as Gaussian kernels with variances σ^2 running over $(0, \infty)$. The multikernel regularization scheme associated with K_Σ is defined as

$$f_{\mathbf{z},\gamma,\Sigma} = \operatorname*{arginf}_{\sigma \in \Sigma} \inf_{f \in \mathcal{H}_{K_\sigma}} \left\{ \frac{1}{m} \sum_{i=1}^{m} V\left(y_i, f(x_i)\right) + \gamma \|f\|_{K_\sigma}^2 \right\}.$$

Here $V : \mathbb{R}^2 \to \mathbb{R}_+$ is a general loss function. In [30] SVMs with multiple parameters are investigated. In [76, 104] mixture-density estimation is considered and Gaussian kernels with variance σ^2 varying on an interval $[\sigma_1^2, \sigma_2^2]$ with $0 < \sigma_1 < \sigma_2 < +\infty$ are used to derive bounds. Multitask learning algorithms involve kernels from a convex hull of several Mercer kernels and spaces with changing norms (e.g. [49, 62]). The learning of kernel functions is studied in [72, 88, 90].

Another related class of multikernel regularization schemes consists that of schemes generated by polynomial kernels $\{K_d(x, y) = (1 + x \cdot y)^d\}$ with $d \in \mathbb{N}$. In [158] convergence rates in the univariate case ($n = 1$) for multikernel regularized classifiers generated by polynomial kernels are derived.

(iii) Online learning algorithms. These algorithms improve the efficiency of learning methods when the sample size m is very large. Their convergence is investigated in [28, 51, 52, 68, 134], and their error with respect to the step size has been analyzed for the least squares regression in [112] and for regularized classification with a general classifying loss in [151]. Error analysis for online schemes with varying regularization parameters is performed in [127] and [149].

References

[1] R.A. Adams. *Sobolev Spaces*. Academic Press, 1975.

[2] C.A. Aliprantis and O. Burkinshaw. *Principles of Real Analysis*. Academic Press, 3rd edition, 1998.

[3] F. Alizadeh and D. Goldfarb. Second-order cone programming. *Math. Program.*, 95:3–51, 2003.

[4] E.L. Allwein, R.E. Schapire, and Y. Singer. Reducing multiclass to binary: a unifying approach for margin classifiers. *J. Mach. Learn. Res.*, 1:113–141, 2000.

[5] N. Alon, S. Ben-David, N. Cesa-Bianchi, and D. Haussler. Scale-sensitive dimensions, uniform convergence and learnability. *J. ACM*, 44:615–631, 1997.

[6] M. Anthony and P. Bartlett. *Neural Network Learning: Theoretical Foundations*. Cambridge University Press, 1999.

[7] M. Anthony and N. Biggs. *Computational Learning Theory*. Cambridge University Press, 1992.

[8] M. Anthony and J. Shawe-Taylor. A result of Vapnik with applications. *Discrete Appl. Math.*, 47:207–217, 1993.

[9] N. Aronszajn. Theory of reproducing kernels. *Trans. Amer. Math. Soc.*, 68:337–404, 1950.

[10] A.R. Barron. Complexity regularization with applications to artificial neural networks. In G. Roussas, editor, *Nonparametric Functional Estimation*, pages 561–576. Kluwer Academic Publishers 1990.

[11] R.G. Bartle. *The Elements of Real Analysis*. John Wiley & Sons, 2nd edition, 1976.

[12] P.L. Bartlett. The sample complexity of pattern classification with neural networks: the size of the weights is more important than the size of the network. *IEEE Trans. Inform. Theory*, 44:525–536, 1998.

[13] P.L. Bartlett, O. Bousquet, and S. Mendelson. Local Rademacher complexities. *Ann. Stat.*, 33:1497–1537, 2005.

[14] P.L. Bartlett, M.I. Jordan, and J.D. McAuliffe. Convexity, classification, and risk bounds. *J. Amer. Stat. Ass.*, 101:138–156, 2006.

[15] M. Belkin and P. Niyogi. Semi-supervised learning on Riemannian manifolds. *Mach. Learn.*, 56:209–239, 2004.

[16] J. Bergh and J. Löfström. *Interpolation Spaces: An Introduction.* Springer-Verlag, 1976.

[17] P. Binev, A. Cohen, W. Dahmen, R. DeVore, and V. Temlyakov. Universal algorithms for learning theory. Part I: piecewise constant functions. *J. Mach. Learn. Res.*, 6:1297–1321, 2005.

[18] C.M. Bishop. *Neural Networks for Pattern Recognition.* Cambridge University Press, 1995.

[19] L. Blum, F. Cucker, M. Shub, and S. Smale. *Complexity and Real Computation.* Springer-Verlag, 1998.

[20] B.E. Boser, I. Guyon, and V. Vapnik. A training algorithm for optimal margin classifiers. In *Proceedings of the Fifth Annual Workshop of Computational Learning Theory*, pages 144–152. Association for Computing Machinery, New York, 1992.

[21] S. Boucheron, O. Bousquet, and G. Lugosi. Concentration inequalities. In O. Bousquet, U. von Luxburg, and G. Rátsch, editors, *Advanced Lectures in Machine Learning*, pages 208–240. Springer-Verlag, 2004.

[22] S. Boucheron, G. Lugosi, and P. Massart. A sharp concentration inequality with applications in random combinatorics and learning. *Random Struct. Algorithms*, 16:277–292, 2000.

[23] O. Bousquet, O. Chapelle, and M. Hein. Measure based regularizations. In S. Thrun, L.K. Saul, and B. Schölkopf, editors, *Advances in Neural Information Processing Systems*, volume 16, pages 1221–1228. MIT Press, 2004.

[24] O. Bousquet and A. Elisseeff. Stability and generalization. *J. Mach. Learn. Res.*, 2:499–526, 2002.

[25] S. Boyd and L. Vandenberghe. *Convex Optimization.* Cambridge University Press, 2004.

[26] P.S. Bradley and O.L. Mangasarian. Massive data discrimination via linear support vector machines. *Optimi. Methods and Softw.*, 13:1–10, 2000.

[27] A. Caponnetto and S. Smale. Risk bounds for random regression graphs. To appear at *Found. Comput. Math.*

[28] N. Cesa-Bianchi, P.M. Long, and M.K. Warmuth. Worst-case quadratic loss bounds for prediction using linear functions and gradient descent. *IEEE Trans. Neural Networks*, 7:604–619, 1996.

[29] N. Cesa-Bianchi and G. Lugosi. *Prediction, Learning, and Games.* Cambridge University Press, 2006.

[30] O. Chapelle, V. Vapnik, O. Bousquet, and S. Mukherjee. Choosing multiple parameters for support vector machines. *Mach. Learn.*, 46:131–159, 2002.

[31] D.R. Chen, Q. Wu, Y. Ying, and D.X. Zhou. Support vector machine soft margin classifiers: error analysis. *J. Mach. Learn. Res.*, 5:1143–1175, 2004.

[32] D.R. Chen and D.H. Xiang. The consistency of multicategory support vector machines. *Adv. Comput. Math.*, 24:155–169, 2006.

[33] D.A. Cohn, Z. Ghahramani, and M.I. Jordan. Active learning with statistical models. *J. Artif. Intell. Res.*, 4:129–145, 1996.

[34] R.R. Coifman, S. Lafon, A.B. Lee, M. Maggioni, B. Nadler, F. Warner, and S.W. Zucker. Geometric diffusions as a tool for harmonic analysis and structure definition of data: diffusion maps. *Proc. Natl. Acad. Sci.*, 102:7426–7431, 2005.

[35] C. Cortes and V. Vapnik. Support-vector networks. *Mach. Learn.*, 20:273–297, 1995.

[36] D.D. Cox. Approximation of least squares regression on nested subspaces. *Ann. Stat.*, 16:713–732, 1988.

[37] N. Cristianini and J. Shawe-Taylor. *An Introduction to Support Vector Machines.* Cambridge University Press, 2000.

[38] F. Cucker and S. Smale. Best choices for regularization parameters in learning theory. *Found. Comput. Math.*, 2:413–428, 2002.

[39] F. Cucker and S. Smale. On the mathematical foundations of learning. *Bull. Amer. Math. Soc.*, 39:1–49, 2002.

[40] L. Debnath and P. Mikusiński. *Introduction to Hilbert Spaces with Applications.* Academic Press, 2nd edition, 1999.

[41] C. de Boor, K. Höllig, and S. Riemenschneider. *Box Splines.* Springer-Verlag, 1993.

[42] E. De Vito, A. Caponnetto, and L. Rosasco. Model selection for regularized least-squares algorithm in learning theory. *Found. Comput. Math.*, 5:59–85, 2005.

[43] E. De Vito, L. Rosasco, A. Caponnetto, U. de Giovannini, and F. Odone. Learning from examples as an inverse problem. *J. Mach. Learn. Res.*, 6:883–904, 2005.

[44] L. Devroye, L. Györfi, and G. Lugosi. *A Probabilistic Theory of Pattern Recognition.* Springer-Verlag, 1996.

[45] D.L. Donoho and C. Grimes. Hessian eigenmaps: locally linear embedding techniques for high-dimensional data. *Proc. Natl. Acad. Sci.*, 100:5591–5596, 2003.

[46] R.M. Dudley, E. Giné, and J. Zinn. Uniform and universal Glivenko–Cantelli classes. *J. Theor. Prob.*, 4:485–510, 1991.

[47] D.E. Edmunds and H. Triebel. *Function Spaces, Entropy Numbers, Differential Operators.* Cambridge University Press, 1996.

[48] H.W. Engl, M. Hanke, and A. Neubauer. *Regularization of Inverse Problems*, volume 375 of *Mathematics and Its Applications.* Kluwer, 1996.

[49] T. Evgeniou and M. Pontil. Regularized multi-task learning. In C.E. Brodley, editor, *Proc. 17th SIGKDD Conf. Knowledge Discovery and Data Mining*, Association for Computing Machinery, New York, 2004.

[50] T. Evgeniou, M. Pontil, and T. Poggio. Regularization networks and support vector machines. *Adv. Comput. Math.*, 13:1–50, 2000.

[51] J. Forster and M.K. Warmuth. Relative expected instantaneous loss bounds. *J. Comput. Syst. Sci.*, 64:76–102, 2002.

[52] Y. Freund and R.E. Shapire. A decision-theoretic generalization of on-line learning and an application to boosting. *J. Comput. Syst. Sci.*, 55:119–139, 1997.

[53] F. Girosi, M. Jones, and T. Poggio. Regularization theory and neural networks architectures. *Neural Comp.*, 7:219–269, 1995.

[54] G. Golub, M. Heat, and G. Wahba. Generalized cross-validation as a method for choosing a good ridge parameter. *Technometrics*, 21:215–223, 1979.

[55] I. Guyon and A. Ellisseeff. An introduction to variable and feature selection. *J. Mach. Learn. Res.*, 3:1157–1182, 2003.

[56] I. Guyon, J. Weston, S. Barnhill, and V. Vapnik. Gene selection for cancer classification using support vector machines. *Mach. Learn.*, 46:389–422, 2002.

[57] L. Györfi, M. Kohler, A. Krzyżak, and H. Walk. *A Distribution-Free Theory of Nonparametric Regression.* Springer-Verlag, 2002.

[58] D. Hardin, I. Tsamardinos, and C.F. Aliferis. A theoretical characterization of linear SVM-based feature selection. In *Proc. 21st Int. Conf. Machine Learning,* 2004.

[59] T. Hastie, R.J. Tibshirani, and J.H. Friedman. *The Elements of Statistical Learning.* Springer-Verlag, 2001.

[60] D. Haussler. Decision theoretic generalizations of the PAC model for neural net and other learning applications. *Inform. and Comput.,* 100:78–150, 1992.

[61] R. Herbrich. *Learning Kernel Classifiers: Theory and Algorithms.* MIT Press, 2002.

[62] M. Herbster. Relative loss bounds and polynomial-time predictions for the k-lms-net algorithm. In S. Ben-David, J. Case, and A. Maruoka, editors, *Proc. 15th Int. Conf. Algorithmic Learning Theory,* Springer 2004.

[63] H. Hochstadt. *Integral Equations.* John Wiley & Sons, 1973.

[64] V.V. Ivanov. *The Theory of Approximate Methods and Their Application to the Numerical Solution of Singular Integral Equations.* Nordhoff International, 1976.

[65] T. Jaakkola, M. Meila, and T. Jebara. Maximum entropy discrimination. In S.A. Solla, T.K. Leen, and K.-R. Müller, editors, *Advances in Neural Information Processing Systems,* volume 12, pages 470–476. MIT Press, 2000.

[66] K. Jetter, J. Stöckler, and J.D. Ward. Error estimates for scattered data interpolation on spheres. *Math. Comp.,* 68:733–747, 1999.

[67] M.J. Kearns and U.V. Vazirani. *An Introduction to Computational Learning Theory.* MIT Press, 1994.

[68] J. Kivinen, A.J. Smola, and R.C. Williamson. Online learning with kernels. *IEEE Trans. Signal Processing,* 52:2165–2176, 2004.

[69] R. Kohavi and G. John. Wrappers for feature subset selection. *Artif. Intell.,* 97:273–324, 1997.

[70] A.N. Kolmogorov and S.V. Fomin. *Introductory Real Analysis.* Dover Publications, 1975.

[71] V. Koltchinskii and D. Panchenko. Empirical margin distributions and bounding the generalization error of combined classifiers. *Ann. Stat.,* 30:1–50, 2002.

[72] G.R.G. Lanckriet, N. Cristianini, P. Bartlett, L. El Ghaoui, and M.I. Jordan. Learning the kernel matrix with semidefinite programming. *J. Mach. Learn. Res.,* 5:27–72, 2004.

[73] P. Lax. *Functional Analysis.* John Wiley & Sons, 2002.

[74] W.-S. Lee, P. Bartlett, and R. Williamson. The importance of convexity in learning with squared loss. *IEEE Trans. Inform. Theory,* 44:1974–1980, 1998.

[75] Y. Lee, Y. Lin, and G. Wahba. Multicategory support vector machines, theory, and application to the classification of microarray data and satellite radiance data. *J. Amer. Stat. Ass.,* 99:67–81, 2004.

[76] J. Li and A. Barron. Mixture density estimation. In S.A. Solla, T.K. Leen, and K.R. Müller, editors, *Advances in Neural Information Processing Systems,* volume 12, pages 279–285. Morgan Kaufmann Publishers, 1999.

[77] Y. Lin. Support vector machines and the Bayes rule in classification. *Data Min. Knowl. Discov.,* 6:259–275, 2002.

[78] G.G. Lorentz. *Approximation of Functions.* Holt, Rinehart and Winston, 1966.

[79] F. Lu and H. Sun. Positive definite dot product kernels in learning theory. *Adv. Comput. Math.*, 22:181–198, 2005.

[80] G. Lugosi and N. Vayatis. On the Bayes-risk consistency of regularized boosting methods. *Ann. Stat.*, 32:30–55, 2004.

[81] D.J.C. Mackay. Information-based objective functions for active data selection. *Neural Comp.*, 4:590–604, 1992.

[82] W.R. Madych and S.A. Nelson. Bounds on multivariate polynomials and exponential error estimates for multiquadric interpolation. *J. Approx. Theory*, 70:94–114, 1992.

[83] C. McDiarmid. Concentration. In M. Habib et al., editors, *Probabilistic Methods for Algorithmic Discrete Mathematics*, pages 195–248. Springer-Verlag, 1998.

[84] S. Mendelson. Improving the sample complexity using global data. *IEEE Trans. Inform. Theory*, 48:1977–1991, 2002.

[85] J. Mercer. Functions of positive and negative type and their connection with the theory of integral equations. *Philos. Trans. Roy. Soc. London Ser. A*, 209:415–446, 1909.

[86] C.A. Micchelli. Interpolation of scattered data: distance matrices and conditionally positive definite functions. *Constr. Approx.*, 2:11–22, 1986.

[87] C.A. Micchelli and A. Pinkus. Variational problems arising from balancing several error criteria. *Rend. Math. Appl.*, 14:37–86, 1994.

[88] C.A. Micchelli and M. Pontil. Learning the kernel function via regularization. *J. Mach. Learn. Res.*, 6:1099–1125, 2005.

[89] C.A. Micchelli and M. Pontil. On learning vector-valued functions. *Neural Comp.*, 17:177–204, 2005.

[90] C.A. Micchelli, M. Pontil, Q. Wu, and D.X. Zhou. Error bounds for learning the kernel. Preprint, 2006.

[91] M. Mignotte. *Mathematics for Computer Algebra*. Springer-Verlag, 1992.

[92] T.M. Mitchell. *Machine Learning*. McGraw-Hill, 1997.

[93] S. Mukherjee and D.X. Zhou. Learning coordinate covariances via gradients. *J. Mach. Learn. Res.*, 7:519–549, 2006.

[94] F.J. Narcowich, J.D. Ward, and H. Wendland. Refined error estimates for radial basis function interpolation. *Constr. Approx.*, 19:541–564, 2003.

[95] P. Niyogi. *The Informational Complexity of Learning*. Kluwer Academic Publishers, 1998.

[96] P. Niyogi and F. Girosi. On the relationship between generalization error, hypothesis complexity and sample complexity for radial basis functions. *Neural Comput.*, 8:819–842, 1996.

[97] P. Niyogi, S. Smale, and S. Weinberger. Finding the homology of submanifolds with high confidence from random samples. Preprint, 2004.

[98] J.P. Pedroso and N. Murata. Support vector machines with different norms: motivation, formulations and results. *Pattern Recognit. Lett.*, 22:1263–1272, 2001.

[99] I. Pinelis. Optimum bounds for the distributions of martingales in Banach spaces. *Ann. Probab.*, 22:1679–1706, 1994.

[100] A. Pinkus. *N-widths in Approximation Theory*. Springer-Verlag, 1996.

[101] A. Pinkus. Strictly positive definite kernels on a real inner product space. *Adv. Comput. Math.*, 20:263–271, 2004.

[102] T. Poggio, V. Torre, and C. Koch. Computational vision and regularization theory. *Nature*, 317:314–319, 1985.

[103] D. Pollard. *Convergence of Stochastic Processes*. Springer-Verlag, 1984.

[104] A. Rakhlin, D. Panchenko, and S. Mukherjee. Risk bounds for mixture density estimation. *ESAIM: Prob. Stat.*, 9:220–229, 2005.

[105] R. Schaback. Reconstruction of multivariate functions from scattered data. Manuscript, 1997.

[106] I.J. Schoenberg. Metric spaces and completely monotone functions. *Ann. Math.*, 39:811–841, 1938.

[107] B. Schölkopf and A.J. Smola. *Learning with Kernels: Support Vector Machines, Regularization, Optimization, and Beyond*. MIT Press, 2002.

[108] B. Schölkopf, A.J. Smola, R.C. Williamson, and P.L. Bartlett. New support vector algorithms. *Neural Comp.*, 12:1207–1245, 2000.

[109] I.R. Shafarevich. *Basic Algebraic Geometry. 1: Varieties in Projective Space*. Springer-Verlag, 2nd edition, 1994.

[110] J. Shawe-Taylor, P.L. Bartlet, R.C. Williamson, and M. Anthony. Structural risk minimization over data dependent hierarchies. *IEEE Trans. Inform. Theory*, 44:1926–1940, 1998.

[111] J. Shawe-Taylor and N. Cristianini. *Kernel Methods for Pattern Analysis*. Cambridge University Press, 2004.

[112] S. Smale and Y. Yao. Online learning algorithms. *Found. Comput. Math.*, 6:145–170, 2006.

[113] S. Smale and D.X. Zhou. Estimating the approximation error in learning theory. *Anal. Appl.*, 1:17–41, 2003.

[114] S. Smale and D.X. Zhou. Shannon sampling and function reconstruction from point values. *Bull. Amer. Math. Soc.*, 41:279–305, 2004.

[115] S. Smale and D.X. Zhou. Shannon sampling II: Connections to learning theory. *Appl. Comput. Harmonic Anal.*, 19:285–302, 2005.

[116] S. Smale and D.X. Zhou. Learning theory estimates via integral operators and their approximations. To appear in *Constr. Approx.*

[117] A. Smola, B. Schölkopf, and R. Herbricht. A generalized representer theorem. *Comput. Learn. Theory*, 14:416–426, 2001.

[118] A. Smola, B. Schölkopf, and K.R. Müller. The connection between regularization operators and support vector kernels. *Neural Networks*, 11:637–649, 1998.

[119] M. Sousa Lobo, L. Vandenberghe, S. Boyd, and H. Lebret. Applications of second-order cone programming. *Linear Algebra Appl.*, 284:193–228, 1998.

[120] E.M. Stein. *Singular Integrals and Differentiability Properties of Functions*. Princeton University Press, 1970.

[121] I. Steinwart. Support vector machines are universally consistent. *J. Complexity*, 18:768–791, 2002.

[122] I. Steinwart and C. Scovel. Fast rates for support vector machines. In P. Auer and R. Meir, editors, *Proc. 18th Ann. Conf. Learn. Theory*, pages 279–294, Springer 2005.

[123] H.W. Sun. Mercer theorem for RKHS on noncompact sets. *J. Complexity*, 21:337–349, 2005.

[124] R.S. Sutton and A.G. Barto. *Reinforcement Learning: An Introduction*. MIT Press, 1998.

[125] J.A.K. Suykens, T. Van Gestel, J. De Brabanter, B. De Moor, and J. Vandewalle. *Least Squares Support Vector Machines*. World Scientific, 2002.

[126] M. Talagrand. New concentration inequalities in product spaces. *Invent. Math.*, 126:505–563, 1996.

[127] P. Tarrès and Y. Yao. Online learning as stochastic approximations of regularization paths. Preprint, 2005.

[128] D.M.J. Tax and R.P.W. Duin. Support vector domain description. *Pattern Recognit. Lett.*, 20:1191–1199, 1999.

[129] M.E. Taylor. *Partial Differential Equations I: Basic Theory*, volume 115 of *Applied Mathematical Sciences*. Springer-Verlag, 1996.

[130] A.N. Tikhonov and V.Y. Arsenin. *Solutions of Ill-Posed Problems*. W.H. Winston, 1977.

[131] A.B. Tsybakov. Optimal aggregation of classifiers in statistical learning. *Ann. Stat.*, 32:135–166, 2004.

[132] A.W. van der Vaart and J.A. Wellner. *Weak Convergence and Empirical Processes*. Springer-Verlag, 1996.

[133] V.N. Vapnik. *Estimation of Dependences Based on Empirical Data*. Springer-Verlag, 1982.

[134] V. Vapnik. *Statistical Learning Theory*. John Wiley & Sons, 1998.

[135] V. N. Vapnik and A. Ya. Chervonenkis. On the uniform convergence of relative frequencies of events to their probabilities. *Theory Prob. Appl.*, 16:264–280, 1971.

[136] M. Vidyasagar. *Learning and Generalization*. Springer-Verlag, 2003.

[137] G. Wahba. *Spline Models for Observational Data*. SIAM, 1990.

[138] G. Wahba. Support vector machines, reproducing kernel Hilbert spaces and the randomized GACV. In B. Schölkopf, C. Burges, and A. Smola, editors, *Advances in Kernel Methods – Support Vector Learning*, pages 69–88. MIT Press, 1999.

[139] J. Weston and C. Watkins. Multi-class support vector machines. Technical Report CSD-TR-98-04, Department of Computer Science, Royal Holloway, University of London, 1998.

[140] Q. Wu. Classification and regularization in learning theory. *PhD thesis*, City University of Hong Kong, 2005.

[141] Q. Wu, Y. Ying, and D.X. Zhou. Learning theory: from regression to classification. In K. Jetter, M. Buhmann, W. Haussmann, R. Schaback, and J. Stoeckler, editors, *Topics in Multivariate Approximation and Interpolation*, volume 12 of *Studies in Computational Mathematics*, pages 257–290. Elsevier, 2006.

[142] Q. Wu, Y. Ying, and D.X. Zhou. Multi-kernel regularized classifiers. To appear in *J. Complexity*.

[143] Q. Wu, Y. Ying, and D.X. Zhou. Learning rates of least-square regularized regression. *Found. Comput. Math.*, 6:171–192, 2006.

[144] Q. Wu and D.X. Zhou. SVM soft margin classifiers: linear programming versus quadratic programming. *Neural Comp.*, 17:1160–1187, 2005.

[145] Q. Wu and D.X. Zhou. Analysis of support vector machine classification. *J. Comput. Anal. Appl.*, 8:99–119, 2006.

[146] Q. Wu and D.X. Zhou. Learning with sample dependent hypothesis spaces. Preprint, 2006.

[147] Z. Wu and R. Schaback. Local error estimates for radial basis function interpolation of scattered data. *IMA J. Numer. Anal.*, 13:13–27, 1993.

[148] Y. Yang and A. R. Barron. Information-theoretic determination of minimax rates of convergence. *Ann. Stat.*, 27:1564–1599, 1999.

[149] G.B. Ye and D.X. Zhou. Fully online classification by regularization. To appear at *Appl. Comput. Harmonic Anal.*

[150] Y. Ying and D.X. Zhou. Learnability of Gaussians with flexible variances. To appear at *J. Mach. Learn. Res.*

[151] Y. Ying and D.X. Zhou. Online regularized classification algorithms. To appear in *IEEE Trans. Inform. Theory*, 52:4775–4788, 2006.

[152] T. Zhang. On the dual formulation of regularized linear systems with convex risks. *Machine Learning*, 46:91–129, 2002.

[153] T. Zhang. Leave-one-out bounds for kernel methods. *Neural Comp.*, 15:1397–1437, 2003.

[154] T. Zhang. Statistical behavior and consistency of classification methods based on convex risk minimization. *Ann. Stat.*, 32:56–85, 2004.

[155] D.X. Zhou. The covering number in learning theory. *J. Complexity*, 18:739–767, 2002.

[156] D.X. Zhou. Capacity of reproducing kernel spaces in learning theory. *IEEE Trans. Inform. Theory*, 49:1743–1752, 2003.

[157] D.X. Zhou. Density problem and approximation error in learning theory. Preprint, 2006.

[158] D.X. Zhou and K. Jetter. Approximation with polynomial kernels and SVM classifiers. *Adv. Comput. Math.*, 25:323–344, 2006.

Index

.

Printed in the United States
By Bookmasters